D1714089

WAVES AND INTERACTIONS
IN SOLID STATE PLASMAS

SOLID STATE PHYSICS

Advances in
Research and Applications

Editors

Henry Ehrenreich
*Division of Engineering and Applied Physics, Harvard University
Cambridge, Massachusetts*

Frederick Seitz
Rockefeller University, New York, New York

David Turnbull
*Division of Engineering and Applied Physics, Harvard University
Cambridge, Massachusetts*

The following monographs are published within the framework of the series:

1. T. P. Das and E. L. Hahn, *Nuclear Quadrupole Resonance Spectroscopy*, 1958
2. William Low, *Paramagnetic Resonance in Solids*, 1960
3. A. A. Maradudin, E. W. Montroll, G. H. Weiss, and I. P. Ipatova, *Theory of Lattice Dynamics in the Harmonic Approximation*, Second Edition, 1971
4. Albert C. Beer, *Galvanomagnetic Effects in Semiconductors*, 1963
5. Robert S. Knox, *Theory of Excitons*, 1963
6. S. Amelinckx, *The Direct Observation of Dislocations*, 1964
7. James W. Corbett, *Electron Radiation Damage in Semiconductors and Metals*, 1966
8. Jordan J. Markham, *F-Centers in Alkali Halides*, 1966
9. Esther M. Conwell, *High Field Transport in Semiconductors*, 1967
10. C. B. Duke, *Tunneling in Solids*, 1969
11. M. Cardona, *Modulation Spectroscopy*, 1969
12. A. A. Abrikosov, *Introduction to the Theory of Normal Metals*, 1972
13. P. M. Platzman and P. A. Wolff, *Waves and Interactions in Solid State Plasmas*, 1972

WAVES AND INTERACTIONS IN SOLID STATE PLASMAS

P. M. PLATZMAN

Bell Telephone Laboratories
Murray Hill, New Jersey

P. A. WOLFF

Department of Physics
Massachusetts Institute of Technology
Cambridge, Massachusetts

ACADEMIC PRESS · New York and London · **1973**

COPYRIGHT © 1973, BY BELL TELEPHONE LABORATORIES,
INCORPORATED, AND P. A. WOLFF
ALL RIGHTS RESERVED.
NO PART OF THIS PUBLICATION MAY BE REPRODUCED OR
TRANSMITTED IN ANY FORM OR BY ANY MEANS, ELECTRONIC
OR MECHANICAL, INCLUDING PHOTOCOPY, RECORDING, OR ANY
INFORMATION STORAGE AND RETRIEVAL SYSTEM, WITHOUT
PERMISSION IN WRITING FROM THE PUBLISHER.

ACADEMIC PRESS, INC.
111 Fifth Avenue, New York, New York 10003

United Kingdom Edition published by
ACADEMIC PRESS, INC. (LONDON) LTD.
24/28 Oval Road, London NW1

LIBRARY OF CONGRESS CATALOG CARD NUMBER: 55-12299

PRINTED IN THE UNITED STATES OF AMERICA

Contents

PREFACE ix

I. Characteristics of Solid State Plasmas

1. What Is a Plasma? 1
2. Conditions for Weak Coupling 3
3. Fermi Liquid Theory—Introductory Remarks 5
4. Coulomb Energy in a Plasma 7
5. What Parameters Characterize a Plasma?—Screening Lengths 8
6. The Characteristic Frequency 11
7. Other Plasma Parameters 15
   References 19

II. Theory of Important Plasma Experiments

8. Qualitative Discussion of Scattering Experiments 20
9. Analysis of Electron Scattering Experiments 23
10. Scattering from a Noninteracting Electron System 26
11. Light Scattering—The Classical, Single-Component Case . 27
12. Transmission Experiments: Introductory Remarks 31
13. Response Functions 33
14. Response to the Total Field 37
15. The Plasma Dielectric Constant 40
16. Response to Transverse Fields 43
17. The Fluctuation–Dissipation Theorem 47
    References 48

III. The Random Phase Approximation

18. The Self-Consistent Field Method 49
19. Discussion of the Random Phase Approximation 52
20. Static Properties of a Degenerate Plasma 55
21. High-Frequency Properties of a Degenerate Plasma . . . 57
22. Screening in Classical Plasmas 59
23. Plasma Waves in Maxwellian Plasmas 60
    References 62

IV. Comparison with Experiment—Degenerate Plasmas

24. RPA Formula for the Spectrum 64
25. Integrated Spectrum for Characteristic Energy Loss Experiments 65
26. Mean Free Path For Plasmon Emission 66
27. Surface Plasmons 67
28. Angular Variation of Characteristic Energy Loss Spectra . . 68
29. Exhange Corrections—The Generalized Random Phase Approximation 71
30. The Cutoff Angle and Decay of Long-Wavelength Plasmons . 75
31. Effects of Band Structure 77
32. Plasmons in Doped Semiconductors 80
    References 81

V. Light Scattering Experiments

33. Introduction 82
34. Plasmon Light Scattering and Plasmon–Phonon Coupling . 85
35. The Magnetoplasma and Light Scattering Experiments Involving Magnetized Plasmas 90
36. Multicomponent Plasmas 99
37. Light Scattering by Multicomponent Plasmas 107
38. Band Structure Effects in Light Scattering 111
    References 117

VI. Waves in Metals

39. Introductory Remarks 119
40. General Formulation of the Wave Propagation Problem . . 123
    References 125

VII. The Local Limit

41. The Dispersion Relation for Helicons 127
42. The Dielectric Tensor 131
43. Alfven Waves 142
44. Helicon–Phonon Interaction 144
    References 150

VIII. Nonlocal Effects in a Noninteracting Gas

45. The Solution of the Boltzmann–Vlasov Equation 151
46. Helicons in the Nonlocal Regime 157

47. Helicon Propagation at an Angle to the Magnetic Field . . 161
48. Cyclotron Wave Propagation in a Noninteracting Gas . . . 164
49. The "Problem" of Boundaries 191
   References 202

IX. Conduction Electron Spin Resonance in an Electron Gas

50. Hydrodynamic or Long-Wavelength Treatment of CESR . 204
51. CESR in a Bounded Medium 212
   References 219

X. The Interacting Electron Liquid

52. The Phenomenological Formulation of Fermi-Liquid Theory 220
53. Equilibrium Properties of the Interacting Electron Liquid . 227
54. Transport Equations for the Interacting Fermi Liquid . . 232
55. The Electrical Transport Equation for Fermi Liquids . . . 241
56. CW in the Presence of Interactions 245
57. CW Propagating Parallel to the Magnetic Field 252
58. Spin Waves in the Interacting Electron Liquid 256
   References 271

XI. The Microscopic Theory of Fermi Liquids

59. Introduction to the Microscopic Theory of Fermi Liquids . 273
60. The General Formulation of the Microscopic Theory . . . 274
61. Coulomb Interactions and the Landau Scattering Function . 283
62. The Electron–Phonon Interaction and the Landau Scattering Function 287
63. Comparison of Numerical Estimates and Experimental Results 290
   References 294

ADDENDUM: ALKALI METAL PARAMETERS 295
AUTHOR INDEX 297
SUBJECT INDEX 301

Preface

Solid state plasma (SSP) is a phrase used to describe any collection of mobile charge carriers in a solid. Such plasmas occur in metals, semimetals, and semiconductors. In the case of metals, the electron plasma plays a crucial role in determining all properties of the crystal. Such vital features as cohesive energy, crystal structure, electrical properties, and phonon frequencies are controlled by the plasma-like behavior of the conduction electrons. An understanding of such plasmas is a prerequisite, therefore, to any understanding of metallic behavior. Recently, the point of view of plasma physics has increasingly been applied to the problem of conduction electron dynamics in metals. In this approach one studies wavelike, collective excitations of the electronic medium, and seeks to relate their properties to the fundamental features of the interacting electron system. The primary aim of this monograph is to elucidate this relationship—to show how plasma experiments in metals can be used to obtain information about the dense, strongly interacting electron liquid which they contain. For this purpose we will discuss several theories of quantum plasmas in detail, and apply them to a wide variety of experimental situations. These theories are of two sorts—microscopic and phenomenological. The phenomenological (Fermi-liquid) theory is in excellent quantitative agreement with wave experiments. Indeed, it can well be argued that SSP work provides the best proof of its validity for metallic systems. Microscopic SSP theories, on the other hand, are somewhat less satisfactory. Accurate, first principle calculations can only be performed for weakly interacting plasmas —most metallic plasmas do not meet this condition. Nevertheless, the weak coupling model provides a remarkably good qualitative (and in some cases quantitative) description of metallic plasma behavior. The reasons for this success are not entirely clear—the model often works well when considerations of the relative strength of the electron-electron interaction would lead one to expect it to fail. In short, we can explain many experiments with a theory that we do not entirely trust. This is a field in which a new theoretical viewpoint is needed to understand why the weak coupling model works as well as it does, and to understand where and how it ultimately breaks down.

Another, rather different, area of investigation in SSP work is that of semiconductor and semimetal plasmas. These plasmas are quite dilute compared to those found in metals, hence they play almost no role in determining the overall structure of the solid. A Ge crystal, for example, is essentially unchanged (except for electrical properties) when doped from intrinsic to quite high carrier densities. One does not, therefore, study semiconductor plasmas to learn about the structure of such crystals, but as systems of intrinsic interest in their own right. Such systems occur in enormous variety and with ranges of parameters that far exceed those attainable in metals. Moreover, they are the only SSP in which all parameters of interest—plasma frequency, cyclotron frequency, Fermi energy, phonon energy, and band gap—can be made comparable. This combination of properties makes them a particularly flexible medium for testing a wide range of plasma phenomena. They have been extensively studied for this purpose. Since the theoretical methods used to investigate wave propagation in semiconductor plasmas are similar to those used for metals, it is natural to discuss these phenomena, as well, in a monograph of this type.

The name *solid state plasma* suggests that these media are similar to gaseous plasmas. This statement is qualitatively correct, but it should be recognized that there are large *quantitative* differences between the two types of systems. For example, in gaseous plasmas the mean free paths of the various particles are often comparable to the dimensions of the container, while in solid state plasmas this is rarely true. As a consequence the solid state plasma is effectively an infinite medium where boundary conditions play little role, whereas in gaseous plasmas they can be crucial. This difference dooms most attempts to model gas plasma devices (such as fusion machines) by solid state experiments.

Another, related, difference between gaseous and solid state plasmas concerns the ease with which the two types of systems can be driven out of thermal equilibrium. The SSP is a uniform, equilibrium system, closely coupled to the lattice, which can only be strongly excited by relatively drastic perturbations. Small mean free paths for electron–lattice scattering are responsible for this state of affairs. On the other hand, the gaseous plasma is a highly excited (and excitable) state of matter in which particle collisions are infrequent. Nonequilibrium velocity distributions can easily be generated in such plasmas which, in turn, give rise to a host of instabilities.

These distinctions account for an important difference in emphasis between the present work and most others in SSP physics. The authors believe that the fundamental advances in SSP have been concerned with near equilibrium phenomena, particularly the wavelike excitations mentioned earlier. Our monograph is concerned with this aspect of the subject and does not discuss a considerable body of research dealing with instabilities in SSP. The instability studies are inherently quite interesting, but have yet to achieve the interplay between theory and experiment that characterizes the wave work. Moreover, they have been discussed at length in several books and review articles. The situation in SSP is rather different from that which obtains in gas plasma physics. In the latter medium, instabilities are easily generated and have been extensively studied (they are, of course, the central problem in the development of fusion power). Extraordinary care is required to produce laboratory gaseous plasmas that are sufficiently uniform and quiescent to permit the study of small amplitude wave propagation. On the other hand, in SSP instabilities are complicated and difficult to excite. Furthermore, there is no overriding practical reason for wishing to understand their properties. Thus, though both media are termed plasmas, they are actually best viewed as complementary to one another. The SSP is a naturally occurring, uniform system almost ideally adapted to the study of wave propagations; gas plasmas are inherently unstable, often nonuniform systems, best suited to the study of the instabilities to which they are so susceptible.

The ideas discussed in this manuscript have largely been developed during the past decade. The authors have been especially fortunate, during that period, to have been members of a large and active research group concerned with SSP at Bell Telephone Laboratories. Many of the viewpoints presented here evolved in the collaborations of those years, and this manuscript could not have been written without the criticism and encouragement of our colleagues there, and elsewhere. We owe a particular debt of gratitude to G. A. Baraff, S. J. Buchsbaum, D. Fredkin, C. C. Grimes, J. M. Luttinger, C. K. N. Patel, S. Schultz, and W. M. Walsh. They, and others too numerous to mention, have helped bring a measure of unity to the subjects we will discuss.

P. M. PLATZMAN
P. A. WOLFF

I. Characteristics of Solid State Plasmas

1. What Is a Plasma?

A plasma is a collection of relatively mobile, charged particles that interact with one another via Coulomb forces. Intuitively, one thinks of such a system as an electrified gas or fluid, characterized by its great ability to respond to electric and magnetic perturbations. The responsiveness of a plasma is a direct consequence of the mobility of its constituents, and occurs only when the charged particles which comprise it are relatively free. Gas discharges, electrons and holes in semiconductors, and electrons in metals are examples of systems that show plasma-like behavior. However, not all charged-particle systems have this character. A sodium chloride crystal, for instance, is made up of positively and negatively charged (Na^+ and Cl^-) ions, but these are fixed in a crystal lattice and the system shows none of the typical plasma behavior. In this case the ions are too tightly bound to one another by electrostatic forces to move about freely.

One may well ask, when are the particles sufficiently free for plasmalike behavior to occur in a charged-particle system? A precise answer to this question is not known. For the NaCl case one might expect, on intuitive grounds, that the ions would become mobile when their vibrational (thermal) energy was an appreciable fraction of the Coulomb energy binding them in the lattice. This example suggests that the condition for plasmalike behavior should be of the form

$$V/K \lesssim 1, \tag{1.1}$$

where V is the average potential energy of interaction between particles, and K their kinetic energy. We will see later that the ratio V/K *is* the crucial parameter which determines the properties of a plasma. Actually, in the NaCl case the crystal melts (and the ions become mobile) for values of V/K well above unity. Similarly, in metals, plasmalike behavior is observed when V/K is considerably greater than one. Experience, therefore, suggests that the condition for plasmalike behavior should be more like $V/K \gtrsim 10$.

As the preceding examples indicate, charged-particle systems have quite different behavior in the regimes $V/K \ll 10$ and $V/K \gg 10$.

If $V/K \ll 10$, the system is a plasma. Its constituents move freely relative to one another, and there is *no long-range order*. On the other hand, if $V/K \gg 10$ the particles *order*, i.e., crystallize, into a lattice. An ionic crystal, such as NaCl, is one example of this behavior. Another is the dilute, free electron gas, neutralized by a uniform positive background charge. Wigner (*1*) has shown that in the limit $V/K \gg 1$ this system also forms a crystal—in which electrons, rather than ions, are the constituents of the lattice. The essential point, in both cases, is that the Coulomb energy dominates when $V/K \gg 1$. In this limit the problem becomes classical, and the gound state is obtained by seeking a particle configuration that minimizes the electrostatic energy. This configuration is invariably one in which the charged particles are arranged in a periodic array.

These ideas are particularly well illustrated by another example —that of the alkali metals. A sodium crystal, for instance, contains two types of charged particles—mobile conduction electrons, and Na$^+$ ions which are fixed in a lattice. In Na metal the electron gas is degenerate, and therefore has considerable kinetic energy. The ratio V/K for electrons is near unity (actually somewhat greater) in the alkalis. As a consequence, the electron system in these metals is expected to be plasmalike, as observed. The ions, on the other hand, because of their much larger mass, are nondegenerate and have little kinetic energy. For them, the ratio V/K is large. Hence they form a lattice. Moreover, this lattice is sufficiently open that the ion cores do not touch, and it can fairly be said that Coulomb energy is responsible for the crystallization. We see, therefore, that in a single material it is possible to observe both limits of the behavior of charged-particle systems.

In this book our main interest is in *plasmas*, i.e., in charged-particle systems for which $V/K \gtrsim 10$. Solid state plasmas, in fact, span the range from $V/K \ll 1$ to $V/K \simeq 5$. In principle, one would like to understand them all in detail. At present, however, this is not possible. So far, the only plasma problems which have been completely solved are those in which, in some sense, the particle interactions are a small perturbation, i.e., systems for which $V/K \ll 1$. Fortunately, many of the general features of such weakly coupled plasmas seem, also, to occur in more strongly interacting systems ($V/K \simeq 1$).

The condition $V/K \ll 1$ is met in two important cases. The first is that of a dilute, high-temperature plasma, in which the thermal energy ($\tfrac{3}{2}kT$) of the charged particles is large compared to their average energy of interaction via the Coulomb forces. Most gas plasmas are of this type, as well as many plasmas in semiconductors. The second case is that of a cold, dense plasma in which the electron gas is so concentrated as to be degenerate. Here, though there is no thermal energy, electrons acquire kinetic energy because of the exclusion principle. At sufficiently high densities this energy becomes large compared to the Coulomb energy, and it is again legitimate to treat the latter as a perturbation. This situation is approximated (though not particularly well) in metals. Rather surprisingly, it can also be achieved in semimetals and some heavily doped semiconductors. In such materials the Coulomb interaction between carriers is reduced by the large dielectric constant due to polarization of the core electrons, and the Fermi energy enhanced by small effective masses.

Much of this book will be concerned with one or the other of these limiting cases. There are many solid state plasmas that satisfy the condition $V/K < 1$. For others, which do not, the weak-coupling theories will be useful in providing qualitative understanding of plasma properties. These theories illustrate most of the important features of plasma behavior, and are the only ones for which detailed *microscopic* calculations can easily be done. It should be kept in mind, however, that the weak-coupling theories are only approximately valid for plasmas in which $V/K \simeq 1$. There is a definite need for a more rigorous theory of strongly interacting plasmas and, in particular, for a theory of the transition between the plasmalike and crystalline states of charged-particle systems.

2. Conditions for Weak Coupling

For orientation, we commence by estimating when these two ideal cases are achieved. To reach the dilute, high-temperature limit, it is required that the average Coulomb energy of charged particles in a nondegenerate plasma be small compared to kT. The Coulomb energy is roughly $e^2/\epsilon_0 r_0$, where ϵ_0 is the dielectric constant of the medium through which the particles move, and r_0 the mean interparticle spacing. The quantity r_0 is about equal to $n_0^{-1/3}$, where n is the particle density. Thus, the weak-coupling criterion can be written in the form

$$V/K \simeq e^2 n_0^{1/3}/(\epsilon_0 kT) \ll 1. \tag{2.1}$$

In gas plasmas, electron densities are relatively low and temperatures are fairly high. Typical values for an active discharge are $n_0 \simeq 10^{12}$ cm^{-3} and $kT \simeq 5$ eV. The ratio V/K is about 2×10^{-4} in such a medium, so the weak-coupling condition, Eq. (2.1), is satisfied. This is true of almost all gas plasmas.

The situation is somewhat different in solids. The important, nondegenerate solid state plasmas are those which occur in semiconductors. In such crystals, electron densities are usually much higher than those found in gas plasmas (a value $n_0 \simeq 10^{18}$ cm^{-3} is not unusual), and experiments are commonly done at room temperature, or below. The higher densities and lower temperatures combine to increase the ratio V/K in semiconductor plasmas. These effects are to some extent counterbalanced by the fact that in the semiconductors which are commonly used in solid state plasma work, the static dielectric constant ϵ_0 is of order ten. In such materials, at room temperature, Eq. (2.1) is satisfied until n_0 approaches 10^{17} electrons cm^{-3}. Many semiconductors contain fewer carriers than this and are good plasmas in the sense of Eq. (2.1). On the other hand, the condition is more stringent at low temperatures, and most semiconductor plasmas violate it. However, it should be remembered that electron gases in semiconductors usually become degenerate as the temperature is lowered, and thus may qualify as weakly coupled plasmas of the cold, dense type. Several materials actually show such behavior. In any event, it is clear that plasmas in semiconductors span the range from fairly good ones to quite poor ones.

A criterion similar to Eq. (2.1) applies to the cold, degenerate plasma. To reach the weak-coupling limit in this case one requires that the Fermi energy of the electrons (rather than kT) be large compared to their average Coulomb energy. Since the Fermi energy varies as $n^{2/3}$, as compared to the $n_0^{1/3}$ variation of the Coulomb energy, it is clear that any plasma can be driven to the degenerate, weak-coupling limit by compressing it sufficiently. The condition which is the analog, of Eq. (2.1) is

$$e^2 n_0^{1/3}/\epsilon_0 E_F \ll 1, \tag{2.2}$$

where E_F is the Fermi energy. In a free electron gas, which is a good model for the behavior of electrons in the simpler metals, $E_F = \hbar^2 k_F^2/2m^*$ where $k_F = (3\pi^2 n_0)^{1/3}$ is the Fermi wave vector and m^* is the effective mass. It is convenient to rewrite Eq. (2.2) in terms of an effective Bohr radius $a_0^* = \hbar^2 \epsilon_0/m^* e^2$ and another length r_0 which determines the average volume per electron via the relation $(4\pi r_0^3/3) n_0 = 1$. The criterion for weak coupling then becomes (ignoring numerical factors of order unity)

$$r_s \equiv r_0/a_0^* \ll 1. \tag{2.3}$$

the dimensionless parameter r_s is often used to characterize the strength of electron–electron interactions in degenerate plasmas. It is important to notice that the dependence of Eqs. (2.2) and (2.3) on density is exactly the reverse of that of Eq. (2.1). The high-temperature plasma is weakly coupled when it is dilute, whereas in the degenerate plasma this limit is reached at high density.

In most metals the parameter r_s is *not* small compared to unity. It usually lies in the range $2 \leqslant r_s \leqslant 6$. As a consequence, one cannot expect the theory of the dense, weakly coupled plasma to predict quantitatively the behavior of real metals. It is, however, a model whose properties are well understood, and which, in fact, comes closer to describing metallic plasmas than one might suspect from the values of r_s quoted above. It is also the only model of a degenerate plasma whose properties can be calculated in detail—for all frequencies and wave vectors. For this reason, it plays a central role in our understanding of degenerate plasmas. We will study it in considerable detail, and often use it as a guide (though necessarily a qualitative one) in interpreting the behavior of real metals.

Before concluding this discussion of degenerate plasmas, a word should also be said about semimetals and heavily doped, degenerate semiconductors. In such materials electron densities are low (generally in the range 10^{18}–10^{20} cm^{-3}) compared to those found in metals. At first glance, one might expect the coupling parameter r_s in these systems to be large compared to unity. This is not always true however. Such crystals often have big dielectric constants and low effective masses, leading to effective Bohr radii which are large. As a consequence, it is possible to satisfy the weak-coupling condition $r_0/a_0{}^* < 1$ in materials with electron densities of the order of 10^{18}–10^{20} cm^{-3}. Certain semiconductors (InSb, PbTe, etc.) easily satisfy this condition. The semimetal bismuth is an even more striking case. For it, $n_0 \simeq 10^{18}$ cm^{-3}, $\epsilon_0 \simeq 100$, and $m^* \simeq 0.1 m_0$ (the mass given is a rough average of the extremely anisotropic mass tensor). These figures lead to the value $r_s \simeq 0.1$. Thus, the electrons and holes in Bi constitute, at low temperatures, a degenerate plasma which is quite a good one according to the criterion of Eq. (2.3). It is not surprising, therefore, that a great deal of solid state plasma work has been done in this material (2).

3. FERMI LIQUID THEORY—INTRODUCTORY REMARKS

The weak–coupling model, though only approximately valid at the electron densities found in metals, has the virtue of being a rather universal theory in the sense that it permits one to calculate the response of the plasma to perturbations having *arbitrary* variation in space or time. There is another theory of the degenerate plasma which has almost the opposite features. This is the Landau–Silin Fermi-liquid theory (3) which will be discussed in the latter part of the book. The Fermi liquid theory, in contrast to the weak-coupling model, is believed to be exact for all "normal" (systems which have not undergone some type of second order phase transition) Fermion systems. It is derived by a formal summation, to infinite order, of the perturbation series in the parameter r_s. This theory, however, can only be used to calculate the response of the electron liquid to perturbations of low frequency and long wavelength. It is a phenomenological theory which contains unknown functions that measure the strength (and form) of electron–electron and electron–phonon interactions in the degenerate plasma. Some attempts have been made to calculate these functions from first principles. These calculations, though somewhat beyond the scope of this book, are briefly discussed in Chapter XI. In the final analysis, however, most of the Fermi-liquid parameters must be determined by experiment.

Until recently, the experimental determination of the parameters which characterize the Fermi liquid in a metal has been beset with difficulties. The early work dealt with equilibrium properties of the electron gas, such as spin susceptibility, electronic specific heat, and so forth. These quantities are finite in the absence of electron–electron interactions but, according to the Fermi-liquid theory, are modified when interactions are taken into account. Usually, however, these modifications are relatively small—typically about 20% in alkali metals. They are difficult to measure and the experiments suffer from numerous uncertainties. Moreover, the number of effects that can be studied in this way is small. It is difficult, therefore, to use the equilibrium experiments to provide a convincing verification of the theory.

Recently, it has become clear that certain types of plasma waves—in particular, the electromagnetic waves mentioned in the preface—are a far better tool for this purpose. These disturbances are low-frequency, long-wavelength excitations that are naturally described in terms of the Fermi-liquid theory. In some cases (paramagnetic spin waves) the very existence of the waves depends upon explicit electron–electron interaction. Thus, when one measures their characteristics, one directly measures a quantity which does not exist in the absence of interaction, rather than a small correction to a quantity which is finite even in the noninteracting electron gas. These waves have the further advantage of being an essentially spectroscopic tool (with all the precision this implies) for measuring the properties of interacting electrons in metals. They provide the most detailed proof of the validity of the Fermi-liquid theory for metal systems. As a result of recent work with these electromagnetic waves, it is now possible to give a fairly complete specification of the Fermi-liquid parameters in simple metals such as Na or K. This is a new and exciting direction for solid state plasma research which will be described in the second half of the book.

4. Coulomb Energy in a Plasma

Equations (2.1) and (2.3) are sufficient conditions for classical and degenerate charged-particle systems to behave as plasmas. The physics of the two criteria is, of course, the same, namely that the average interaction energy of the particles be small compared to their kinetic energy. When this condition is satisfied, the individual particle–particle interactions are relatively small perturbations and it is possible to calculate the properties of the plasma in detail. It is important to realize, however, that the weakness of *individual* particle–particle interactions does not imply that the *overall* effect of the Coulomb interactions in a plasma need be small. Because of the peculiar, long-range nature of the Coulomb potential, it is possible to generate sizable electric fields in plasmas by perturbing the medium over a large volume, and summing the individually weak contributions of many particles. To achieve such a situation, however, it is necessary that the individual contributions be in phase with one another or, in other words, that the plasma be perturbed in a *coherent* way.

We may illustrate this point by considering the expression for the potential energy of a charged particle system, which is

$$V = \frac{1}{2} \int \frac{\rho(\mathbf{r}) \rho(\mathbf{r}') d^3r \, d^3r'}{\epsilon_0 |\mathbf{r} - \mathbf{r}'|}, \tag{4.1}$$

where $\rho(\mathbf{r})$ is the charge density at position $\mathbf{r}$. We assume that the quiescent plasma is neutral, so that in it $\rho(\mathbf{r}) = 0$. Now let us imagine that by some unspecified means we produce a charge density fluctuation of the form

$$\rho(\mathbf{r}) = \rho_\mathbf{k} e^{i\mathbf{k}\cdot\mathbf{r}} + \text{c.c.} \tag{4.2}$$

The potential energy (per unit volume) can easily be calculated from Eq. (4.1). It is

$$V_\mathbf{k} = |\rho_\mathbf{k}|^2 (4\pi/\epsilon_0 k^2). \tag{4.3}$$

The crucial feature of this formula is its dependence on the wave vector $\mathbf{k}$ and, in particular, the *divergence* of $V_\mathbf{k}$ as $k \to 0$. This divergence is a direct consequence of the long-range nature of the Coulomb interaction and is absent with a potential of finite range. Physically, the divergence represents the Coulomb interaction energy of many particles (an infinite number in the limit $k \to 0$) spread over a large region of space. The occurrence of the divergence has nothing to do with the *strength* of the individual particle–particle interactions, but is a result of its *form*. Thus, even in plasmas that are weakly coupled according to either Eq. (2.1) or (2.3), there is a range of wave vectors (or wavelengths) within which Coulomb energies are large and dominate the behavior of the plasma. It is this special feature of the Coulomb force that gives a plasma its unique properties and makes its study a particularly interesting one.

5. What Parameters Characterize a Plasma?— Screening Lengths

The considerations outlined above suggest—and this viewpoint will be borne out by later detailed calculations—that a plasma is characterized by two quite different modes of behavior, depending upon the wavelength of the probe with which it is studied. When it is perturbed by a short-wavelength (large k) disturbance the Coulomb energy, according to Eq. (4.3), is relatively small. In this limit the weakly coupled plasma behaves as a collection of noninteracting particles. Its response to the perturbation can be calculated by solving the individual, free particle equations of motion, and then summing the result over the velocity distribution. On the other hand, if the perturbation has a sufficiently long wavelength (small k), the Coulomb energy is dominant and forces the plasma to respond in a "collective" way. In this limit the particlelike aspects of the medium are suppressed, and it acts like an electrified fluid or jelly. The properties of this jelly are, in large part, characterized by a dispersion relation whose features are determined by Coulomb fources rather than the particle dynamics.

The graininess which distinguishes a particle system from a continuum also tends to disappear in the long-wavelength limit. Particle density fluctuations, which are a measure of this graininess, are strongly suppressed by the Coulomb interaction. Fluctuations that would be statistically probable in uncharged-particle systems are energetically unfavorable in plasmas. The net effect is that the Coulomb force holds the charge density much nearer to its average value than statistics would dictate, thereby making the medium behave as a uniform continuum rather than a collection of discrete particles.

The two domains of behavior of a plasma are separated by a critical length. The plasma responds collectively to wavelengths greater than the critical length; shorter-wavelength perturbations produce a particle-like response. The critical length is known as the *Debye length* in a classical plasma. In a degenerate plasma it is termed the *Fermi–Thomas length*. These lengths are among the most important parameters which characterize plasmas.

5. PARAMETERS CHARACTERIZING A PLASMA

We may estimate the critical length of a plasma by considering the effect of Coulomb interactions on charge density fluctuations. Particlelike behavior occurs in regimes where statistics play the major role in determining the size of these fluctuations, "collective" behavior in regimes where their size is controlled by the Coulomb energy. To simplify matters, we consider the case of a single-component, classical (Maxwellian) plasma which is neutralized by a stationary, positive background charge. Plasmas of this type occur in doped semiconductors such as n-type GaAs or InSb. Imagine a small subvolume δV of the plasma and consider the probability, P_N, of finding N electrons in δV. If the average number of particles ($\bar{N} = n_0 \delta V$) is large, and if P_N is determined by statistics alone (ignoring Coulomb forces for a moment), this distribution is Gaussian, i.e.,

$$P_N \sim \exp[-(N - \bar{N})^2 / 2\bar{N}]. \tag{5.1}$$

On the other hand, when interactions are included, the P_N's must be modified by the Coulomb energy of the fluctuation. This energy is approximately given by the square of the total charge fluctuation, divided by its radius. Since the plasma is neutral in its equilibrium state ($N = \bar{N}$), we have the following expression for the Coulomb energy of the fluctuations:

$$V_N \simeq e^2 (N - \bar{N})^2 / r, \tag{5.2}$$

where $4\pi r^3 / 3 \simeq \delta V$. At temperature T, the modified probability of finding N particles in δV is

$$P_N \sim \exp\left\{ -\frac{(N - \bar{N})^2}{2\bar{N}} \left[1 + \frac{2\bar{N}e^2}{kTr\epsilon_0} \right] \right\}. \tag{5.3}$$

Here it should be realized that the second term in the exponential, which represents the effect of Coulomb forces on the charge density fluctuation, is given only approximately by this calculation. Correction factors of order unity are to be expected from a more exact treatment. The Coulomb term can be rewritten in the form

$$2\bar{N}e^2 / kTr\epsilon_0 \simeq (8\pi/3)(n_0 e^2 / kT\epsilon_0) r^2 \simeq r^2 / \lambda_D^2, \tag{5.4}$$

where we have introduced the characteristic length λ_D defined by the relation

$$\lambda_D^2 = kT\epsilon_0 / 4\pi n_0 e^2. \tag{5.5}$$

λ_D is the *Debye length*. In subvolumes whose radii are small compared to λ_D, the Coulomb correction to Eq. (5.3) is negligible, and the particle density fluctuations are purely statistical. On the other hand, in the opposite limit $(r \gg \lambda_D)$ the Coulomb term dominates and suppresses density fluctuations that would be statistically probable.

The length defined by Eq. (5.5) has another equally important meaning. It is the *screening length* of the plasma, i.e., the distance to which an external electrostatic field will penetrate it before being counterbalanced by induced fields due to polarization of the medium. To see how this screening occurs let us imagine that a small external charge density, $\rho_{ext}(\mathbf{r})$, is immersed in the electron gas. This is a formal device that is often used to investigate the response of plasmas. The external charge (assumed positive) attracts the electrons in its vicinity and polarizes the plasma as shown in Fig. I-1. As a consequence, the *total*

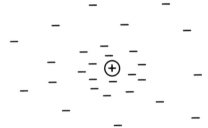

FIG. I-1. Screening of an external positive charge in a plasma.

charge density consists of two parts: the external charge and an induced charge $\rho_{ind}(\mathbf{r})$ due to the departure of the plasma from uniformity. These charge densities combine to generate an electrostatic potential which can be calculated from Poisson's equation:

$$-\nabla^2 \phi = (4\pi/\epsilon_0)(\rho_{ext} + \rho_{ind}). \tag{5.6}$$

Equation (5.6) contains two unknown functions, ϕ and ρ_{ind}. However, in thermal equilibrium, the induced electron density can be related to the potential via the Boltzmann factor. Thus, in a single-component plasma, one has

$$\rho_{ind} = en_0(e^{-e\phi/kT} - 1). \tag{5.7}$$

Substitution of this result into Eq. (5.6) leads to a nonlinear differential equation for ϕ. However, if ρ_{ext} is small (as we have assumed), the induced potential will also be small $(e\phi/kT \ll 1)$, and one may approximate Eq. (5.7) by the simpler relation

$$\rho_{ind} \simeq -n_0 e^2 \phi/kT. \tag{5.8}$$

Combining Eq. (5.8) with Eq. (5.6) yields the linearized equation

$$-\nabla^2 \phi + (4\pi n_0 e^2/kT\epsilon_0)\phi = 4\pi \rho_{\text{ext}}/\epsilon_0. \tag{5.9}$$

Equation (5.9) is a Yukawa equation containing the same characteristic length λ_D as was obtained from our earlier fluctuation argument. This length was first defined by Debye and Hückel (4) in connection with the theory of electrolytic solutions. The solution of Eq. (5.9) is

$$\phi = \int \frac{\exp(-\kappa_D |\mathbf{r} - \mathbf{r}'|)}{|\mathbf{r} - \mathbf{r}'| \epsilon_0} \rho_{\text{ext}}(\mathbf{r}') \, d^3r', \tag{5.10}$$

where $\kappa_D = \lambda_D^{-1}$. Note that the polarization of the plasma converts the bare Coulomb interaction $(1/|\mathbf{r} - \mathbf{r}'| \epsilon_0)$ into a finite range interaction of the form

$$[\exp(-\kappa_D |\mathbf{r} - \mathbf{r}'|)]/|\mathbf{r} - \mathbf{r}'| \epsilon_0. \tag{5.11}$$

This is the famous plasma screening. All plasmas show it. Indeed, it could well be said that a medium is not a plasma unless it *does* screen external, electrostatic fields.

It should be emphasized that Eq. (5.11) applies to the Maxwellian case. Screening also occurs in degenerate and nonequilibrium plasmas; but the screening function may have a quite different form from that of Eq. (5.11). In degenerate plasmas, the length which characterizes the screened potential (5) is the Fermi–Thomas length, defined by

$$(1/\lambda_{\text{FT}})^2 \equiv \kappa_{\text{FT}}^2 = 6\pi n_0 e^2/\epsilon_0 E_F. \tag{5.12}$$

Equation (5.12) is a natural generalization of Eq. (5.5) obtained by making the replacement $\frac{3}{2}kT \to E_F$. For values of $|\mathbf{r} - \mathbf{r}'|$ comparable to λ_{FT}, the screened potential in a degenerate plasma falls off roughly exponentially. However, for large values of $|\mathbf{r} - \mathbf{r}'|$, the potential has a long, nonexponential, oscillatory tail. This rather surprising feature of the degenerate plasma is a consequence of the exclusion principle, and has no counterpart in the classical case.

Numerically, screening lengths in doped semiconductors (λ_D, or λ_{FT} when that is appropriate) lie in the range 10^{-4}–10^{-6} cm, depending upon the impurity concentration. In metals, the plasma is much more dense and invariably degenerate. The typical screening length for a metal is $\lambda_{\text{FT}} \simeq 10^{-8}$ cm.

6. THE CHARACTERISTIC FREQUENCY

The characteristic lengths discussed above separate two wavelength regions. In one of them ($\lambda > \lambda_D$) a plasma responds collectively; in the other ($\lambda < \lambda_D$) its behavior is single-particle like. One may well ask whether there is a comparable separation in the time (or frequency) domain. This is indeed the case. Every plasma has a characteristic frequency—the *plasma frequency*—which sets the scale of its response to time-varying perturbations. The plasma frequency is the frequency of the basic collective mode of a charged medium, which is a longitudinal, charge-density wave. In the limit $\lambda \gg \lambda_D$ (or $\lambda \gg \lambda_{FT}$) the properties of this mode can be calculated quite simply.

Imagine an electron gas of density n_0, which is neutralized by a fixed background charge. This model is a good one for the simpler metals, such as Al, Na, or K, as well as many doped semiconductors. Now let us consider the behavior of a small-amplitude density fluctuation in this plasma. We may write the electron density as

$$n(\mathbf{r}, t) = n_0 + n_1(\mathbf{r}, t), \tag{6.1}$$

where $n(\mathbf{r}, t)$ is the density at position $\mathbf{r}$ and time t, and $|n_1|/n_0 \ll 1$. The perturbation n_1 produces electrostatic fields that cause currents to flow in the plasma. The particle current $\mathbf{\Gamma}(\mathbf{r}, t)$ must obey the continuity equation

$$\partial n_1/\partial t + \nabla \cdot \mathbf{\Gamma} = 0. \tag{6.2}$$

In addition, the electrostatic field satisfies Poisson's equation

$$\nabla \cdot \mathbf{E} = 4\pi e n_1. \tag{6.3}$$

Finally, we need a constitutive equation relating $\mathbf{\Gamma}$ to $\mathbf{E}$. This is the difficult part of the problem and the point at which, for the present, we will make a simplifying assumption. Consider, in particular, the limit of long wavelengths. An electron in the plasma then sees a uniform, oscillating field due to the plasma wave and its equation of motion is

$$m\, d\mathbf{v}/dt \equiv m\dot{\mathbf{v}} = e\mathbf{E}. \tag{6.4}$$

Since $\mathbf{\Gamma} \simeq n_0 \mathbf{v}$, the constitutive equation in this limit becomes

$$\partial \mathbf{\Gamma}/\partial t = n_0 e \mathbf{E}/m. \tag{6.5}$$

Combining Eqs. (6.2)–(6.5) yields a harmonic oscillator equation,

$$\partial^2 n_1/\partial t^2 + \omega_p^2 n_1 = 0, \tag{6.6}$$

describing the electron density fluctuation in a long-wavelength plasma oscillation. The characteristic frequency (6) is

$$\omega_p = (4\pi n_0 e^2/m)^{1/2}. \tag{6.7}$$

Equation (6.6) has been derived in a simple way but, in the long-wavelength limit, can be shown to be rigorously correct even if one takes into account electron–electron interactions (which are ignored above). The physical origin of the oscillation is also fairly clear from this derivation. A density perturbation in the plasma generates electrostatic fields [Eq. (6.3)]. The fields, in turn, cause currents to flow [Eq. (6.5)] in such a direction as to smooth out the original perturbation. However, because of the electrons' inertia [Eq. (6.4)], this process tends to "overshoot"—regions which were originally negatively charged become positively charged, and vice versa. The process then reverses and the net result is the oscillation described by Eq. (6.6). From this description we see that the spring constant of this oscillator arises from the Coulomb forces between electrons, and its inertia is just their mass density.

The frequency ω_p separates high- and low-frequency domains, within which a plasma has quite different behavior. Electrostatic fields of frequency lower than ω_p are screened, whereas the plasma has little effect on higher-frequency fields. There is a corresponding behavior in the response to transverse fields—a plasma reflects electromagnetic fields with frequency below ω_p, but transmits higher-frequency ones.

It is interesting to estimate values of ω_p for solid state plasmas. In metals electron densities are high, and ω_p is very large—typically about 10^{16} rads sec^{-1}. The "quantum" of plasma oscillation ($\hbar\omega_p$) is usually, about 10 eV (15.8 eV in Al, e.g.). This energy is far above thermal energy so plasma waves (often called plasmons) are not normally excited in metals. To study them, one must use a probe sufficiently energetic to excite them. High-energy electrons are an almost ideal tool for this purpose. In a typical experiment of this type, an electron beam is fired through a thin foil (usually less than 500Å thick) of the metal to be investigated. The beam is monoenergetic to within a fraction of an electron volt, and has an energy of several kilovolts or greater. After the electrons have passed through the foil their energy spectrum is measured with high resolution. A typical characteristic energy loss spectrum (number of electrons versus energy lost) is shown in Fig. I-2 for the case of the metal aluminum. It is clear from this figure that the electrons have lost energy in integral multiples of a basic quantum, which in Al and other simple metals is almost exactly $\hbar\omega_p$ [Eq. (6.7)]. Thus, there seems little question that what is observed in the characteristic energy loss experiments is multiple excitation of plasma quanta in the metal film by the fast electron beam.

Plasmons in metals were first observed (7) about twenty years ago—this was really the start of solid state plasma work—in such an experiment.

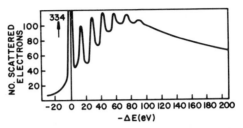

Fig. I-2. The characteristic energy loss spectrum of Al [after Ruthemann and Lang (7)].

The experiment has since been repeated many times and, in recent years, has been greatly refined to make possible the measurement of plasmon dispersion relations and damping rates (8). We will discuss these experiments, which have played an important role in supporting the currently accepted picture of the interacting electron gas, in Chapter IV.

Plasma waves also occur in semiconductors. In fact, in doped semiconductors, there are two sorts of plasma modes—a high-frequency oscillation, involving all the valence electrons; and a low-frequency mode, in which only the conduction electrons participate. The high-frequency mode has been studied by characteristic energy loss experiments, just as in metals. For example, in silicon the characteristic energy loss spectrum (9) is similar to that observed in Al (Fig. I-2). The measured plasma frequency agrees with that calculated from Eq. (6.7), assuming *four* electrons per silicon atom. The energy ($\hbar\omega_p$) of such a wave is well above the band gaps of the silicon crystal. At such frequencies the valence electrons are essentially free, and hardly feel the effect of the periodic crystal potential. This statement accounts for the similarity of the characteristic energy loss spectra of metals and semiconductors, despite the fact that, at low frequencies, these materials have quite different electrical properties. We will discuss this situation in more detail in Chapter IV.

In a semiconductor with a single, isotropic conduction band (such as InSb or GaAs), the expression for the *low-frequency plasma mode* takes the form

$$\omega_p^2 = 4\pi n_0 e^2 / m^* \epsilon_0 , \qquad (6.8)$$

where m^* is the effective mass of the carriers, n_0 is the electron density, and ϵ_0 the static dielectric constant of the undoped crystal. This formula is valid when ω_p is below the reststrahl frequency of the crystal. (We will see later what happens when this is not the case.) The frequency defined by Eq. (6.8) varies, of course, with doping level n_0 in the crystal.

Typically, the low-frequency plasma oscillation has energy in the range $\hbar\omega_p \lesssim 0.01$ eV. Such energies are too small to be conveniently detected in an electron scattering (characteristic energy loss) experiment. Instead, light scattering is often used to study these excitations. Semiconductors are relatively transparent to light waves whose frequency ω lies in the range $\omega_p < \omega < E_G/\hbar$, where E_G is the energy gap of the crystal. This band of frequencies generally extends into the portion of the infrared spectrum where powerful laser sources are available. As a consequence, light scattering, using such sources as the Nd:YAG and CO_2 lasers, has become a powerful tool for probing the structure of conduction electron plasmas in semiconductors. Such experiments will be discussed in detail in Chapters II and V. They are the analog, for low-frequency plasma waves, of the characteristic energy loss experiments which are used to study high-frequency plasma waves. The light scattering technique has been used to study a number of low-frequency collective modes in semiconductor plasmas, as well as to measure electron velocity distributions. These semiconductor plasmas are particularly interesting because they occur in enormous variety, and because their properties can be strongly influenced by applied fields or quantum-mechanical effects.

7. Other Plasma Parameters

In the preceding sections, we have discussed the lengths and frequencies that characterize the response of a plasma to electrostatic perturbations. Another important frequency enters the problem when the plasma is placed in a static magnetic field $\mathbf{H}_0$. This frequency is the cyclotron frequency, given by

$$\omega_c = eH_0/m^*c. \tag{7.1}$$

It is the angular frequency of the electrons helical motion about the magnetic lines of force. For fields of the order of 10kG and $m^* \simeq m$, it is a frequency that lies in the microwave range i.e., $\omega_c \simeq 3 \times 10^{10}$ Hz.

The presence of a magnetic field greatly alters the response of a plasma to low-frequency ($\omega \lesssim \omega_c$) perturbations. This is a dramatic effect which makes it possible for a magnetized plasma to support low-frequency, propagating electromagnetic waves, whereas an unmagnetized one cannot. To understand the difference between the two cases, it is convenient to return to Eq. (6.5) which (in the long-wavelength limit) determines the current induced in an unmagnetized plasma by a perturbing electric field. If the field has frequency ω, the current is

$$\mathbf{j} = e\mathbf{\Gamma} = in_0 e^2 \mathbf{E}/m\omega. \tag{7.2}$$

The corresponding plasma polarization **P** defined as

$$\mathbf{P} = ie\mathbf{\Gamma}/\omega \tag{7.3}$$

is then

$$\mathbf{P} = -n_0 e^2 \mathbf{E}/m\omega^2. \tag{7.4}$$

We may define a plasma conductivity σ and dielectric constant ϵ by the relations

$$\sigma/i\omega = \lim_{E \to 0} (P/E) = -n_0 e^2/m\omega^2 \tag{7.5}$$

and

$$\epsilon = 1 + 4\pi\sigma/i\omega = 1 - \omega_p^2/\omega^2. \tag{7.6}$$

The important point is that, in the absence of a magnetic field, $\sigma/i\omega$ is negative, indicating that the electron's displacement is 180° out of phase with the driving field. As a result, for $\omega < \omega_p$, ϵ is also negative. Electromagnetic waves cannot propagate in a medium for which $\epsilon < 0$, but are totally reflected by it (the waves are evanescent). These facts account for the high reflectivity of most metals in the visible and near infrared ranges.

The quantity $\sigma/i\omega$ is always negative in unmagnetized plasmas because the electrons in such a medium are completely free. As soon as the electrons are constrained, however, their response becomes more complicated. A harmonically bound electron, for example, has a characteristic frequency; it responds *in phase* to perturbations below this frequency, and *out of phase* to higher-frequency ones. With this idea, we can begin to understand why a magnetic field has such a drastic effect on the properties of a plasma. The essential point is that the field confines the electron's motion in the plane perpendicular to it. There is a restoring force and a characteristic frequency ω_c. The electron responds in phase (and thereby generates a positive contribution to $\sigma/i\omega$) to frequencies below ω_c. Clearly, however, this effect can only take place if the electric field has an appreciable component in the plane perpendicular to $\mathbf{H}_0$. We may anticipate, therefore, that the polarizability of a magnetized plasma is highly anisotropic. Many of the peculiarities of such media arise from this fact.

Later in the book we will thoroughly discuss the problem of wave propagation in magnetized plasmas. To some extent, this can be considered the main theme of the book. The problem is a complicated one, however, and we do not wish to enter into details now. It will suffice to quote from the later work a single formula which illustrates the point we have made. This is the expression for the effective dielectric constant of a plasma when it supports a circularly polarized electromagnetic wave propagating parallel to H_0 [see Eq. (42.15) of Chapter VII]:

$$\epsilon_+ = 1 - [\omega_p^2/\omega(\omega - \omega_c)] \qquad (7.7)$$

Notice that ϵ_+ is positive if $\omega < \omega_c$, i.e., if the applied frequency is below the electron's characteristic frequency. In this range the plasma can support a circularly polarized, propagating electromagnetic wave. These waves are known as helicons. They are the solid state analog of the "whistler" modes observed in the ionosphere. Such waves can be visualized as torsional oscillations of the magnetic lines of force in the plasma. They have been studied in a wide variety of solid state plasmas, and their properties are now well understood.

The helicon wave is just one example (probably the simplest) of a hierarchy of propagating electromagnetic modes that occur in magnetized plasmas. These excitations are all transverse in nature, and owe their existence to the fact that the magnetic restoring forces on an electron can cause its motion to be *in phase* with an applied electric field. In the past few years, a variety of these modes have been identified in solid state plasmas, and are now being used as probes of their structure.

It is important to realize that the electromagnetic modes of a (magnetized) plasma are quite different in character from the plasma waves discussed previously. The electromagnetic excitations can only propagate at low frequencies ($\omega \lesssim \omega_c$); they are transverse waves; and their properties depend strongly upon the direction of propagation relative to the dc magnetic field. By contrast, the plasma waves (at least in metals) have very high frequencies and are longitudinal. Their dispersion relations are essentially isotropic and are determined by Poisson's equation (plus the constitutive relations), whereas the complete set of Maxwell equations is needed to describe the electromagnetic modes.

These facts force one to use entirely different experimental methods in studying the two types of modes. The low-frequency, electromagnetic waves have been investigated with microwave or radio-frequency measurements. These experiments have much in common with the nuclear magnetic resonance and electron spin resonance techniques. The plasma waves, on the other hand, can only be excited with an energetic probe. Moreover, they are longitudinal in nature and for this reason do not directly couple (except in rather special circumstances) to electromagnetic perturbations. Their behavior is best studied via electron or light scattering experiments.

In the derivation of Eqs. (7.5) or equivalently Eq. (7.7), it was tacitly assumed that the electrons in the plasma were completely free, i.e., that in the absence of external perturbations, they would move in straight lines or helices forever. This is seldom true in real plasmas, particularly those that occur in solids. Electrons in such media are moving through a crystal lattice which inevitably contains defects and thermal disorder (phonons). These imperfections scatter the electrons, thereby destroying the current induced in the plasma by external fields. The effect is particularly bad when one is trying to study the propagation "windows" which are created by a static magnetic field. The windows, as we have seen, have their origin in the fact that the electron motion has a characteristic frequency ω_c. This frequency can be well defined only if the electron does not scatter before making several turns of the helix—a criterion which may be written in the form

$$\omega_c \tau \gg 1. \tag{7.8}$$

Here τ is the collision time of electrons with defects or impurities. This condition is a prerequisite for most low-frequency solid state plasma experiments. It introduces another characteristic frequency τ^{-1} into our problem.

From the practical point of view, the experimentalist almost invariably tries to makes the collision frequency τ^{-1} as small as possible. In metals, this aim is achieved by using pure samples at low temperatures. Low temperatures eliminate phonon scattering, and purity reduces the number of defects. Electron–lattice scattering times as long as 10^{-9} sec are regularly achieved by these methods. The corresponding values of $\omega_c \tau$ can be as large as 50–100. Such plasmas are nearly lossless, and are well described by a theory in which collisions are ignored.

The situation is less favorable in doped semiconductors. In such materials, impurities are required to produce the plasma and these inevitably give rise to scattering. The scattering is particularly strong at low temperatures. Electron–lattice collision times in doped semiconductors usually are about 10^{-12} sec, and seldom can be made longer than 10^{-11} sec. Fortunately, in some crystals there is a nearly corresponding reduction in the effective mass of the carriers. For example, in InSb the electron effective mass $m^* \simeq 0.015m$. For a given magnetic field, the cyclotron frequency is scaled up by nearly a factor of one hundred. As a consequence, one can still achieve the condition $\omega_c \tau \gg 1$ in this and several other doped semiconductors.

There is one final point that should be mentioned in connection with the scattering problem. This concerns the effect of electron–electron collisions, which were ignored in the estimates of collision times made above. In degenerate plasmas electron–electron scatterings are, to a large extent, frozen out by the exclusion principle. Only a tiny fraction of the electrons ($\sim kT/E_\mathrm{F}$) can partake in such collisions. But more important—and this comment applies to both classical and degenerate plasmas—electron–electron collisions conserve the total current and thus do *not* appear as damping terms in equations of motion, such as Eq. (6.5). Thus, even though the electron density is exceedingly high in solid state plasmas, electron–electron collisions play a relatively minor role in determining their properties. This is an important point, to which we will return later.

References

1. E. P. Wigner, *Phys. Rev.* **46**, 1002 (1934); *Trans. Faraday Soc.* **34**, 678 (1938); see also D. Pines, "Elementary Excitations in Solids," p. 91. Benjamin, New York, 1963.
2. The properties of bismuth are reviewed in an article by L. A. Fal'kovskii, *Sov. Phys.— Usp.* **11**, 1 (1968). Another recent article is that of R. T. Isaacson and G. A. Williams, *Phys. Rev.* **177**, 738 (1969). These papers contain extensive bibliographies.
3. An excellent discussion of the Fermi liquid theory is given by D. Pines and P. Nozières, "The Theory of Quantum Liquids," Vol. I, Chapters 1 and 3. Benjamin, New York, 1966. This reference contains a detailed bibliography.
4. P. P. Debye and E. Hückel, *Phys. Z.* **24**, 185 (1923).
5. Screening in degenerate plasmas is discussed in terms of the Fermi-Thomas theory by N. F. Mott and H. Jones, "Theory of Metals and Alloys." Dover, New York, 1958. However, the detailed form of the screening function was not worked out until relatively recently (see Chapter III).
6. The plasma frequency was first defined by L. Tonks and I. Langmuir, *Phys. Rev.* **33**, 195 (1929). Their work concerned electron density oscillations in gas discharges. The realization that similar oscillations could occur in crystals came much later. Key papers in this development are: D. Bohm and D. Pines, *Phys. Rev.* **85**, 338 (1951); **92**, 609 (1953); D. Pines, *ibid.* **92**, 626 (1953).
7. G. Ruthemann, *Ann. Phys.* **2**, 113 (1948); W. Lang, *Optik (Stuttgart)* **3**, 233 (1948).
8. Work on characteristic energy losses is reviewed in an article by H. Raether, *in* "Springer Tracts in Modern Physics," Vol. 38, p. 84. Springer-Verlag, Berlin and New York, 1965. More detailed references are given in Chapter 4.
9. L. Marton, L. B. Leder, and H. Mendlowitz, *Advan. Electron. Electron Phys.* **7**, 183 (1955); H. Dimigen, *Z. Phys.* **165**, 53 (1961).

II. Theory of Important Plasma Experiments

In Chapter I plasmas were described in a qualitative way, and two experimental techniques that are used to study their properties—scattering and transmission of low-frequency electromagnetic waves—were briefly discussed. These experiments are the primary tools for investigating solid state plasmas. It is important, therefore, to understand exactly which features of the many-body system they measure. To answer this question, we next consider the theory of such experiments.

8. Qualitative Discussion of Scattering Experiments

In the scattering experiments that are commonly used to study solid state plasmas, the probing beam may consist of fast electrons or electromagnetic radiation of various frequencies (IR, visible, X rays). For our immediate purpose, which is to investigate the kinematics of this process, the nature of the radiation is not important. The geometry of a typical scattering experiment is illustrated in Fig. II-1. Here a primary beam,

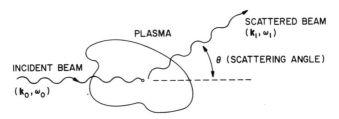

Fig. II-1. Schematic diagram of a typical scattering experiment.

of wave vector $\mathbf{k}_0$ and frequency ω_0, impinges on the plasma; a small fraction of the incident energy is scattered in other directions. The scattered radiation is characterized by its own wave vector $\mathbf{k}_1$ and frequency ω_1. Usually, the frequency ω_1 is different from ω_0. Scattering is caused by fluctuations in the many-body system. These fluctuations are moving; hence, there is a Doppler shift $\omega = \omega_0 - \omega_1$ of the scattered radiation. The fluctuations can be ascribed to particlelike or collective modes of the plasma. In either case, there is a spectrum of scattered

8. DISCUSSION OF SCATTERING EXPERIMENTS

frequencies which is rich in information concerning the structure of the many-body system.

If the primary beam interacts relatively weakly with the plasma, the scattering process can be discussed with the Born approximation; i.e., the coupling between the probe beam and the many-body system can be treated in lowest-order perturbation theory. These statements are true irrespective of the strength of the interactions *within* the many-body system. Thus, the scattering technique can equally well be used to study weakly or strongly interacting systems. Moreover, when the Born approximation is valid, it can be shown that the scattering experiment determines an important property of the many-body system, *the pair correlation function*. The ability to measure directly such a central feature of an interacting system accounts for the power and generality of the scattering technique. Needless to say, in most applications experimentalists try to do their measurements in such a way that the Born approximation is valid. This condition will be satisfied in all of the scattering experiments we discuss. In the case of the characteristic energy loss experiments, it is achieved by using energetic electron beams (as high as 50 keV in recent work) and thin targets. Light scattering experiments must be performed in such a way that the plasma transmits the primary wave, without strongly scattering or absorbing it. In semiconductor plasmas these condition are generally equivalent to the restrictions

$$\omega_\mathrm{p} < \omega_0 < E_\mathrm{G}/\hbar \tag{8.1}$$

that were mentioned in Chapter I.

With this background, we can now ask how a plasma scatters radiation for various values of k. In most cases the scattering is nearly elastic ($|\mathbf{k}_1| \simeq |\mathbf{k}_0|$) so (see Fig. II-1)

$$k \simeq 2k_0 \sin(\theta/2). \tag{8.2}$$

Thus, an equivalent question is to ask how the cross section varies with scattering angle θ. By varying k (or θ), one varies the spatial "resolution" with which the plasma is probed. Scattering with wave vector transfer k is caused by fluctuations of size $1/k$. Its intensity determines the amplitude of these fluctuations in the medium.

As may be anticipated from the discussion of Chapter I, there are two different regimes in plasma scattering experiments, depending upon whether $k\lambda_\mathrm{D}$ (or $k\lambda_\mathrm{FT}$ in the degenerate case) is large or small compared to unity. In the large k limit, waves scattered from different particles in the many-body system *do not interfere with one another*. Each particle scatters independently of the others; the result is a spectrum (called the *single-particle spectrum*) which reflects the individual particle motions. On the other hand, when k is small there is interference between waves scattered from different particles. The amplitude of the total scattered wave is then very sensitive to the relative positioning of the particles, i.e., to correlation in the many-body system. This limit is known as the *collective regime*.

The form of the single-particle spectrum ($k\lambda_D \gg 1$ limit) can be understood by utilizing the idea that the electrons in the plasma scatter independently. An electron of velocity **v** scatters radiation with a Doppler shift $\omega \simeq \mathbf{k} \cdot \mathbf{v}$. The complete spectrum is obtained by summing the intensities (not amplitudes) scattered by individual electrons over the velocity distribution. As an example, we show in Fig. II-2 the sort of

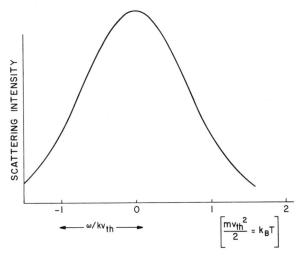

Fig. II-2. The single-particle spectrum, via light scattering, for a single-component, Maxwellian plasma.

spectrum one observes, via light scattering, from a Maxwellian plasma in the limit $k\lambda_D \gg 1$. Note that the scattered intensity $I(\omega)$ peaks at $\omega = 0$ and has a width $\Delta\omega \simeq kv_{th}$, where v_{th} is the thermal velocity of the electrons. A measurement of the width determines the electron temperature. In nonequilibrium situations, the shape of the $I(\omega)$ vs. ω curve can be used to deduce the electron velocity distribution.

In the opposite limit ($k\lambda_D \ll 1$) the spectrum is dominated by collective effects. Single-particle scattering is suppressed (at least in a single-component plasma) as a result of the phase interference mentioned above. Instead of a Doppler-broadened peak centered at $\omega = 0$, the spectrum now consists of two fairly sharp lines, corresponding to Stokes and anti-Stokes Raman scattering from plasma waves in the electron gas (Fig. II-3). The position of these lines determines the frequency $\omega(\mathbf{k})$ of plasma waves of momentum $\mathbf{k}$.

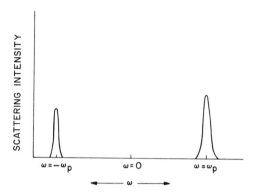

FIG. II-3. The collective spectrum for a single-component plasma.

It is important to realize that the two types of spectra illustrated above (Figs. II-2 and II-3) reveal quite different aspects of the many-body system. For example, the single-particle spectrum is determined by the electron velocity distribution, and can be used to measure it, whereas the collective spectrum is nearly independent of the velocity distribution. Plasma waves, on the other hand, show up as distinct features in the collective regime, but do not affect the single-particle spectra.

9. Analysis of Electron Scattering Experiments

If the coupling between the probing beam and the target is weak, the scattering rate can be calculated in Born approximation, using wave functions which are products of many-body wave functions for the scattering system, and plane waves for the incoming and outgoing particles. Thus, the rate of scattering[†] from an initial state $\psi_0 e^{i\mathbf{k}_0 \cdot \mathbf{r}}$ to a final state $\psi_f e^{i\mathbf{k}_1 \cdot \mathbf{r}}$ is

$$R_{0f} = 2\pi \{|\langle \psi_f e^{i\mathbf{k}_1 \cdot \mathbf{r}} | V(\mathbf{r}) | \psi_0 e^{i\mathbf{k}_0 \cdot \mathbf{r}}\rangle|^2 \, \delta(E_0 + \omega_0 - E_f - \omega_1)\}. \quad (9.1)$$

Here E_0, E_f and ψ_0, ψ_f are energies and wave functions of the many-body system; ω_0 and ω_1 are the energies of the scattering particle. $V(\mathbf{r})$ is the

---

[†] We use units such that $\hbar = 1$.

coupling between the probing beam and the plasma. Its form depends, of course, upon the nature of the probe. One of the simplest interactions is that of a fast electron beam. To be specific, we will develop the theory for this case and later modify it to discuss other processes, especially light scattering.

The coupling between an energetic (but nonrelativistic) electron beam and a plasma takes place via the Coulomb interaction (1). In a single-component plasma—which is a good model for the behavior of many solid state plasmas—it has the form

$$V = \sum_i e^2/|\mathbf{r} - \mathbf{r}_i|, \tag{9.2}$$

where the $\mathbf{r}_i$ label positions of electrons in the plasma and $\mathbf{r}$ is the position of the incoming high energy electron. The latter is moving rapidly compared to the other electrons so we can treat it as a distinguishable particle, ignoring the possibility of exchange collisions. This is an excellent approximation (1) for the small-angle scatterings which are of primary interest in determining the structure of the plasma.

One can now perform the $\mathbf{r}$ integration in Eq. (9.1). The result is the Fourier transform,

$$V_k \equiv 4\pi e^2/k^2 = \int e^{i\mathbf{k}\cdot\mathbf{r}}(e^2/r)\,d^3r, \tag{9.3}$$

of the potential. Here $\mathbf{k} = (\mathbf{k}_0 - \mathbf{k}_1)$ is the momentum transfer to the plasma. With the aid of Eq. (9.3), we rewrite Eq. (9.1) in the form

$$R_{0f} = 2\pi \left\{ (4\pi e^2/k^2)^2 \left| \langle \psi_f | \sum_i (e^{i\mathbf{k}\cdot\mathbf{r}_i}) | \psi_0 \rangle \right|^2 \delta(E_0 + \omega - E_f) \right\}, \tag{9.4}$$

where $\omega = \omega_0 - \omega_1$. The operator whose matrix element appears in Eq. (9.4) is the Fourier transform of the *electron density operator* for the plasma:

$$n_\mathbf{k} \equiv \sum_i (e^{-i\mathbf{k}\cdot\mathbf{r}_i})$$

$$= \int e^{-i\mathbf{k}\cdot\mathbf{r}} \sum_i [\delta(\mathbf{r} - \mathbf{r}_i)]\, d^3r$$

$$\equiv \int e^{-i\mathbf{k}\cdot\mathbf{r}_n} n(\mathbf{r})\, d^3r. \tag{9.5}$$

Note that Eq. (9.4) depends upon the variables $\mathbf{k} = \mathbf{k}_0 - \mathbf{k}_1$ and $\omega = \omega_0 - \omega_1$, rather than $\mathbf{k}_0$, $\mathbf{k}_1$, ω_0, and ω_1 separately. This is often the case when scattering can be treated in the Born approximation. In

9. ANALYSIS OF SCATTERING EXPERIMENTS

any event, ω and $\mathbf{k}$ are the important variables in the problem since they determine the momentum ($\hbar\mathbf{k}$) and energy ($\hbar\omega$) of whatever excitation is created in the plasma.

In most experiments, one does not measure R_{0f} but the *total transition rate* consistent with a given final energy for the scattered particle. This quantity is

$$R = 2\pi(4\pi e^2/k^2)^2 \sum_f \{|\langle\psi_f | n_{-\mathbf{k}} | \psi_0\rangle|^2 \,\delta(E_0 + \omega - E_f)\}. \tag{9.6}$$

The sum over final states f appearing in Eq. (9.6) may be carried out formally by using the representation

$$\delta(E_0 + \omega - E_f) = (2\pi)^{-1} \int_{-\infty}^{\infty} e^{i(E_0+\omega-E_f)t} \, dt. \tag{9.7}$$

Thus

$$R = (4\pi e^2/k^2)^2 \int_{-\infty}^{\infty} dt \, e^{i\omega t} \sum_f [\langle\psi_0 | n_{\mathbf{k}} | \psi_f\rangle e^{i(E_0-E_f)t} \langle\psi_f | n_{-\mathbf{k}} | \psi_0\rangle]$$

$$= (4\pi e^2/k^2)^2 \int_{-\infty}^{\infty} dt \, e^{i\omega t} \sum_f [\langle\psi_0 | e^{iHt} n_{\mathbf{k}} e^{-iHt} | \psi_f\rangle\langle\psi_f | n_{-\mathbf{k}} | \psi_0\rangle]$$

$$= (4\pi e^2/k^2)^2 \int_{-\infty}^{\infty} dt \, e^{i\omega t} \langle\psi_0 | n_{\mathbf{k}}(t) \, n_{-\mathbf{k}}(0) | \psi_0\rangle, \tag{9.8}$$

where H is the Hamiltonian of the many-body system and

$$n_{\mathbf{k}}(t) \equiv e^{iHt} n_{\mathbf{k}} e^{-iHt}. \tag{9.9}$$

Finally, if there is a distribution of initial states, we obtain the total scattering rate by taking an ensemble average of Eq. (9.8). The result is

$$R_{\text{total}} = (4\pi e^2/k^2)^2 \int_{-\infty}^{\infty} dt \, e^{i\omega t} \langle n_{\mathbf{k}}(t) \, n_{-\mathbf{k}}(0)\rangle, \tag{9.10}$$

where the angular brackets indicate a weighted average over initial states. Equation (9.10) is an important result, that we will often use in the subsequent discussion.

The quantity

$$S(\mathbf{k}, \omega) \equiv (2\pi)^{-1} \int_{-\infty}^{\infty} dt \, e^{i\omega t} \langle n_{\mathbf{k}}(t) \, n_{-\mathbf{k}}(0)\rangle \tag{9.11}$$

which appears in Eq. (9.10) is called the *structure factor* (2). It is the space–time Fourier transform of the electron density–density correlation function:

$$S(\mathbf{k}, \omega) = (2\pi)^{-1} \int_{-\infty}^{\infty} e^{i\omega t} \, dt \int\int e^{-i\mathbf{k}\cdot(\mathbf{r}-\mathbf{r}')} \langle n(\mathbf{r}, t) \, n(\mathbf{r}', 0) \rangle \, d^3r \, d^3r'. \qquad (9.12)$$

This fact indicates that *electron density fluctuations* are responsible for the scattering of an electron beam. More generally, in a multicomponent plasma, it is *charge density fluctuations* that cause the scattering.

Equation (9.10) is an elegant result that was first derived by Van Hove (3). However, the idea that correlation functions determine scattering cross sections is an old one, dating back to Einstein's (4) work in 1910. Equation (9.10) is the product of two factors: the structure factor, which determines the structure of the many-body system; and the Fourier transform $(4\pi e^2/k^2)^2$, of the interparticle potential. Such a separation usually occurs in scattering processes which can be described by the Born approximation.

10. Scattering from a Noninteracting Electron System

Though Eq. (9.10) appears a bit formidable, one can easily evaluate it in a simple case—that of a *noninteracting* electron gas. In second quantized notation, the electron density operator has the form

$$n_\mathbf{k} = \sum_\mathbf{p} (c_\mathbf{p}^+ c_{\mathbf{p}+\mathbf{k}}). \qquad (10.1)$$

Here the c's are the usual annihilation and creation operators. The Hamiltonian of the system is

$$H_0 = \sum_\mathbf{p} [c_\mathbf{p}^+ c_\mathbf{p} E(p)], \qquad (10.2)$$

where $E(p) \equiv p^2/2m$ is the single-particle energy. Since

$$[c_\mathbf{p}^+, c_{\mathbf{p}'}]_+ = \delta(\mathbf{p} - \mathbf{p}'), \qquad (10.3)$$

one can show that

$$c_\mathbf{p}(t) = e^{iH_0 t} c_\mathbf{p} e^{-iH_0 t} = e^{-iE(p)t} c_\mathbf{p}. \qquad (10.4)$$

From this result, it follows that

$$n_\mathbf{k}(t) = \sum_\mathbf{p} c_\mathbf{p}^+ c_{\mathbf{p}+\mathbf{k}} \exp\left\{ i \left[\frac{p^2}{2m} - \frac{(\mathbf{p}+\mathbf{k})^2}{2m} \right] t \right\} \qquad (10.5)$$

and

$$S^0(\mathbf{k}, \omega) = \sum_{\mathbf{p}} \left\{ n(\mathbf{p}) [1 - n(\mathbf{p} + \mathbf{k})] \delta \left[\omega + \frac{p^2}{2m} - \frac{(\mathbf{p} + \mathbf{k})^2}{2m} \right] \right\}. \quad (10.6)$$

Here

$$n(\mathbf{p}) = \langle c_\mathbf{p}^+ c_\mathbf{p} \rangle. \quad (10.7)$$

Equation (10.6) represents the sum of all possible scattering events in which an electron goes from a filled state $\mathbf{p}$ to an empty state $\mathbf{p} + \mathbf{k}$ with conservation of energy and momentum between the target and the incident particle. Conservation is ensured by the delta-function in Eq. (10.6). Thus, in the noninteracting case $S(\mathbf{k}, \omega)$ has a quite pedestrian interpretation. We will see later, however, that this expression for the structure factor actually has a far wider range of applicability. It also applies to an interacting plasma (5) provided k satisfies the condition $k\lambda_D \gg 1$. This condition is just that for being in the single-particle regime.

Despite its simplicity, Eq. (10.6) is a result of great interest because it suggests a method of directly measuring electron velocity distributions in plasmas. To date, experiments of this kind have been done in semiconductor plasmas and gas plasmas, using light rather than an electron beam as the probe, and in metals using X rays. Thus, we are led to ask how Eqs. (9.10) and (10.6) must be modified to describe light scattering from plasmas. Generally speaking, this is not an easy question to answer. In solid state plasmas, the form of the effective electron–photon interaction is determined by the band structure—and can be different from that of a free electron. We will later consider these band structure effects in detail, since they are of considerable interest and importance. Here, we restrict our attention to the one case in which there is a simple connection between the light and electron scattering experiments—the problem of light scattering from a single-component, classical plasma. By the word "classical," we mean a plasma whose constituents have an energy–momentum relation of the form $E = p^2/2m^*$. Many plasmas in solids, such as those in simple semiconductors (n-GaAs, n-InAs, etc.), closely approximate this condition. For such media, the light scattering spectrum is directly related to that determined by the electron scattering experiments. Both experiments measure the density–density correlation function [Eq. (9.11)] of the medium.

11. Light Scattering—The Classical, Single-Component Case

To discuss light scattering from a single-component, classical plasma, we consider the Hamiltonian of the many-body system, including its coupling to the electromagnetic field $\mathbf{A}$:

II. THEORY OF PLASMA EXPERIMENTS

$$H_{\text{total}} = \sum_i [\mathbf{p}_i - (e/c)\mathbf{A}_i]^2/2m^* + \tfrac{1}{2}\sum_{i\neq j} e^2/\epsilon_0 r_{ij}$$
$$\equiv H - (e/m^*c)\sum_i \mathbf{p}_i \cdot \mathbf{A}_i + (e^2/2m^*c^2)\sum_i A_i^2. \tag{11.1}$$

We have written H_{total} in terms of an effective mass m^* to indicate that, under appropriate conditions (6), this Hamiltonian can be used to describe semiconductors. Note that there are two types of electron–photon coupling terms in Eq. (11.1): linear terms, of the form $\mathbf{p}_i \cdot \mathbf{A}_i$; and quadratic terms, of the form A_i^2. Since light scattering is a two-photon process, the former will contribute to the matrix element in second order, while the latter makes a first-order contribution. We will show that, in a classical plasma, the A_i^2 terms are the dominant ones. This statement can easily be verified for the case of light scattering from a single free electron. In that problem (nonrelativistic Compton scattering) the matrix element for light scattering consists of three terms: two second-order terms due to the $\mathbf{p} \cdot \mathbf{A}$ interaction, and a first-order term coming from A^2,

$$M = \left(\frac{e^2}{m^*c^2}\right)\left(\frac{2\pi c^2}{\omega_0}\right)^{1/2}\left(\frac{2\pi c^2}{\omega_1}\right)^{1/2}\left[\frac{[\mathbf{p}\cdot\boldsymbol{\epsilon}_0][(\mathbf{p}+\mathbf{k}_0)\cdot\boldsymbol{\epsilon}_1]}{m^*[\mathbf{p}^2/2m^* + \omega_0 - (\mathbf{p}+\mathbf{k}_0)^2/2m^*]}\right.$$
$$\left. + \frac{(\mathbf{p}\cdot\boldsymbol{\epsilon}_1)[(\mathbf{p}-\mathbf{k}_1)\cdot\boldsymbol{\epsilon}_0]}{m^*[\mathbf{p}^2/2m^* - \omega_1 - (\mathbf{p}-\mathbf{k}_1)^2/2m^*]} + (\boldsymbol{\epsilon}_0\cdot\boldsymbol{\epsilon}_1)\right]. \tag{11.2}$$

Here $\mathbf{p}$ is the electron momentum; $\boldsymbol{\epsilon}_0$ and $\boldsymbol{\epsilon}_1$ are the polarization vectors of the light waves. The first two terms in Eq. (11.2) come from the $\mathbf{p}\cdot\mathbf{A}$ interaction and correspond to virtual processes in which (i) the electron absorbs a photon of frequency ω_0, then emits a photon of frequency ω_1 and (ii) emits ω_1 first, then absorbs ω_0. The third corresponds to the simultaneous emission and absorption of two photons via the A^2 interaction. Graphically, these three processes are conveniently pictured as in Fig. II-4. If the electron from which the light scatters is

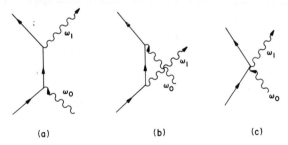

FIG. II-4. Diagrams representing the three possible contributions to the light scattering matrix element.

at rest, the first and second terms of Eq. (11.2) vanish. Even when p is finite, they cancel one another to order v/c. Hence, one can ignore all but the last term in Eq. (11.2). The expression for the scattering cross section then becomes

$$d\sigma/d\Omega = (e^2/m^*c^2)^2 \, (\boldsymbol{\epsilon}_0 \cdot \boldsymbol{\epsilon}_1)^2. \tag{11.3}$$

Equation (11.3) is the well-known Thomson formula which can be derived by a variety of semiclassical arguments.

Equation (11.3), and the arguments which lead up to it, apply to the case of light scattering by a single, nonrelativistic electron. They strongly suggest, however, that in a plasma as well, light scattering will occur via the A_i^2 terms in the electron–photon coupling [see Eq. (11.1)]. This is indeed the case (7). To see why it is so, we consider the matrix elements of the $\mathbf{p} \cdot \mathbf{A}$ interaction when many electrons are present. For light of a given wave vector $\mathbf{k}_0$ this interaction takes the form

$$\sum_i (\mathbf{p}_i \cdot \mathbf{A}_i) = (2\pi c^2/\omega)^{1/2} \sum_i [(\mathbf{p}_i \cdot \boldsymbol{\epsilon}) \, e^{i\mathbf{k}_0 \cdot \mathbf{r}_i}]. \tag{11.4}$$

In the limit $k_0 \rightarrow 0$, Eq. (11.4) becomes

$$(2\pi c^2/\omega)^{1/2} \sum_i (\mathbf{p}_i \cdot \boldsymbol{\epsilon}) = (2\pi c^2/\omega)^{1/2} (\mathbf{P} \cdot \boldsymbol{\epsilon}), \tag{11.5}$$

where $\mathbf{P}$ is the *total* momentum of the electron system. For a single-component plasma, this quantity is a constant of the motion, which is zero in the plasma rest frame. For a multicomponent plasma, the total momentum of *all* species is conserved. However, if collisions between different species are relatively weak (this is always true in a good plasma), the momentum of each is approximately conserved (and again can be taken to be zero). These arguments suggests that in many cases the matrix element of the $\mathbf{p} \cdot \mathbf{A}$ interaction [Eq. (11.4)] has a finite value only because of the recoil momentum $\mathbf{k}_0$ picked up from the light beam. This result leads to the following estimate of the ratio of the $\mathbf{p} \cdot \mathbf{A}$ and A^2 contributions to the light scattering matrix element:

$$\mathscr{R} = \frac{M_{\mathbf{p} \cdot \mathbf{A}}}{M_{A^2}} \sim \frac{\langle \sum_i (\mathbf{p}_i \cdot \mathbf{A}_i) \rangle \langle \sum_i (\mathbf{p}_i \cdot \mathbf{A}_i) \rangle}{m \, \Delta E \langle \sum_i A_i^2 \rangle}$$

$$\approx (k^2/m \, \Delta E), \tag{11.6}$$

where ΔE is the energy denominator of the virtual, second-order process, and the angular brackets indicate matrix elements with respect to exact wave functions of the many-body system. For cases in which the optical energies are large compared to important excitation energies of the many-body system (most light scattering experiments are performed in this regime), $\Delta E \simeq \omega_0$. The ratio R is then about

$$\mathscr{R} \simeq \hbar\omega_0/mc^2 \ll 1. \tag{11.7}$$

This argument ignores possible cancellation of the two $\mathbf{p} \cdot \mathbf{A}$ terms in the matrix element. For nearly elastic scattering, this cancellation causes a further reduction in $\mathscr{R}$. We conclude, therefore, that in most situations one is justified in neglecting the $\mathbf{p}_i \cdot \mathbf{A}_i$ terms in calculating the light scattering spectrum from a classical plasma.

For a process in which light scatters from $(\mathbf{k}_0, \omega_0)$ to $(\mathbf{k}_1, \omega_1)$, the A_i^2 terms take the form

$$\frac{e^2}{2m^*c^2} \sum_i A_i^2 = \frac{2\pi c^2}{(\omega_0\omega_1)^{1/2}} \left(\frac{e^2}{m^*c^2}\right) \sum_i (\boldsymbol{\epsilon}_0 \cdot \boldsymbol{\epsilon}_1)\, e^{i\mathbf{k}\cdot\mathbf{r}_i}. \tag{11.8}$$

Equation (11.8) indicates that, in the classical, single-component plasma, light couples to electron density fluctuations just as does the fast electron beam. From this conclusion, it is easy to derive an expression for the spectrum. The calculation exactly parallels that we performed to obtain Eq. (9.11). One finds

$$\frac{d^2\sigma}{d\Omega\, d\omega} = (e^2/m^*c^2)^2\, (\omega_1/\omega_0)(\boldsymbol{\epsilon}_0 \cdot \boldsymbol{\epsilon}_1)^2\, S(\mathbf{k}, \omega), \tag{11.9}$$

where $d^2\sigma/d\Omega\, d\omega$ is the differential cross section for scattering into solid angle $d\Omega$ and frequency interval $d\omega$, $S(\mathbf{k}, \omega)$ is the structure factor defined by Eq. (9.11). Thus, in a classical plasma, the characteristic energy loss experiments [Eq. (9.10)] and light scattering experiments [Eq. (11.9)] measure the same feature of the many-body system. However, it is important to notice that in electron scattering small values of k are strongly weighted (by a factor k^{-4}), whereas they are not in light scattering. Because of this fact, electron scattering experiments are primarily useful for studying collective modes of solid state plasmas ($k\lambda_{\text{FT}} \ll 1$ limit). In particular, the spectrum of all scattered electrons [obtained by integrating Eq. (9.10) over solid angle] reflects the small k collective modes of the plasma.

The parallelism between the characteristic energy loss and light scattering experiments [Eqs. (9.10) and (11.9)] only holds in the case of a single-component, classical plasma. In multicomponent plasmas, an electron beam couples to the total charge density:

$$\rho_\mathbf{k} = \sum_I \rho_\mathbf{k}^I, \tag{11.10}$$

where $\rho_k{}^I$ is the charge density of the Ith species in the plasma. On the other hand, light couples (via the A^2 terms in the classical Hamiltonian) as follows to the multicomponent, classical plasma:

$$[2\pi c^2/(\omega_0\omega_1)^{1/2}] \sum_I (e_I{}^2/m_I c^2)\, n_k{}^I. \tag{11.11}$$

Here e_I, m_I, and n^I are the charge, mass, and density of the Ith species. This formula is an obvious generalization of Eq. (11.8). Note that in both Eqs. (11.10) and (11.11) the probes interact with particle density fluctuations. However, the fluctuations of the various species are weighted differently in the two cases. This fact has important consequences. For example, a multicomponent plasma can sometimes support a nearly neutral, collective mode. The densities of the individual species fluctuate in this excitation, but in such a way as to produce little net charge. Electrons couple weakly to such modes, and one cannot hope to observe them via electron scattering. On the other hand, if the masses of the various species are appreciably different from one another, the light scattering cross sections can be sizable.

Equation (11.9) also fails in the case of a quantum-mechanical plasma, i.e., a solid state plasma in which band structure effects play a real role. In a crystal, the operator $\mathbf{p} \cdot \mathbf{A}$ has *interband* matrix elements; and their contributions to the matrix element for light scattering do not cancel. Under certain circumstances, the interband matrix elements of $\mathbf{p} \cdot \mathbf{A}$ may be removed by a canonical transformation (8) (the effective mass transformation), yielding a classical Hamiltonian of the form indicated in Eq. (11.1). The theory of light scattering from a classical plasma then applies [Eq. (11.9)]. However, such a procedure is at best an approximation, and completely fails when the optical frequencies approach the band gap. In these cases one must calculate, in detail, the interband matrix elements of $\mathbf{p} \cdot \mathbf{A}$ to determine the light scattering cross sections. We will discuss such situations in Chapter V. They are of considerable interest because the interband matrix elements enable light to scatter from other sorts of fluctuations in the plasma—such as spin density or energy density fluctuations. In this sense, the light scattering experiment is somewhat richer than its fast electron counterpart; though, with relativistic beams, it may ultimately be possible to study spin densities via electron scattering as well (*1*).

12. Transmission Experiments: Introductory Remarks

As discussed in Chapter I, when a low-loss plasma is placed in a dc magnetic field, it can support a variety of propagating, electromagnetic waves. These waves, besides having great intrinsic interest, are powerful probes of the structure of the many-body system. Their study is now a major topic in solid state plasma work. The primary technique for studying waves is a transmission experiment of the sort illustrated in Fig. II-5. Later, we will discuss such experiments in detail. Here, we

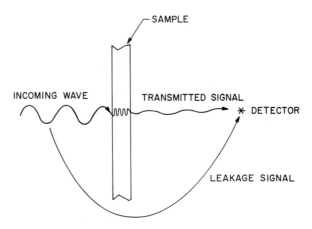

FIG. II-5. Schematic of a typical transmission experiment.

wish to consider their features in a general way, with the aim of isolating the important quantities that the theory must determine.

In a typical transmission experiment a relatively low-frequency (rf or microwave) electromagnetic wave impinges on a slab-shaped sample which contains the plasma. The sample is most often a metal, though transmission experiments have also been done in doped semiconductors and semimetals. For optimal results, it should be pure and cold to reduce scattering losses. Usually the sample thickness is 0.1–1.0 mm. These values are orders of magnitude greater than the skin depth of a pure metal. Radiation transmitted through such a sample is either carried by some sort of *propagating* excitation, or by free electrons moving on ballistic trajectories. To identify and study the wavelike excitations, one measures the amplitude and phase of the transmitted field as a function of the strength (and sometimes the direction) of the dc magnetic field. Such measurements may conveniently be done by beating the transmitted wave against a portion of the primary radiation. (See Chapter VI for details.) From the phase and amplitude of the transmitted signal one can, in many cases, infer the effective index of refraction of the plasma for the wave under study. When it works, this technique is an elegant one since the information of interest

—the plasma dielectric constant— is obtained directly, without the solution of a boundary value problem. In this respect, transmission experiments are vastly simpler than measurements of surface impedance. To interpret the latter, one must solve a boundary value problem which, in the anomalous skin effect range, becomes exceedingly complicated. Some discussion of the boundary value problem is presented in Chapter VIII, but in general one tries to avoid it.

Any wave that propagates in a plasma is a solution of Maxwell's equations. In particular, it is a solution of those Maxwell equations which contain the constitutive relations of the plasma. Thus, in comparing the results of a transmission experiment with theory, two problem arise. The first is that of determining the constitutive relations. This amounts to a calculation of the plasma conductivity and susceptibility. The second problem is that of solving Maxwell's equations in the bounded medium. Here, however, a complete solution is often not necessary. The wave one studies in transmission experiments has traveled a relatively large distance through the plasma. Its properties are mainly given (particularly as to phase) by the infinite-medium dispersion relation. This relation can be calculated quite simply, and suffices to determine the gross features of the transmitted wave. One thereby avoids a complicated (and often insoluble) boundary value problem. A solution of the **bound**ary value problem is usually required only if one is interested in detailed line shapes or intensities.

13. Response Functions

The preceding discussion indicates that the central theoretical problem in the interpretation of transmission experiments is that of calculating the constitutive relations. These relations describe the *linear* response of the plasma to (assumed weak) electric and magnetic fields. If the plasma properties are independent of time, the most general relations of this type have the form

$$j_\alpha(\mathbf{r}, \omega) = \int d^3r'\, \sigma_{\alpha\beta}(\mathbf{r}, \mathbf{r}'; \omega)\, E_\beta(\mathbf{r}', \omega)$$

$$M_\alpha(\mathbf{r}, \omega) = \int d^3r'\, \chi_{\alpha\beta}(\mathbf{r}, \mathbf{r}'; \omega)\, H_\beta(\mathbf{r}', \omega)$$

(13.1)

where $j_\alpha(\mathbf{r}, \omega)$ and $M_\alpha(\mathbf{r}, \omega)$ are the current and magnetization induced in the plasma, at position $\mathbf{r}$ and frequency ω, by the fields $E_\beta(\mathbf{r}', \omega)$ and $H_\beta(\mathbf{r}', \omega)$. The coefficients which relate these quantities are *response functions* of the plasma, in this case the conductivity and susceptibility tensors $\sigma_{\alpha\beta}(\mathbf{r}, \mathbf{r}'; \omega)$ and $\chi_{\alpha\beta}(\mathbf{r}, \mathbf{r}'; \omega)$. These functions are important examples of a general class of functions which describe the response of a plasma to weak perturbations of arbitrary types. Response functions play an important role in our understanding of the properties of many-body systems, and are closely related to the correlation functions which determine scattering cross sections and spectra. We will discuss their general features at this point. Later, for specific examples, these functions will be computed in detail.

In analyzing response functions, the crucial assumption that we will make is that the plasma interacts *weakly* with the external perturbation. This approximation will enable us to relate the response function to an appropriate property of the unperturbed plasma. Let us imagine that the plasma is described by the Hamiltonian

$$H = H_0 + H_1, \tag{13.2}$$

where H_0 is the Hamiltonian of the isolated many-body system, and H_1 is its coupling to a perturbation, which may be a function of both position and time. We wish to calculate the *first-order* change in the density matrix $\boldsymbol{\rho}$ of the system due to this perturbation (9). The equation of motion of the density matrix is

$$i\,\partial\boldsymbol{\rho}/\partial t = [(H_0 + H_1), \boldsymbol{\rho}] + i(\boldsymbol{\rho}_0 - \boldsymbol{\rho})/\tau. \tag{13.3}$$

Except for the last term, this is a standard formula. The final term is a phenomenological one, describing the relaxation of the perturbed density matrix (10) $\boldsymbol{\rho}$ to its equilibrium value $\boldsymbol{\rho}_0$. It is important to include such a term—otherwise, the long-time application of even a feeble perturbation will gradually change (by heating) the state of the system. This problem is avoided by using sufficiently weak probes (small H_1) that the natural damping processes hold the system near its equilibrium state. In the following analysis, we will assume that H_1 is small enough that this criterion has been met. It must also, of course, be satisfied in experiments that are used to measure response functions.

Equation (13.3) can most easily be solved in the interaction representation, where it becomes

$$i\,\partial\boldsymbol{\rho}_1/\partial t + i\boldsymbol{\rho}_1/\tau = e^{iH_0 t}[H_1, \boldsymbol{\rho}]\,e^{-iH_0 t}, \tag{13.4}$$

with $\boldsymbol{\rho}_1 = e^{iH_0 t}(\boldsymbol{\rho} - \boldsymbol{\rho}_0)\,e^{-iH_0 t}$. The formal solution is

$$\boldsymbol{\rho}_1 = (e^{-t/\tau}/i) \int_{-\infty}^{t} e^{t'/\tau} e^{iH_0 t'}[H_1(t'), \boldsymbol{\rho}(t')]\,e^{-iH_0 t'}\,dt'. \tag{13.5}$$

13. RESPONSE FUNCTIONS

To first order in H_1, Eq. (13.5) reduces to

$$\boldsymbol{\rho}_1 \simeq (e^{-t/\tau}/i) \int_{-\infty}^{t} e^{t'/\tau} e^{iH_0 t'} [H_1(t'), \boldsymbol{\rho}_0] e^{-iH_0 t'} \, dt'. \tag{13.6}$$

Correspondingly,

$$\boldsymbol{\rho} \simeq \boldsymbol{\rho}_0 + (1/i) \int_{-\infty}^{t} \{e^{-(t-t')/\tau} e^{-iH_0(t-t')} [H_1(t'), \boldsymbol{\rho}_0] e^{iH_0(t-t')}\} \, dt'. \tag{13.7}$$

With the aid of Eq. (13.7), we can now calculate the induced value (to first order in H_1) of any dynamical variable $\mathcal{O}$. One finds

$$\langle \mathcal{O}(t) \rangle \equiv \mathrm{tr}(\boldsymbol{\rho}\mathcal{O})$$

$$\simeq \mathrm{tr}(\boldsymbol{\rho}_0 \mathcal{O}) + (1/i) \int_{-\infty}^{t} e^{-(t-t')/\tau} \mathrm{tr}\{\mathcal{O} e^{-iH_0(t-t')} [H_1(t'), \rho_0] e^{iH_0(t-t')}\} \, dt'. \tag{13.8}$$

After some rearrangements, this expression takes the form

$$\langle \mathcal{O}(t) \rangle \simeq \langle \mathcal{O} \rangle + (1/i) \int_{-\infty}^{t} e^{-(t-t')/\tau} \langle [\mathcal{O}, e^{-iH_0(t-t')} H_1(t') e^{iH_0(t-t')}] \rangle \, dt', \tag{13.9}$$

where the notation

$$\langle \mathcal{O} \rangle = \mathrm{tr}(\mathcal{O} \boldsymbol{\rho}_0)$$

has been used. Finally, we Fourier transform the time dependence of H_1 (which is that of the external field):

$$H_1(t) = \int e^{-i\omega t} H_1(\omega) \, d\omega. \tag{13.10}$$

The response at frequency ω then becomes

$$\langle \mathcal{O}(\omega) \rangle = (1/i) \int_{0}^{\infty} e^{i(\omega + i/\tau)t} \langle [\mathcal{O}(t), H_1(\omega)] \rangle \, dt, \tag{13.11}$$

where $\mathcal{O}(t) \equiv e^{iH_0 t} \mathcal{O} e^{-iH_0 t}$.

Equation (13.11), which involves operators and wave functions that refer to the *unperturbed medium*, is a general formula for the response of a system to a weak, external perturbation. Equations of this sort were first developed by Kubo (*11*). Though simple to derive, they have far-reaching consequences. They immediately indicate, for a given type of perturbation, how to express the corresponding response function for the many-body system. They also contain causality information that determines the analytic behavior, in the frequency plane, of this function.

The fact that the t' integration in Eq. (13.8) is restricted to values $t' \leq t$ indicates that the signal $\langle \mathcal{O}(t) \rangle$ cannot precede its stimulus. This result leads in a familiar way to the Kramers–Kronig (*12*) relations. Finally, Eq. (13.11) is useful because it suggests which quantities one should calculate to understand the properties of a many-body system. The commutator average $\langle [\mathcal{O}(t), H(\omega)] \rangle$ that appears in Eq. (13.11) is typical of all quantum-mechanical response functions. Moreover, its value at $t = 0$ is known from commutation rules. These facts suggest that one should seek equations of motion for the commutator averages, and integrate them directly. One important approach to many-body problems is based on this idea (*13*).

In applying Eq. (13.11), we will sometimes pass to the limit $H_1 \to 0$, $1/\tau \equiv \delta \to 0$ to determine the response function of an undamped many-body system. Since the energy transferred to the system is proportional to H_1^2, this procedure gives meaningful results if the limit is taken in such a way that $H_1^2/\delta \to 0$. In practice, this goal can usually be achieved by setting $\tau = \infty$ ($\delta = 0$) at the end of the calculation.

Finally, let us consider the particular response function which is most closely related to the density–density correlation function discussed in the preceding section. This function determines the *charge density* induced in a plasma by an external, electrostatic potential. It is one of the simplest, yet most important, response functions which characterize these media. Several basic experiments can be interpreted with its aid. Nevertheless, it should be clearly understood that this function does not completely characterize the response of an electron system to electromagnetic perturbations—but only its response to *longitudinal* ones. Later, we will consider the behavior of plasmas in *transverse* fields, and discuss the response functions which are relevant in such cases.

To determine the induced charge in a plasma, we imagine that the system is perturbed by an external, electrostatic potential which couples to the electron charge density via the interaction

$$H_1 = \int \rho(\mathbf{r}) \phi_{\text{ext}}(\mathbf{r}, t) \, d^3r. \tag{13.12}$$

Such a potential might be produced by a fast particle passing through the plasma (as in the characteristic energy loss experiments) or by a static charge (impurity) imbedded in it. To calculate the induced electron density, we use Eq. (13.11). The result is

$$\langle \rho(\mathbf{r}, \omega) \rangle \equiv (2\pi)^{-1} \int_{-\infty}^{+\infty} e^{i\omega t} \langle \rho(\mathbf{r}, t) \rangle \, dt$$

$$= \int F(\mathbf{r}, \mathbf{r}'; \omega) \phi_{\text{ext}}(\mathbf{r}', \omega) \, d^3r', \tag{13.13}$$

where the response function F is given by

$$F(\mathbf{r}, \mathbf{r}'; \omega) = -i \int_0^\infty e^{i(\omega+i\delta)t} \langle [\rho(\mathbf{r}, t), \rho(\mathbf{r}', 0)] \rangle \, dt. \tag{13.14}$$

An important feature of Eq. (13.13) is the fact that the induced charge density can be *nonlocally* related to the driving field ϕ_{ext}. This is actually the case in all plasmas. In a *translationally invariant* plasma, F must be a function of the coordinate difference $(\mathbf{r} - \mathbf{r}')$. It is then convenient to Fourier transform Eqs. (13.13) and (13.14) to obtain the algebraic relations

$$\rho_{\text{ind}}(\mathbf{k}, \omega) = F(\mathbf{k}, \omega) \phi_{\text{ext}}(\mathbf{k}, \omega), \tag{13.15}$$

with

$$F(\mathbf{k}, \omega) = -i \int_0^\infty e^{i(\omega+i\delta)t} \langle [\rho_\mathbf{k}(t), \rho_{-\mathbf{k}}(0)] \rangle \, dt. \tag{13.16}$$

These equations indicate that, in translationally invariant systems, the charge density perturbation has the same wave vector and frequency as the driving potential.

14. Response to the Total Field

In the preceding sections we have discussed the structure factor $[S(\mathbf{k}, \omega)]$ and the corresponding response function $[F(\mathbf{k}, \omega)]$ for a many-body system. The latter can be used to calculate the response of a plasma to an *arbitrary* electrostatic perturbation. We will see presently (Section 17) that there is a close relation between these two functions, so that $S(\mathbf{k}, \omega)$ can also be used to determine the response to electrostatic perturbations. However, neither function is the most convenient for this purpose, since they relate the induced electron density to the external perturbation ϕ_{ext}. In practice, we know that when a plasma is perturbed there is generally an induced potential ϕ_{ind} in addition to ϕ_{ext} (see the discussion of Section 5). The response of the plasma to the total potential, $\phi_{\text{total}} = \phi_{\text{ext}} + \phi_{\text{ind}}$, is a considerably more physical quantity than its response to ϕ_{ext}. This thinking leads one to define a new sort of response function, relating the induced charged density to the *total electrostatic potential*:

$$\rho_{\text{ind}}(\mathbf{r}, \omega) \equiv \int \alpha(\mathbf{r}, \mathbf{r}'; \omega) \phi_{\text{total}}(\mathbf{r}', \omega) \, d^3r'. \tag{14.1}$$

The function $\alpha(\mathbf{r}, \mathbf{r}', \omega)$ is the analog, for an electrostatic perturbation, of the conductivity tensor [Eq. (13.1)] which determines the response of the plasma to fields of arbitrary polarization. In both cases, these functions relate the response of the medium to the *total* perturbation, rather than the external one.

A relation between $\alpha(\mathbf{r}, \mathbf{r}', \omega)$ and $F(\mathbf{r}, \mathbf{r}'; \omega)$ can be derived by making use of the fact that ϕ_{ind} is a solution of Poisson's equation:

$$\phi_{\text{ind}}(\mathbf{r}, \omega) = \int \frac{\rho_{\text{ind}}(\mathbf{r}', \omega) \, d^3 r'}{|\mathbf{r} - \mathbf{r}'|}. \tag{14.2}$$

By combining Eqs. (13.13), (14.1), and (14.2), one finds

$$\rho_{\text{ind}}(\mathbf{r}, \omega) = \int F(\mathbf{r}, \mathbf{r}'; \omega) \phi_{\text{ext}}(\mathbf{r}', \omega) \, d^3 r'$$

$$= \int \alpha(\mathbf{r}, \mathbf{r}'; \omega) \phi_{\text{ext}}(\mathbf{r}', \omega) \, d^3 r'$$

$$+ \iiint \alpha(\mathbf{r}, \mathbf{r}', \omega) \frac{1}{|\mathbf{r}' - \mathbf{r}''|} F(\mathbf{r}'', \mathbf{r}'''; \omega) \phi_{\text{ext}}(\mathbf{r}''', \omega) \, d^3 r' \, d^3 r'' \, d^3 r'''. \tag{14.3}$$

Since $\phi_{\text{ext}}(\mathbf{r}, \omega)$ is an arbitrary function of $\mathbf{r}$, it follows that

$$\alpha(\mathbf{r}, \mathbf{r}'; \omega) = F(\mathbf{r}, \mathbf{r}'; \omega) - \iint \alpha(\mathbf{r}, \mathbf{r}''; \omega) \frac{1}{|\mathbf{r}'' - \mathbf{r}'''|} F(\mathbf{r}''', \mathbf{r}'; \omega) \, d^3 r'' \, d^3 r'''. \tag{14.4}$$

This is an *integral* equation relating $\alpha(\mathbf{r}, \mathbf{r}'; \omega)$ to $F(\mathbf{r}, \mathbf{r}'; \omega)$. Its formal solution is

$$\alpha = F \left[\frac{1}{1 - 4\pi (1/\nabla^2) F} \right], \tag{14.5}$$

but to evaluate this expression one must solve Eq. (14.4). Fortunately, in the most important case—that of a translationally invariant system—the solution is trivial. In such a medium $F(\mathbf{r}, \mathbf{r}'; \omega) \equiv F(\mathbf{r} - \mathbf{r}'; \omega)$ and $\alpha(\mathbf{r}, \mathbf{r}'; \omega) \equiv \alpha(\mathbf{r} - \mathbf{r}'; \omega)$. Equations (14.4) and (14.5) then reduce to *algebraic* relations between the Fourier transforms of F and α, i.e., Eq. (14.5) becomes

$$\alpha(\mathbf{k}, \omega) = \frac{F(\mathbf{k}, \omega)}{[1 + (4\pi/k^2) F(\mathbf{k}, \omega)]}. \tag{14.6}$$

Equation (14.6) is an important formula in the theory of solid state plasmas.

The reader may wonder how a result that is derived for a translationally invariant system can be applied to crystals, which certainly do not have this property. In practice, it appears that plasma effects in solids (at least in the simpler materials) are largely unaffected by lattice periodicity. In subsequent chapters we will see, for specific examples, why this is the case. These calculations show that plasma properties are mainly determined by the many-body (Coulomb) interactions, with relatively minor perturbations due to band structure and lattice periodicity. Such a viewpoint is also supported by many experiments (see Chapter IV), and explains why the interacting (but translationally invariant) electron gas is such an important theoretical model.

Equation (14.6) is an interesting and illuminating relation. Let us imagine, in particular, that we are dealing with a weakly coupled plasma, as defined by Eqs. (2.1) or (2.2). One then expects that, in some sense, the response of the plasma to an external perturbation might be calculated by treating the system as a *noninteracting* electron gas. But now the problem arises, should one use the noninteracting approximation to calculate F or α? Equation (14.6) shows that, in the small k limit, the results of these two approaches will be entirely different from one another, so the question is an important one to answer. Let us first see what happens if we calculate F for noninteracting electrons [from Eq. (13.16)]. After a simple calculation, one finds $F \simeq e^2 Q(\mathbf{k}, \omega)$, where

$$Q(\mathbf{k}, \omega) \equiv \sum_{\mathbf{p}} \frac{n(\mathbf{p}) - n(\mathbf{p} + \mathbf{k})}{\omega + (p^2/2m) - [(\mathbf{p} + \mathbf{k})^2/2m] + i\delta}. \tag{14.7}$$

The function Q is the response function of a *noninteracting* electron gas, and will be used repeatedly in the subsequent analysis. The approximation $F \simeq e^2 Q$ may be tested by calculating the induced potential in the plasma. Using Eqs. (13.15) and (14.2) one finds

$$\phi_{\text{ind}}(\mathbf{k}, \omega) = (4\pi/k^2) F(\mathbf{k}, \omega) \phi_{\text{ext}}(\mathbf{k}, \omega)$$

$$\simeq (4\pi e^2/k^2) Q(\mathbf{k}, \omega) \phi_{\text{ext}}(\mathbf{k}, \omega). \tag{14.8}$$

It is easily verified from Eq. (14.7) that Q remains finite as $\omega \to 0$, $k \to 0$. From this fact, we see that $\phi_{\text{ind}}(\mathbf{k}, 0) \to \infty$ as $k \to 0$, if $\phi_{\text{ext}}(0, 0)$ is finite. An induced potential that becomes infinite is certainly not physically reasonable, so we are forced to the conclusion that in the small k limit a free electron approximation to F is very bad. This is indeed the case.

More sensible results are obtained by making the approximation $\alpha \simeq e^2 Q$. In terms of α, the induced and external potentials are related by the equation

$$\phi_{\text{ind}}(\mathbf{k}, \omega) = \frac{(4\pi/k^2) \alpha(\mathbf{k}, \omega) \phi_{\text{ext}}(\mathbf{k}, \omega)}{1 - (4\pi/k^2) \alpha(\mathbf{k}, \omega)}. \tag{14.9}$$

Now in the limit $k = 0$, $\phi_{\text{ind}} = -\phi_{\text{ext}}$ if we approximate α by $e^2 Q$. This is a reasonable and satisfying result. ϕ_{ind} is finite at $k = 0$ and exactly cancels ϕ_{ext}; i.e., the plasma completely screens a long-wavelength perturbation. The lesson to be learned from this discussion is that one should approximate to α, rather than F. In retrospect, this is not surprising. Since a plasma screens long-wavelength perturbations, its response is quite different from that of a noninteracting electron gas. On the other hand, it is possible (in a weakly coupled plasma) for the response to the *total potential* to be close to that of the noninteracting system.

We may also look at this question from an analytic point of view, by considering the inverse equation to Eq. (14.6):

$$F(\mathbf{k}, \omega) = \frac{\alpha(\mathbf{k}, \omega)}{1 - (4\pi/k^2)\,\alpha(\mathbf{k}, \omega)}. \tag{14.10}$$

Let us now imagine that one could vary the particle interactions by changing the coupling constant e^2. In particular, we are interested in the limit of weak coupling ($e^2 \to 0$) and long wavelengths ($k \to 0$), where plasmalike effects are most pronounced. If $\alpha \simeq e^2 Q$, it follows from Eq. (14.10) that $F(\mathbf{k}, \omega)$ is singular as $k \to 0$, in the sense that the value of F depends upon how the limit $e^2 \to 0$, $k \to 0$ is taken. It is also clear from Eq. (14.10) that it is *not* possible to make an expansion of F in powers of e^2 near the point $k = 0$. This is a general feature of plasmas, whether they are weakly coupled or not. Even in the weak coupling limit, many of their properties *cannot* be expressed as a power series in the coupling constant V/K.

15. The Plasma Dielectric Constant

It is often convenient to describe the properties of a plasma in terms of a *frequency and wave-vector-dependent dielectric constant*. For translationally invariant systems, where the induced charge has the same frequency and wave vector as the perturbing potential, the definition of the dielectric constant is a straightforward generalization of the usual formula of electrostatics:

$$1/\epsilon(\mathbf{k}, \omega) \equiv E(\mathbf{k}, \omega)/D(\mathbf{k}, \omega) = \phi_{\text{total}}(\mathbf{k}, \omega)/\phi_{\text{ext}}(\mathbf{k}, \omega). \tag{15.1}$$

By combining this definition with Eqs. (13.15), (14.2), and (14.6), one finds that

$$\frac{1}{\epsilon(\mathbf{k}, \omega)} = 1 + \frac{4\pi}{k^2} F(\mathbf{k}, \omega) = \frac{1}{1 - (4\pi/k^2)\,\alpha(\mathbf{k}, \omega)}. \tag{15.2}$$

These rather simple relations apply to translationally invariant systems.

In crystals, on the other hand, a perturbation of wave vector $\mathbf{k}$ induces a response at all wave vectors $(\mathbf{k} + \mathbf{K})$—where $\mathbf{K}$ is any reciprocal lattice. Generally speaking, however, one is interested in the *macroscopic* (14) fields generated in the system, since these are easily measurable. Such fields are determined by the average, over a region which contains many unit cells but is small compared to the wavelength of the perturbation, of the plasma response, i.e., by the $K = 0$ component of the induced potential. This line of thinking leads us to define a *macroscopic* dielectric constant in the crystalline case by the formula

$$\frac{1}{\epsilon_M(\mathbf{k}, \omega)} \equiv \frac{\int \langle \phi_{\text{total}}(\mathbf{r}, \omega) \rangle e^{-i\mathbf{k} \cdot \mathbf{r}} d^3r}{\phi_{\text{ext}}(\mathbf{k}, \omega)}. \quad (15.3)$$

Here the angular brackets indicate a spatial average over many unit cells, and it is assumed that $k \ll K$. To relate $\epsilon_M(\mathbf{k}, \omega)$ to the response functions defined previously, we combine Eqs. (13.13), (14.2), and (15.3). The result is

$$\frac{1}{\epsilon_M(\mathbf{k}, \omega)} = 1 + \frac{(4\pi/k^2) \iint \langle F(\mathbf{r}, \mathbf{r}'; \omega) \phi_{\text{ext}}(\mathbf{r}', \omega) \rangle e^{-i\mathbf{k} \cdot \mathbf{r}} d^3r \, d^3r'}{\phi_{\text{ext}}(\mathbf{k}, \omega)}. \quad (15.4)$$

This expression can be simplified by noting that, because of the lattice periodicity, F can be expanded as follows (14):

$$F(\mathbf{r}, \mathbf{r}'; \omega) = \sum_{\mathbf{k}; \mathbf{K}, \mathbf{K}'} \{F(\mathbf{k} + \mathbf{K}, \mathbf{k} + \mathbf{K}'; \omega) e^{i[(\mathbf{k}+\mathbf{K}) \cdot \mathbf{r} - (\mathbf{k}+\mathbf{K}') \cdot \mathbf{r}']}\}, \quad (15.5)$$

where the sum on $\mathbf{k}$ extends over the first Brillouin zone. Substitution into Eq. (15.4) yields the result

$$1/\epsilon_M(\mathbf{k}, \omega) = 1 + (4\pi/k^2) F(\mathbf{k}, \mathbf{k}; \omega). \quad (15.6)$$

This formula is a natural generalization of the first part of Eq. (15.2). However, the second part of this equation—the relation between ϵ and α—does not carry over in as simple a way to the crystalline case. The complication has its origin in the fact that the relation between α and F is no longer an algebraic one. To study this problem, we use Eq. (14.1) to relate ϕ_{total} to ϕ_{ext}:

$$\rho_{\text{ind}}(\mathbf{r}, \omega) = \int \alpha(\mathbf{r}, \mathbf{r}'; \omega) \phi_{\text{total}}(\mathbf{r}', \omega) d^3r'$$

$$= -(\nabla^2/4\pi) [\phi_{\text{total}}(\mathbf{r}, \omega) - \phi_{\text{ext}}(\mathbf{r}, \omega)], \quad (15.7)$$

hence

$$\phi_{\text{ext}}(\mathbf{r}, \omega) = \phi_{\text{total}}(\mathbf{r}, \omega) - \iint |\mathbf{r} - \mathbf{r}'|^{-1} \alpha(\mathbf{r}', \mathbf{r}'', \omega) \phi_{\text{total}}(\mathbf{r}'', \omega) \, d^3r' \, d^3r''.$$
(15.8)

This expression can formally be inverted to give the results

$$\phi_{\text{total}} = [1 + (4\pi/\nabla^2)\alpha]^{-1} \phi_{\text{ext}} \tag{15.9}$$

and

$$\frac{1}{\epsilon_M(\mathbf{k}, \omega)} \equiv \int \left\langle \frac{1}{1 + (4\pi/\nabla^2)\alpha} \right\rangle e^{-i\mathbf{k}\cdot(\mathbf{r}-\mathbf{r}')} d^3r \, d^3r'. \tag{15.10}$$

To evaluate Eq. (15.10), however, one must realize that α is a *matrix* in **K** space, with an expansion similar to that of Eq. (15.5). The matrix

$$1 - (4\pi/|\mathbf{k} + \mathbf{K}|^2) \alpha(\mathbf{k} + \mathbf{K}, \mathbf{k} + \mathbf{K}'; \omega)$$

must be inverted to determine $\epsilon_M(\mathbf{k}, \omega)$. Generally speaking,

$$\frac{1}{\epsilon_M(\mathbf{k}, \omega)} \neq \frac{1}{1 - (4\pi/k^2) \alpha(\mathbf{k}, \mathbf{k}; \omega)}, \tag{15.11}$$

i.e., the simplest generalization of Eq. (15.2) fails. The differences between the two sides of Eq. (15.11) are due to local field effects (*15*). Fortunately, in most solid state plasma applications these effects are relatively unimportant, and the approximation

$$\frac{1}{\epsilon_M(\mathbf{k}, \omega)} \simeq \frac{1}{1 - (4\pi/k^2) \alpha(\mathbf{k}, \mathbf{k}; \omega)} \tag{15.12}$$

can be used.

The importance of the dielectric function is that it directly relates the total, macroscopic potential in a plasma to the external one via Eq. (15.1) or (15.3). In the limit $\omega = 0$, Eq. (15.1) describes the screening of a static perturbation in a plasma. Zeros of the dielectric function, on the other hand, correspond to points at which it is possible to have charge density perturbations in the medium, without external driving forces. These roots are the collective modes. Their dispersion relation is given by the equation

$$\epsilon(\mathbf{k}, \omega) = 0, \tag{15.13}$$

which is an important formula in many-body theory.

The functions F, α, and ϵ describe the response of a plasma to longitudinal perturbations. Thus $\epsilon(\mathbf{k}, \omega)$, as defined by (15.1), is often called the *longitudinal dielectric function*. In writing the various equations which connect these functions, we have tacitly assumed that a longitudinal perturbation can only induce longitudinal currents in the plasma. Generally speaking, this is not the case. If magnetic fields are present, or if the crystal containing the plasma is anisotropic, longitudinal fields will give rise to both longitudinal and transverse currents. Under such conditions, the plasma response functions become tensors. These effects are particularly important at low frequencies ($\omega \lesssim \omega_c$), and play a crucial role in determining the properties of electromagnetic waves in the plasma.

On the other hand, at frequencies well above the cyclotron frequency ($\omega \gg \omega_c$), the magnetic field induced anisotropy of the plasma is small. In cubic crystals, the conductivity and dielectric tensors then become diagonal. These conditions are satisfied in many important solid state plasmas, such as the simple metals Na, K, etc. In such media, the response of the plasma to longitudinal, high-frequency ($\omega \gg \omega_c$) perturbations can be described by a single scalar function such as the plasma dielectric constant.

16. Response to Transverse Fields

To discuss the response of a many-body system to transverse, electromagnetic perturbations, we must generalize Eq. (13.12) to include coupling to a vector potential. To lowest order in the fields, this perturbation takes the form

$$H_1 = \int \rho(\mathbf{r}) \phi_{\text{ext}}(\mathbf{r}, t) \, d^3r - \frac{1}{c} \int \mathbf{j}(\mathbf{r}) \cdot \mathbf{A}_{\text{ext}}(\mathbf{r}, t) \, d^3r, \quad (16.1)$$

where $\mathbf{A}_{\text{ext}}(\mathbf{r}, t)$ is the vector potential, and

$$\mathbf{j}(\mathbf{r}) = (e/2m) \sum_i [\mathbf{p}_i \, \delta(\mathbf{r} - \mathbf{r}_i) + \delta(\mathbf{r} - \mathbf{r}_i) \, \mathbf{p}_i] \quad (16.2)$$

is the current density operator for the electron system. A straightforward application of Eq. (13.11) yields the following formula for the induced current in the medium

$$\begin{aligned}
\mathbf{j}_{\text{ind}}(\mathbf{r}, \omega) &\equiv \int_{-\infty}^{\infty} e^{i\omega t} \, \text{tr}\{\boldsymbol{\rho}[\mathbf{j}(\mathbf{r}) - (e/mc) \, \rho(\mathbf{r}) \, A_{\text{ext}}(\mathbf{r}, t)]\} \, dt \\
&= -(e/mc) \langle \rho(\mathbf{r}) \rangle \mathbf{A}_{\text{ext}}(\mathbf{r}, \omega) \\
&\quad + \int \mathbf{F}_{j\rho}(\mathbf{r}, \mathbf{r}'; \omega) \, \phi_{\text{ext}}(\mathbf{r}', \omega) \, d^3r' \\
&\quad - (1/c) \int \mathbf{F}_{jj}(\mathbf{r}, \mathbf{r}'; \omega) \cdot \mathbf{A}_{\text{ext}}(\mathbf{r}', \omega) \, d^3r', \quad (16.3)
\end{aligned}$$

where

$$\mathbf{F}_{j\rho}(\mathbf{r}, \mathbf{r}'; \omega) = -i \int_0^\infty e^{i(\omega+i\delta)t} \langle [\mathbf{j}(\mathbf{r}, t), \rho(\mathbf{r}', 0)] \rangle \, dt \quad (16.4)$$

and

$$\mathbf{F}_{jj}(\mathbf{r}, \mathbf{r}'; \omega) = -i \int_0^\infty e^{i(\omega+i\delta)t} \langle [\mathbf{j}(\mathbf{r}, t), \mathbf{j}(\mathbf{r}', 0)] \rangle \, dt. \quad (16.5)$$

The functions defined by Eqs. (16.4) and (16.5) are similar to the density–density response function [which in the present notations would be labeled $F \equiv F_{\rho\rho}(\mathbf{r}, \mathbf{r}'; \omega)$] defined by Eq. (13.14). As in the electrostatic problem, the response to transverse fields is determined by the thermal average of time-dependent commutators— in this case the current–current and current–charge commutators, rather than the charge–charge commutator which determines $F \,(\equiv F_{\rho\rho})$.

Equation (16.3) must be gauge invariant. This requirement imposes a relation between $\mathbf{F}_{j\rho}$ and $\mathbf{F}_{jj}$, which can also be proved by integrating Eq. (16.4) by parts and using the continuity equation to set $\partial \rho / \partial t = -\nabla \cdot \mathbf{j}$. The result is *(14)*

$$\mathbf{F}_{j\rho}(\mathbf{r}, \mathbf{r}'; \omega) = (ie/m\omega) \langle \rho(\mathbf{r}) \rangle \nabla_{\mathbf{r}'} \delta(\mathbf{r} - \mathbf{r}') + (i/\omega)[\nabla_{\mathbf{r}'} \cdot \mathbf{F}_{jj}(\mathbf{r}, \mathbf{r}'; \omega)]. \quad (16.6)$$

With the aid of this formula, Eq. (16.3) can be rewritten in the manifestly gauge-invariant form

$$\mathbf{j}_{\text{ind}}(\mathbf{r}, \omega) = \left(\frac{i}{\omega}\right) \int \left[\mathbf{F}_{jj}(\mathbf{r}, \mathbf{r}'; \omega) + \frac{e\langle\rho(r)\rangle}{m} \delta(\mathbf{r} - \mathbf{r}') \mathbf{1} \right] \cdot \mathbf{E}_{\text{ext}}(\mathbf{r}', \omega) \, d^3r' \quad (16.7)$$

where **1** is the unit tensor. The continuity equation can also be used to relate the response functions $F(\equiv F_{\rho\rho})$ and $\mathbf{F}_{jj}$. One finds

$$F \equiv F_{\rho\rho}(\mathbf{r}, \mathbf{r}'; \omega) = (-i/\omega) \nabla_{\mathbf{r}} \cdot \mathbf{F}_{j\rho}(\mathbf{r}, \mathbf{r}'; \omega)$$
$$= (\nabla_{\mathbf{r}} \nabla_{\mathbf{r}'}/\omega^2): \mathbf{F}_{jj}(\mathbf{r}, \mathbf{r}'; \omega) - (ne^2/m\omega^2) \nabla_{\mathbf{r}}^2 \delta(\mathbf{r} - \mathbf{r}'). \quad (16.8)$$

Equation (16.6) determines the induced current in terms of the *external* field applied to the plasma. As in the electrostatic case, the function which relates $\mathbf{j}_{\text{ind}}(\mathbf{r}, \omega)$ to the *total* field is a considerably **more** convenient one for most purposes. This is the *conductivity tensor*, defined by [see Eq. (13.1)]

$$\mathbf{j}_{\text{ind}}(\mathbf{r}, \omega) = \int \boldsymbol{\sigma}(\mathbf{r}, \mathbf{r}'; \omega) \cdot \mathbf{E}_{\text{total}}(\mathbf{r}', \omega) \, d^3r. \quad (16.9)$$

$\boldsymbol{\sigma}(\mathbf{r}, \mathbf{r}'; \omega)$ is related to $\mathbf{F}_{jj}(\mathbf{r}, \mathbf{r}'; \omega)$ by a complicated integral equation, similar to Eq. (14.4). However, this equation greatly

16. TRANSVERSE FIELDS

simplifies in the case of a translationally invariant system, where $\mathbf{F}_{jj}(\mathbf{r}, \mathbf{r}'; \omega) \equiv \mathbf{F}_{jj}(\mathbf{r} - \mathbf{r}'; \omega)$ and $\boldsymbol{\sigma}(\mathbf{r}, \mathbf{r}'; \omega) \equiv \boldsymbol{\sigma}(\mathbf{r} - \mathbf{r}'; \omega)$. One then finds

$$\mathbf{j}_{\text{ind}}(\mathbf{k}, \omega) = \boldsymbol{\sigma}(\mathbf{k}, \omega) \cdot \mathbf{E}_{\text{total}}(\mathbf{k}, \omega)$$
$$= \boldsymbol{\sigma}(\mathbf{k}, \omega) \cdot [\mathbf{E}_{\text{ext}}(\mathbf{k}, \omega) + \mathbf{E}_{\text{ind}}(\mathbf{k}, \omega)]. \quad (16.10)$$

By inverting the wave equation,

$$(k^2 - \mathbf{k}\mathbf{k} - \omega^2/c^2) \mathbf{E}_{\text{ind}} = (4\pi i \omega/c^2) \mathbf{j}_{\text{ind}}, \quad (16.11)$$

this expression can be rewritten in the form

$$\mathbf{j}_{\text{ind}}(\mathbf{k}, \omega)$$
$$= \boldsymbol{\sigma}(\mathbf{k}, \omega) \cdot \mathbf{E}_{\text{ext}}(\mathbf{k}, \omega)$$
$$+ \boldsymbol{\sigma}(\mathbf{k}, \omega) \left(\frac{4\pi i \omega}{c^2}\right) \left(\frac{1}{k^2 - \mathbf{k}\mathbf{k} - \omega^2/c^2}\right) \left(\frac{i}{\omega}\right) \left(\mathbf{F}_{jj}(\mathbf{k}, \omega) + \frac{n_0 e^2}{m} \mathbf{1}\right) \cdot \mathbf{E}_{\text{ext}}(\mathbf{k}, \omega)$$
$$= \left(\frac{i}{\omega}\right) \left(\mathbf{F}_{jj}(\mathbf{k}, \omega) + \frac{n_0 e^2}{m} \mathbf{1}\right) \cdot \mathbf{E}_{\text{ext}}(\mathbf{k}, \omega). \quad (16.12)$$

Since $\mathbf{E}_{\text{ext}}(\mathbf{k}, \omega)$ is an arbitrary vector, it follows that

$$\boldsymbol{\sigma}(\mathbf{k}, \omega) = \left(\frac{i}{\omega}\right) \left[\mathbf{F}_{jj}(\mathbf{k}, \omega) + \frac{n_0 e^2}{m} \mathbf{1}\right]$$
$$+ 4\pi \boldsymbol{\sigma}(\mathbf{k}, \omega) \left[\frac{1}{c^2(k^2 - \mathbf{k}\mathbf{k}) - \omega^2}\right] \left[\mathbf{F}_{jj}(\mathbf{k}, \omega) + \frac{n_0 e^2}{m} \mathbf{1}\right]. \quad (16.13)$$

This formula is the analog, for transverse response functions, of Eq. (14.6).

In an isotropic, translationally invariant system the conductivity tensor must (from symmetry) have the form

$$\sigma_{ij}(\mathbf{k}, \omega) = \sigma^{(1)}(\mathbf{k}, \omega) \delta_{ij} + (k_i k_j/k^2) \sigma^{(2)}(\mathbf{k}, \omega), \quad (16.14)$$

where $\sigma^{(1)}(\mathbf{k}, \omega)$ and $\sigma^{(2)}(\mathbf{k}, \omega)$ are *scalars*. This formula also applies to the macroscopic conductivity of cubic crystals. It is clear from Eq. (16.10) that the combination $\sigma_L(\mathbf{k}, \omega) \equiv [\sigma^{(1)} + \sigma^{(2)}]$ determines the response of the system to longitudinal perturbations, whereas $\sigma_T \equiv \sigma^{(1)}$ gives the response to transverse ones. In general, $\sigma_L(\mathbf{k}, \omega)$ is *not* equal to $\sigma_T(\mathbf{k}, \omega)$. This difference is hardly surprising since a longitudinal perturbation compresses the electron gas, whereas a transverse one does not. It is interesting to relate $\sigma_L(\mathbf{k}, \omega)$ to $\alpha(\mathbf{k}, \omega)$ [Eq. (14.1)]. For a translationally invariant system, one finds

$$\nabla \cdot \mathbf{j}_{\text{ind}} = -i\omega\rho_{\text{ind}}(\mathbf{r}, \omega) = -i\omega\alpha(\mathbf{k}, \omega)\phi_{\text{total}}(\mathbf{k}, \omega)$$
$$= ik\sigma_L(\mathbf{k}, \omega) E_{\text{total}}(\mathbf{k}, \omega) = k^2\sigma_L(\mathbf{k}, \omega)\phi_{\text{total}}(\mathbf{k}, \omega). \quad (16.15)$$

Thus

$$\sigma_L(\mathbf{k}, \omega) = -i\omega\alpha(\mathbf{k}, \omega)/k^2, \quad (16.16)$$

and [from Eq. (15.2)]

$$\epsilon_L(\mathbf{k}, \omega) = 1 - 4\pi\alpha(\mathbf{k}, \omega)/k^2 = 1 + (4\pi/i\omega)\,\sigma_L(\mathbf{k}, \omega), \quad (16.17)$$

which is the standard relation between dielectric constant and conductivity. One may also derive, from the wave equation, an expression for the *transverse dielectric function*:

$$(k^2 - \omega^2/c^2)\,\mathbf{E}_{\text{trans}}(\mathbf{k}, \omega) = (4\pi i\omega/c^2)\,\mathbf{j}_{\text{trans}}(\mathbf{k}, \omega)$$
$$= (4\pi i\omega/c^2)\,\sigma_T(\mathbf{k}, \omega)\,\mathbf{E}_{\text{trans}}(\mathbf{k}, \omega).$$

Hence,

$$\left[k^2 - \frac{\omega^2}{c^2}\left(1 + \frac{4\pi\sigma_T}{i\omega}\right)\right]\mathbf{E}_{\text{trans}}(\mathbf{k}, \omega) = \left[k^2 - \epsilon_T(\mathbf{k}, \omega)\frac{\omega^2}{c^2}\right]\mathbf{E}_{\text{trans}}(\mathbf{k}, \omega) = 0.$$
$$(16.18)$$

The result

$$\epsilon_T(\mathbf{k}, \omega) = 1 + (4\pi/i\omega)\,\sigma_T(\mathbf{k}, \omega), \quad (16.19)$$

is exactly analogous to Eq. (16.17). Note that for arbitrary $\mathbf{k}$, $\epsilon_L(\mathbf{k}, \omega) \neq \epsilon_T(\mathbf{k}, \omega)$. However, as $k \to 0$, $\sigma_L(\mathbf{k}, \omega) \to \sigma_T(\mathbf{k}, \omega)$ since the electrons cannot differentiate between transverse and longitudinal waves in this limit. Hence $\epsilon_L(0, \omega) = \epsilon_T(0, \omega)$. It is sometimes stated that this equality implies that optical experiments [which determine $\epsilon_T(0, \omega)$] and characteristic energy loss experiments [which determine ϵ_L] yield similar information concerning the many-body system. This connection is only valid if, by "characteristic energy loss experiments," one means those experiments in which the spectrum of *all* scattered electrons is measured. As we have seen in Section 11, this spectrum is sensitive to small k fluctuations and thus measures $S(0, \omega)$ or (equivalently) $\epsilon_L(0, \omega)$. On the other hand, in characteristic energy loss experiments in which the spectrum is measured *versus scattering angle*, the entire function $\epsilon_L(\mathbf{k}, \omega)$ is determined. Such measurements are among the most powerful tools available for probing the structure of the interacting electron gas, and yield far more information than can be derived from optical data. We will discuss them in Chapter IV.

17. The Fluctuation–Dissipation Theorem

From Eqs. (9.11) and (13.16) it is clear that there is a close relation between the response function $F(\mathbf{k}, \omega)$ and the structure factor $S(\mathbf{k}, \omega)$. This relation is based on thermodynamic arguments and therefore can only be proved when the many-body system is in thermal equilibrium. Nevertheless, it is an important result which is often used to determine $S(\mathbf{k}, \omega)$ from $F(\mathbf{k}, \omega)$.

To derive this relation *(16)*, we imagine that we know a complete set of states for the interacting system. In terms of these (and their energies) the expression for $S(\mathbf{k}, \omega)$ [Eq. (9.11)] becomes

$$S(\mathbf{k}, \omega) = (2\pi)^{-1} \int_{-\infty}^{\infty} e^{i\omega t} S(\mathbf{k}, t)\, dt$$

$$= \sum_{m,n} \left\{ (2\pi)^{-1} \int_{-\infty}^{\infty} e^{i\omega t} e^{-\beta E_m} \langle m \mid n_{\mathbf{k}} \mid n \rangle\, e^{i(E_m - E_n)t} \langle n \mid n_{-\mathbf{k}} \mid m \rangle\, dt \right\}$$

$$= \sum_{m,n} \{ e^{-\beta E_m} |\langle m \mid n_{\mathbf{k}} \mid n \rangle|^2\, \delta(\omega + E_m - E_n) \}. \tag{17.1}$$

In a similar way, one can show that the correlation function

$$T(\mathbf{k}, \omega) \equiv (2\pi)^{-1} \int_{-\infty}^{\infty} e^{i\omega t} T(\mathbf{k}, t)\, dt \equiv (2\pi)^{-1} \int_{-\infty}^{\infty} e^{i\omega t} \langle n_{-\mathbf{k}}(0)\, n_{\mathbf{k}}(t) \rangle\, dt$$

$$= \sum_{m,n} \{ e^{-\beta E_m} |\langle m \mid n_{-\mathbf{k}} \mid n \rangle|^2\, \delta(\omega + E_n - E_m) \}. \tag{17.2}$$

Upon interchanging the dummy indices m and n, we see that

$$T(\mathbf{k}, \omega) = e^{-\beta\omega} S(\mathbf{k}, \omega) \tag{17.3}$$

This formula enables one to relate the response function to the structure factor:

$$F(\mathbf{k}, \omega) = -i \int_0^\infty e^{i(\omega + i\delta)t} \langle [\rho_{\mathbf{k}}(t), \rho_{-\mathbf{k}}(0)] \rangle$$

$$= -ie^2 \int_0^\infty e^{i(\omega + i\delta)t} [S(\mathbf{k}, t) - T(\mathbf{k}, t)]\, dt$$

$$= -ie^2 \int_0^\infty e^{i(\omega + i\delta)t} \int_{-\infty}^\infty e^{-i\omega' t} [S(\mathbf{k}, \omega') - T(\mathbf{k}, \omega')]\, d\omega'$$

$$= e^2 \int_{-\infty}^\infty \left[\frac{S(\mathbf{k}, \omega') - T(\mathbf{k}, \omega')}{\omega - \omega' + i\delta} \right] d\omega'$$

$$= e^2 \int_{-\infty}^\infty \frac{(1 - e^{-\beta\omega'}) S(\mathbf{k}, \omega')}{(\omega - \omega' + i\delta)}\, d\omega'. \tag{17.4}$$

Since $S(\mathbf{k}, \omega)$ is a real quantity, it follows that

$$e^2 S(\mathbf{k}, \omega) = [\pi(e^{-\beta\omega} - 1)]^{-1} \text{Im}[F(\mathbf{k}, \omega)]. \tag{17.5}$$

Equation (17.5) relates the longitudinal electric response function to the charge density fluctuation for a system in thermodynamic equilibrium. There are similar equations relating such quantities as the spin susceptibility to the spin density fluctuation, or the transverse response function to current density fluctuations. These formulas are known as fluctuation–dissipation relations, and are derived in a manner which parallels that used to obtain Eq. (17.5).

Finally, we use Eq. (15.2) to relate $S(\mathbf{k}, \omega)$ to the longitudinal dielectric function. The result is

$$S(\mathbf{k}, \omega) = \frac{k^2}{4\pi^2 e^2} \frac{1}{[e^{-\beta\omega} - 1]} \text{Im}\left[\frac{1}{\epsilon(\mathbf{k}, \omega)}\right]. \tag{17.6}$$

This is an important formula which we will have several occasions to use in later work.

REFERENCES

1. Electron-electron scattering is discussed by W. Heitler, "The Quantum Theory of Radiation," p. 231. Oxford Univ. Press, London and New York, 1953. In the nonrelativistic limit, the scattering occurs via the Coulomb interaction. The Mott formula shows that exchange scattering (arising from the antisymmetrization of the electron wave function) is unimportant at small scattering angles. Electron-electron scattering becomes spin-dependent in the relativistic limit, and might then be used to study electron spin densities in solids.
2. The properties of the structure factor are discussed in several standard references. See, for example, D. Pines, "The Many-Body Problem," p. 35. Benjamin, New York, 1962.
3. L. Van Hove, *Phys. Rev.* **95**, 249 (1954).
4. A. Einstein, *Ann. Phys. (Leipzig)* **33**, 1275 (1910).
5. P. M. Platzman and N. Tzoar, *Phys. Rev.* **139**, A410 (1965).
6. P. M. Platzman, *Phys. Rev.* **139**, A379 (1965); A. L. McWhorter, "Physics of Quantum Electronics." McGraw-Hill, New York, 1966.
7. P. M. Platzman and N. Tzoar, *Phys. Rev.* **136**, A11 (1964).
8. W. Kohn and J. M. Luttinger, *Phys. Rev.* **97**, 1721 (1955); **98**, 915 (1955).
9. See, for example, G. H. Wannier, "Statistical Physics," p. 57. Wiley, New York, 1966.
10. M. Lax, *Phys. Rev.* **109**, 1921 (1958).
11. R. Kubo, *Can. J. Phys.* **34**, 1274 (1956); *J. Phys. Soc. Jap.* **12**, 570 (1957).
12. R. Kronig, *J. Opt. Soc. Amer.* **12**, 547 (1926); H. C. Kramers, *Atti Congr. Int. Fis. Como* **2**, 545 (1927); see also "Collected Scientific Papers." North-Holland Publ., Amsterdam.
13. The so-called "equation of motion method" is discussed in Pines (2), p. 44.
14. This averaging procedure is discussed by H. Ehrenreich, *Proc. Int. Sch. Phys. "Enrico Fermi"* **34**, 106 (1966).
15. S. L. Adler, *Phys. Rev.* **126**, 413 (1962); N. Wiser, *ibid.* **129**, 62 (1963).
16. D. N. Zubarev, *Sov. Phys.—Usp.* **3**, 320 (1960).

III. The Random Phase Approximation

In the preceding chapter we have defined several functions (S, F, α, ϵ) which determine the response of a plasma to longitudinal perturbations. It has also been suggested that, in a weakly coupled plasma, a free (noninteracting) electron approximation to the function α gives an accurate description of the system's behavior. This is indeed true. We now wish to explore the meaning of this approximation and apply it to two simple cases. For the time being, we restrict our attention to translationally invariant systems—either degenerate or Maxwellian. The former is a good model for the behavior of electrons in simple metals such as Na, K, and Al; the latter describes electron gases in lightly doped, single-valley semiconductors such as n-type GaAs, InSb, and InAs.

The approximation $\alpha = e^2 Q$ [see Eq. (14.7)] is known, for historical reasons, as the random phase approximation (RPA). It is widely used and has been discussed from a number of different viewpoints. We will present what we believe to be the simplest and most physical approach to this approximation—the *self-consistent field* (SCF) method developed by Ehrenreich and Cohen (*1*) and, independently, by Goldstone and Gottfried (*2*). Readers interested in other approaches to this problem are referred to an extensive and important literature on the subject (*3*). Several of these papers are landmarks in many-body theory, and have played a crucial role in elucidating the structure of the interacting electron system.

18. THE SELF-CONSISTENT FIELD METHOD

The goal of the SCF technique is to calculate the response of an interacting electron system to a weak, electrostatic perturbation, i.e., to determine the (longitudinal) plasma dielectric function. The SCF method attacks this problem directly by seeking approximate equations of motion for the charge density. For this reason, it is often called the equation of motion technique.

In a free electron gas the charge density operator is given by Eq. (10.1), $\rho_\mathbf{k} = e \sum_\mathbf{p} (c_\mathbf{p}^+ c_{\mathbf{p}+\mathbf{k}})$. Thus, one is led to a consideration of the equation of motion for the operator $c_\mathbf{p}^+ c_{\mathbf{p}+\mathbf{k}}$:

III. RANDOM PHASE APPROXIMATION

$$i \partial(c_p^+ c_{p+k})/\partial t = [c_p^+ c_{p+k}, H]. \tag{18.1}$$

Here H is the Hamiltonian of the many electron system, including its interaction with the perturbing electrostatic potential. For a free electron gas (no crystal potential), H has the form

$$H = H_0 + H_1, \tag{18.2}$$

where

$$H_0 = \sum_{p'} (p'^2/2m) c_{p'}^+ c_{p'} + \tfrac{1}{2} \sum_{p',p'',q}{}' [c_{p'+q}^+ c_{p''-q}^+ (4\pi e^2/q^2) c_{p''} c_{p'}], \tag{18.3}$$

and

$$H_1 = e\phi_{\text{ext}}(\mathbf{k}, \omega) e^{-i\omega t} \sum_{p'} (c_{p'+k}^+ c_{p'}). \tag{18.4}$$

The prime on the summation in Eq. (18.3) means that the term $q = 0$ is to be omitted, indicating that the charge of the electron gas is neutralized by a uniform, positive background charge. In Eq. (18.4) it is assumed that the perturbation has a well-defined wave vector $\mathbf{k}$ and frequency ω. The linear response of the system to perturbations of arbitrary spatial and temporal variation can be obtained by integrating our final result, with the appropriate weighting factor, over frequency and wave vector.

With this Hamiltonian the equation of motion we wish to study becomes

$$\begin{aligned} i \frac{\partial}{\partial t}(c_p^+ c_{p+k}) &= \left[\left(\frac{(\mathbf{p}+\mathbf{k})^2}{2m}\right) - \frac{p^2}{2m}\right] c_p^+ c_{p+k} \\ &+ \sum_{p'',q}{}' \left[c_p^+ c_{p''-q}^+ \left(\frac{4\pi e^2}{q^2}\right) c_{p''} c_{p+k-q}\right] \\ &- \sum_{p'',q}{}' \left[c_{p+q}^+ c_{p''-q}^+ \left(\frac{4\pi e^2}{q^2}\right) c_{p''} c_{p+k}\right] \\ &+ e\phi_{\text{ext}}(\mathbf{k},\omega) e^{-i\omega t}[c_p^+ c_p - c_{p+k}^+ c_{p+k}]. \end{aligned} \tag{18.5}$$

Next, we take the expectation value of this equation with respect to the complete wave function of the system—including the effects of H_1. The result is

$$\begin{aligned} i \frac{\partial}{\partial t}\langle c_p^+ c_{p+k}\rangle &= \left[\frac{(\mathbf{p}+\mathbf{k})^2}{2m} - \frac{p^2}{2m}\right]\langle c_p^+ c_{p+k}\rangle \\ &+ \sum_{p'',q}{}' \left[\langle c_p^+ c_{p''-q}^+ c_{p''} c_{p+k-q}\rangle \left(\frac{4\pi e^2}{q^2}\right)\right] \\ &- \sum_{p'',q}{}' \left[\langle c_{p+q}^+ c_{p''-q}^+ c_{p''} c_{p+k}\rangle \left(\frac{4\pi e^2}{q^2}\right)\right] \\ &+ e\phi_{\text{ext}}(\mathbf{k},\omega)e^{-i\omega t}[\langle c_p^+ c_p\rangle - \langle c_{p+k}^+ c_{p+k}\rangle]. \end{aligned} \tag{18.6}$$

18. SELF-CONSISTENT FIELD METHOD

If this formula could be used to calculate $\langle c_p^+ c_{p+k} \rangle$, it would then be possible, via Eq. (13.15), to determine the response functions (F, ϵ) that we are seeking. However, as it stands, the equation is a complicated one which involves unknown four-operator averages, such as $\langle c_{p+q}^+ c_{p''-q}^+ c_{p''} c_{p+k} \rangle$, on the right-hand side. These must be simplified to obtain a tractable result. This simplification is the crucial step in the SCF method.

The four-operator average can be rewritten in the form

$$\langle c_p^+ c_{p''-q}^+ c_{p''} c_{p+k-q} \rangle \equiv \langle c_p^+ c_{p+k-q} \rangle \langle c_{p''-q}^+ c_{p''} \rangle$$
$$+ \{\langle c_p^+ c_{p''-q}^+ c_{p''} c_{p+k-q} \rangle - \langle c_p^+ c_{p+k-q} \rangle \langle c_{p''-q}^+ c_{p''} \rangle\}. \tag{18.7}$$

Here the average of an operator product is expressed as the product of two averages, plus a second term [the bracketed portion of Eq. (18.7)] which represents *fluctuations* about the average values. The key approximation of the SCF method is to neglect the fluctuation terms in Eq. (18.7), i.e., to make the replacements

$$\langle c_p^+ c_{p''-q}^+ c_{p''} c_{p+k-q} \rangle \simeq \langle c_p^+ c_{p+k-q} \rangle \langle c_{p''-q}^+ c_{p''} \rangle$$
$$\langle c_{p+q}^+ c_{p''-q}^+ c_{p''} c_{p+k} \rangle \simeq \langle c_{p+q}^+ c_{p+k} \rangle \langle c_{p''-q}^+ c_{p''} \rangle. \tag{18.8}$$

This approximation is equivalent to using an *effective* Coulomb interaction of the form

$$V = \sum_{p',p'',q}{}' [c_{p'+q}^+ c_{p'} (4\pi e^2/q^2) \langle c_{p''-q}^+ c_{p''} \rangle] \tag{18.9}$$

in determining the equations of motion, rather than the true interaction given in Eq. (18.3). The difference between the two interactions is clear. In Eq. (18.3) electrons respond to the *instantaneous* field generated by other carriers, whereas in Eq. (18.9) they feel the *average* field. From this description we can begin to understand why the RPA is a good approximation in weakly coupled plasmas. In such systems, many electrons are within range of one another ($n_0 \lambda_{FT}^3 \gg 1$), hence the instantaneous field experienced by a typical electron does not deviate much from the average value. This is precisely what is required for Eq. (18.8) to be valid.

With the aid of Eq. (18.8), Eq. (18.6) takes the form

$$i\frac{\partial}{\partial t}\langle c_\mathbf{p}^+ c_{\mathbf{p+k}}\rangle = \left[\frac{(\mathbf{p+k})^2}{2m} - \frac{p^2}{2m}\right]\langle c_\mathbf{p}^+ c_{\mathbf{p+k}}\rangle$$
$$+ \sum_{\mathbf{p}'',\mathbf{q}}{}' \{[\langle c_\mathbf{p}^+ c_{\mathbf{p+k-q}}\rangle\langle c_{\mathbf{p}''-\mathbf{q}}^+ c_{\mathbf{p}''}\rangle - \langle c_{\mathbf{p+q}}^+ c_{\mathbf{p+k}}\rangle\langle c_{\mathbf{p}''-\mathbf{q}}^+ c_{\mathbf{p}''}\rangle](4\pi e^2/q^2)\}$$
$$+ e\phi_{\text{ext}}(\mathbf{k},\omega)\,e^{-i\omega t}[\langle c_\mathbf{p}^+ c_\mathbf{p}\rangle - \langle c_{\mathbf{p+k}}^+ c_{\mathbf{p+k}}\rangle]. \quad (18.10)$$

So far, we have made no use of the fact that $\phi_{\text{ext}}(\mathbf{k},\omega)$ is weak. We now evaluate all quantities in Eq. (18.10) to lowest order in the perturbation. In this approximation a product of averages, such as $\langle c_\mathbf{p}^+ c_{\mathbf{p+k-q}}\rangle\langle c_{\mathbf{p}''-\mathbf{q}}^+ c_{\mathbf{p}''}\rangle$, becomes

$$\langle c_\mathbf{p}^+ c_{\mathbf{p+k-q}}\rangle\langle c_{\mathbf{p}''-\mathbf{q}}^+ c_{\mathbf{p}''}\rangle$$
$$\cong \langle c_\mathbf{p}^+ c_{\mathbf{p+k-q}}\rangle_1\langle c_{\mathbf{p}''-\mathbf{q}}^+ c_{\mathbf{p}''}\rangle_0 + \langle c_\mathbf{p}^+ c_{\mathbf{p+k-q}}\rangle_0\langle c_{\mathbf{p}''-\mathbf{q}}^+ c_{\mathbf{p}''}\rangle_1, \quad (18.11)$$

where the subscripts 0 and 1 indicate the order of the term in question when expanded in powers of ϕ_{ext}. For a uniform plasma, translational invariance ensures that the average $\langle c_{\mathbf{p}''-\mathbf{q}}^+ c_{\mathbf{p}''}\rangle_0$ is zero unless $q = 0$. Thus, since $q = 0$ is excluded, the first term of Eq. (18.11) drops out. The last one only gives a contribution if $\mathbf{k} = \mathbf{q}$. This is a characteristic feature of the RPA—that only the kth Fourier component of the Coulomb potential contributes to the response of the system at wave vector $\mathbf{k}$. After reducing the second Coulomb term in Eq. (18.10) in a similar way, one obtains the final result:

$$i\frac{\partial}{\partial t}\langle c_\mathbf{p}^+ c_{\mathbf{p+k}}\rangle_1 = \left[\frac{(\mathbf{p+k})^2}{2m} - \frac{p^2}{2m}\right]\langle c_\mathbf{p}^+ c_{\mathbf{p+k}}\rangle_1$$
$$+ [n(\mathbf{p}) - n(\mathbf{p+k})]\left(\frac{4\pi e^2}{k^2}\right)\sum_{\mathbf{p}''}[\langle c_{\mathbf{p}''}^+ c_{\mathbf{p}''+\mathbf{k}}\rangle_1]$$
$$+ e\phi_{\text{ext}}(\mathbf{k},\omega)\,e^{-i\omega t}[n(\mathbf{p}) - n(\mathbf{p+k})], \quad (18.12)$$

where
$$n(\mathbf{p}) = \langle c_\mathbf{p}^+ c_\mathbf{p}\rangle_0. \quad (18.13)$$

Equation (18.12) is the basic formula of the random phase approximation.

19. Discussion of the Random Phase Approximation

The preceding discussion raises two important questions. What is the meaning of Eq. (18.12)? And what is the range of validity of the

19. DISCUSSION

approximation used to derive it? The first question is the easier of the two. To answer it, we observe that the Coulomb terms in Eq. (18.12) have been reduced to the form

$$[n(\mathbf{p}) - n(\mathbf{p} + \mathbf{k})](4\pi e^2/k^2) \sum_{\mathbf{p}''} [\langle c_{\mathbf{p}''}^+ c_{\mathbf{p}''+\mathbf{k}} \rangle_1] \tag{19.1}$$

which, with the aid of Eqs. (10.1) and (14.2), can be rewritten as

$$[n(\mathbf{p}) - n(\mathbf{p} + \mathbf{k})] \, e\phi_{\text{ind}}(\mathbf{k}, t). \tag{19.2}$$

Here $\phi_{\text{ind}}(\mathbf{k}, t)$ is the *average, induced potential* in the plasma. Equation (18.12) becomes

$$i \frac{\partial}{\partial t} \langle c_{\mathbf{p}}^+ c_{\mathbf{p}+\mathbf{k}} \rangle_1 = \left[\frac{(\mathbf{p}+\mathbf{k})^2}{2m} - \frac{p^2}{2m} \right] \langle c_{\mathbf{p}}^+ c_{\mathbf{p}+\mathbf{k}} \rangle_1$$
$$+ [n(\mathbf{p}) - n(\mathbf{p} + \mathbf{k})][e\phi_{\text{ind}}(\mathbf{k}, t) + e\phi_{\text{ext}}(\mathbf{k}, \omega) \, e^{-i\omega t}]. \tag{19.3}$$

In this formula the plasma responds as a *noninteracting* electron gas to the total potential, $\phi_{\text{total}} = \phi_{\text{ext}} + \phi_{\text{ind}}$. This is exactly what is implied by the approximation $\alpha = e^2 Q$ discussed earlier. As a further confirmation of this fact one can solve Eq. (18.12)—which is a simple integral equation—for $\langle c_{\mathbf{p}}^+ c_{\mathbf{p}+\mathbf{k}} \rangle_1$, and from it calculate the induced charge density. The resultant expression for $\epsilon(\mathbf{k}, \omega)$ is the same as that obtained from Eq. (15.2) by making the replacement $\alpha(\mathbf{k}, \omega) = e^2 Q(\mathbf{k}, \omega)$. We may summarize this discussion by saying that, in RPA, electrons respond as noninteracting particles to the *average, total* potential in the plasma, but that fluctuations about this average are ignored. It is important to recognize (as was pointed out in the preceding chapter) that this approximation is entirely different from that of neglecting all electron–electron interactions.

It is more difficult to discuss the range of validity of the approximation [Eq. (18.8)] used in deriving the RPA result. Intuitively, one expects that fluctuations should be relatively unimportant if there are many electrons within range of any given one, i.e., if $n_0 \lambda_{\text{FT}}^3 \gg 1$. This condition is certainly the right one, and is equivalent to the weak coupling criterion discussed in Chapter I. However, a detailed proof that the RPA is correct in the limit $r_s \ll 1$ is quite complicated. We will not discuss this question, which would take us far from the main theme of the book, but refer the reader to the literature for details (*3*).

There is one point, however, that deserves further comment. This is the behavior of Eq. (18.12) as $k \to 0$. In a weakly coupled plasma, one might hope to calculate the properties of the medium by a power series expansion in the strength of the electron–electron interaction. In Eq. (18.12), this would be equivalent to solving by iteration in the Coulomb term. Such a procedure fails because the Coulomb term contains a factor $4\pi e^2/k^2$ which diverges as $k \to 0$. This is an example of the fact, mentioned in Chapter I, that in all plasmas the Coulomb interactions dominate the long-wavelength behavior of the system. This divergence plagued early attempts to calculate the behavior of electron gases by perturbation theory. For most properties, each term in a strict perturbation expansion is divergent. The divergences result from the piling up of factors $4\pi e^2/k^2$ and, physically, are a consequence of the long-range nature of the Coulomb potential. It was not until the work of Gell-Mann and Brueckner (3) that this dilemma was fully resolved. They realized that to obtain convergent results via perturbation theory in the Coulomb problem, one must sum an infinite subclass of the perturbation terms. In particular, to obtain RPA one sums the most divergent terms in each order of e^2. This procedure requires justification from the mathematical point of view, but when carried through yields exactly the results predicted by Eq. (18.12).

Finally, it should be mentioned that the RPA is no more than a time-dependent version of the Hartree approximation [this is apparent from Eqs. (14.2) and (19.3)]. Hence, the factorization which is the key step in the analysis [Eq. (18.8)] is sometimes called *Hartree factorization*. It is important to notice that this factorization does not include all possible pairings of the operators involved. For example, a complete pairing would suggest the following approximation to the first line of Eq. (18.8):

$$\langle c_{\mathbf{p}}^+ c_{\mathbf{p}''-\mathbf{q}}^+ c_{\mathbf{p}''} c_{\mathbf{p}+\mathbf{k}-\mathbf{q}} \rangle \simeq \langle c_{\mathbf{p}}^+ c_{\mathbf{p}+\mathbf{k}-\mathbf{q}} \rangle \langle c_{\mathbf{p}''-\mathbf{q}}^+ c_{\mathbf{p}''} \rangle - \langle c_{\mathbf{p}}^+ c_{\mathbf{p}''} \rangle \langle c_{\mathbf{p}''-\mathbf{q}}^+ c_{\mathbf{p}+\mathbf{k}-\mathbf{q}} \rangle. \quad (19.4)$$

The new (last) term in this equation describes exchange interactions among the electrons. We will discuss these terms later, and see that the resultant theory is equivalent to the Hartree–Fock theory. For the time being, however, let us restrict our attention to the simple RPA (which is correct in the limit $r_s \ll 1$) and discuss the properties it predicts for various plasmas. The most useful quantity to investigate is the plasma dielectric function which, in RPA, becomes

$$\epsilon(\mathbf{k}, \omega) \simeq 1 - (4\pi e^2/k^2) Q(\mathbf{k}, \omega) \quad (19.5)$$

with $Q(\mathbf{k}, \omega)$ defined by Eq. (14.7). We will use this formula to study the properties of $\epsilon(\mathbf{k}, \omega)$ for both Fermi–Dirac and Maxwellian plasmas.

20. Static Properties of a Degenerate Plasma

At zero temperature in a Fermi–Dirac electron gas, the integrals appearing in Eq. (14.7) can be explicitly evaluated. The result, however, is complicated and obscures much of the physics. Instead of dealing with the complete expression, we prefer to analyze Eq. (19.5) in various limiting cases. First consider the limit $\omega/kv_F \to 0$ (where v_F is the Fermi velocity) which determines the response of the plasma to a static perturbation. An important example of such a perturbation is that produced in metals by dissolving foreign atoms in them. A typical case is that of Zn dissolved in Cu. Zn has one higher valence than Cu and thus will have a net positive charge relative to the host lattice. The field of the Zn atom is not, in fact, sufficiently weak to be considered as a small perturbation on the conduction electron system of the Cu—a nonlinear calculation would be necessary to completely describe its shielding. Nevertheless, in the limit of a dense electron gas, the perturbation *is* relatively small over most of its range. We will treat it as such. The linear calculation gives the general features of the screening and, in particular, is quite accurate in determining the long-range behavior of the potential.

In wave-vector space the external potential due to a point charge is $4\pi e/k^2$. Equation (15.1) then shows that the total potential is

$$\phi_{\text{total}}(\mathbf{k}) = 4\pi e/k^2 \epsilon(\mathbf{k}, 0). \tag{20.1}$$

If $\epsilon(\mathbf{k}, 0)$ were constant (independent of $\mathbf{k}$), this formula would be that appropriate to an insulator, in which the Coulomb field is uniformly reduced without change of shape. In a plasma, on the other hand, the strong wave-vector dependence of $\epsilon(\mathbf{k}, 0)$ completely modifies the form of the potential. We can study this effect by evaluating $\epsilon(\mathbf{k}, 0)$ from Eqs. (14.7) and (19.5). The result is

$$\epsilon(\mathbf{k}, 0) = 1 + \frac{\kappa_{\text{FT}}^2}{2k^2}\left[1 + \left(\frac{4k_F^2 - k^2}{4kk_F}\right)\ln\left|\frac{k + 2k_F}{k - 2k_F}\right|\right], \tag{20.2}$$

where k_F is the Fermi wave vector, and κ_{FT} the Fermi–Thomas wave vector defined by Eq. (5.12). In metals $\kappa_{\text{FT}} \simeq k_F$, corresponding to a screening length of about 10^{-8} cm. For long wavelengths ($k \ll k_F$), $\epsilon(\mathbf{k}, 0) \to (1 + \kappa_{\text{FT}}^2/k^2)$ and the total potential approaches $4\pi e^2/(k^2 + \kappa_{\text{FT}}^2)$, which is the Fourier transform of the function $(e^{-\kappa_{\text{FT}} r}/r)$. From this result one might be tempted to infer that screening in a degenerate plasma is exponential, as in the Maxwellian case. This is not at all true, however, for there is a quite surprising quantum-mechanical effect due to the short-wavelength features of $\epsilon(\mathbf{k}, 0)$. It is clear from Eq. (20.2) that $\epsilon(\mathbf{k}, 0)$ has an infinite slope at $k = 2k_\text{F}$. It can be shown that this anomaly gives rise to a charge density that falls off (for large r) as

$$A \cos[(2k_\text{F}r + \phi)/r^3], \qquad (20.3)$$

rather than exponentially as in a classical plasma. This remarkable effect was first noticed by Friedel (4), and has since been discussed in detail by Kohn and Vosko (5). It has been observed in a number of nuclear magnetic resonance experiments. In a typical measurement of this kind, one studies the resonance signal of nuclei in a metal into which has been dissolved small percentages of other elements. One system that has been carefully investigated (6) is that of Cu containing small amounts of other elements such as Zn, Ga, Ge, etc. These impurities produce electric fields, according to Eq. (20.3), which interact with the quadrupole moments of neighboring Cu nuclei. For Cu nuclei that are close to an impurity, the interaction is sufficiently large that they no longer contribute to the main Cu resonance. Thus, the nuclear resonance signal is smaller in the alloys than in pure Cu. The extent of this reduction gives a measure of the range of the fields surrounding impurity atoms. Experimentally, the reduction is much bigger than one would expect from exponentially screened impurity potentials. More detailed analysis shows that the resonance data agree, in all respects, with the predictions of Eq. (20.3). This agreement has an important implication. Equation (20.2) shows that the occurrence of an infinite slope in the static dielectric constant is a direct result of the sharp Fermi surface in the degenerate gas (5). If the Fermi surface were diffuse, by an amount Δk, the potential of Eq. (20.3) would be multiplied by a factor $e^{-(\Delta k)r}$. In the Cu resonance experiments the effects of the impurity potentials were observed out to sixth or seventh neighbors. From this result one may conclude that in Cu

$$\Delta k \leqslant 0.1 k_\text{F}. \qquad (20.4)$$

The complete form of the impurity potential can be obtained by Fourier transforming Eq. (20.1). This integration must be done numerically (7). The computations show that $\phi_\text{total}(\mathbf{r})$ has about the same range as the Fermi–Thomas potential, but differs from it in having a finite electron charge density at $\mathbf{r} = 0$, as well as the long, oscillatory tail described by Eq. (20.3). Differences at the origin should not be taken seriously, since there the potential is strong and neither calculation can be correct. At larger r, however, the discrepancy is a real one. The experiments indicate quite definitely that Eq. (20.3) is the more correct form in this range.

21. High-Frequency Properties of a Degenerate Plasma

So far we have considered the static behavior of a degenerate plasma. We now return to Eqs. (14.7) and (19.5) to investigate its high-frequency properties. In the long-wavelength limit, we will see that the behavior of the gas is controlled by the collective modes. Assuming that $\omega/k \gg v_F$ and $k \ll k_F$, one can show that

$$\epsilon(\mathbf{k}, \omega) \simeq 1 - (\omega_p^2/\omega^2)[1 + \tfrac{3}{5}(k^2 v_F^2/\omega^2)], \tag{21.1}$$

where $\omega_p^2 = 4\pi n_0 e^2/m$. It is important to notice that, in the limit $\omega/k \gg v_F$, the denominator of Eq. (14.7) never vanishes within the range of the p integration for a Fermi gas. As a consequence, the dielectric constant is real. $\epsilon(\mathbf{k}, \omega)$ has a zero at the point

$$\omega^2(\mathbf{k}) \simeq \omega_p^2 + \tfrac{3}{5} k^2 v_F^2. \tag{21.2}$$

This root [see Eq. (15.13)] of $\epsilon(\mathbf{k}, \omega)$ is the longitudinal collective mode of the electron gas. Its phase and group velocities are given by the formulas

$$v_\phi \equiv \omega(\mathbf{k})/k \simeq (\omega_p^2 + \tfrac{3}{5} k^2 v_F^2)^{1/2}/k, \tag{21.3}$$

$$v_g \equiv \partial\omega(\mathbf{k})/\partial k \simeq \tfrac{3}{5} k v_F^2/(\omega_p^2 + \tfrac{3}{5} k^2 v_F^2)^{1/2} \tag{21.4}$$

Note that $v_\phi \to \infty$ and $v_g \to 0$ as $k \to 0$. The fact that $\text{Im}(\epsilon) = 0$ indicates that, in RPA, long-wavelength plasma waves are undamped excitations of an electron gas. These waves are, of course, the plasmons observed in characteristic energy loss experiments. By combining Eqs. (17.6) and (21.1), one can show that long-wavelength plasmons contribute a term

$$S(\mathbf{k}, \omega)|_{\text{plasmon}} = (k^2/8\pi e^2)\, \omega(\mathbf{k})\, \delta[\omega - \omega(\mathbf{k})] \tag{21.5}$$

to the structure factor.

We now consider the behavior of plasma waves of shorter wavelength (large k). For small k, the phase velocity of the waves is large—greater than the Fermi velocity. As k increases, the velocity decreases, ultimately reaching the Fermi velocity. At this point, there is an important change in the energy denominators of Eq. (14.7). So long as the phase velocity is larger than the Fermi velocity these denominators are finite; $\epsilon(\mathbf{k}, \omega)$ is purely real; and the plasma waves are undamped. However, at a critical value of k ($k = k_c$) the denominator of Eq. (14.7) vanishes for electrons traveling parallel to the wave at the Fermi velocity.

The integral then acquires a sizable imaginary part. The point $k = k_c$ marks an abrupt change in the behavior of the plasma waves. For $k < k_c$, they are weakly damped (there is no damping in RPA), whereas for $k \geqslant k_c$ the damping is so strong that they no longer are well-defined collective excitations. Damping of this sort is called *Landau damping* (8). Its physical origin can be seen by examining Eq. (14.7). The vanishing of the denominator in this formula—which signals the onset of Landau damping—first occurs when it is possible to satisfy the condition

$$\left[\omega(\mathbf{k}) + \frac{p^2}{2m} - \frac{(\mathbf{p} + \mathbf{k})^2}{2m}\right] = 0 \qquad (21.6)$$

for $p \leqslant k_F$. This condition expresses energy and momentum conservation for a process in which the wave transfers its energy and momentum to a *single* electron in the plasma. Such single-particle decays are impossible for long-wavelength ($\omega_p/kv_F \gg 1$) plasmons. Hence, they are undamped in RPA. In real metals, there is a small residual damping of long-wavelength plasmons caused by a process in which two electrons share the energy of the wave. We will discuss this process, which is beyond the RPA, somewhat later.

Landau damping may also be viewed from a classical point of view. Imagine that a wave with phase velocity $v_\phi = \omega/k$ is propagating through the electron gas, and consider the behavior of an electron whose velocity (parallel to $\mathbf{k}$) is close to v_ϕ. In the wave frame, the electron is almost at rest and sees an essentially time-independent electric field. This field continuously transfers energy from the wave to those electrons whose velocities are matched to it. Such electrons are essentially "surf riding" on the wave. A detailed calculation (9) shows that electrons whose velocities are just below that of the wave draw energy from it, while electrons traveling slightly faster than the wave transfer energy to it. As a consequence, the net rate of energy transfer is proportional to the derivative of the velocity distribution, evaluated at the velocity $v = v_\phi$.

The onset of Landau damping is abrupt in degenerate plasmas because there is a strict upper bound (v_F) on the particle velocities. An approximate formula for the critical wave vector (k_c) at which the damping begins can be derived by matching the phase velocity of the wave to the Fermi velocity:

$$v_F \simeq v_\phi = \omega(k_c)/k_c \simeq \omega_p/k_c \qquad \text{or} \qquad k_c \simeq \omega_p/v_F . \qquad (21.7)$$

This result has the form

$$k_c/k_F \simeq (\text{const})(r_s^{1/2}), \qquad (21.8)$$

where r_s is the strength parameter for a degenerate plasma defined in Eq. (2.3). A more accurate treatment shows that the "constant" varies slightly with r_s. Ferrell (10) has calculated it, and finds values ranging between 0.7 and 1.1 as r_s varies from 2 to 6 (extreme values for metals).

22. Screening in Classical Plasmas

We now consider Maxwellian plasmas. In this case one is usually interested in the weak-coupling limit, where $\epsilon(\mathbf{k}, \omega)$ differs appreciably from unity only if $k \ll mv_{\text{th}}$ (v_{th} is the particle thermal velocity). Thus, one may use a small k approximation in evaluating Q. Dropping the $k^2/2m$ term in the denominator of Eq. (14.7), and replacing $[n(\mathbf{p}) - n(\mathbf{p} + \mathbf{k})]$ by $-\mathbf{k} \cdot \nabla_p n = (-\mathbf{k} \cdot \mathbf{v}) \, \partial n/\partial E$, yields the result

$$Q \simeq -n_0 \int \frac{(\mathbf{v} \cdot \mathbf{k})(\partial f_0/\partial E) \, d^3v}{(\omega - \mathbf{v} \cdot \mathbf{k} + i\delta)} \tag{22.1}$$

which, for a Maxwellian distribution, becomes

$$Q(\mathbf{k}, \omega) = \frac{n_0}{kT} \int \frac{(\mathbf{v} \cdot \mathbf{k}) f_0(\mathbf{v}) \, d^3v}{(\omega - \mathbf{v} \cdot \mathbf{k} + i\delta)}. \tag{22.2}$$

As is customary in the classical case, we have written these formulas in terms of the velocity distribution, $n_0 f_0(\mathbf{v}) = (m/2\pi)^3 \, n(\mathbf{p})$. $f_0(\mathbf{v})$ is normalized so that $\int f_0(\mathbf{v}) \, d^3v = 1$.

Equation (22.2) has been derived quantum mechanically, but the result is essentially a classical one. The formula follows directly from the Boltzmann equation

$$\partial f/\partial t + \mathbf{v} \cdot \nabla f - (e/m) \nabla \phi_{\text{total}} \cdot \nabla_v f = 0. \tag{22.3}$$

Assuming ϕ_{total} is a small perturbation, one may linearize Eq. (22.3) by writing $f \simeq f_0(\mathbf{v}) + f_1(\mathbf{v}, \mathbf{r}, t)$. The resulting equation is the Fourier transform (in the small k limit) of Eq. (19.3). Its solution is

$$f_1 = -\frac{en_0}{m} \frac{[\mathbf{k} \cdot \nabla_v f_0] \, \phi_{\text{total}}(\mathbf{k}, \omega)}{(\omega - \mathbf{k} \cdot \mathbf{v})} \tag{22.4}$$

and the induced electron density is

$$n_{\text{ind}}(\mathbf{k}, \omega) = -\frac{en_0}{m} \phi_{\text{total}}(\mathbf{k}, \omega) \int \frac{\mathbf{k} \cdot \nabla_v f_0 \, d^3v}{\omega - \mathbf{k} \cdot \mathbf{v}}. \tag{22.5}$$

For a Maxwellian $f_0(\mathbf{v})$, Eq. (22.5) takes the form

$$n_{\text{ind}}(\mathbf{k},\omega) = \left[\frac{en_0}{kT}\int \frac{(\mathbf{v}\cdot\mathbf{k})f_0(\mathbf{v})\,d^3v}{\omega - \mathbf{k}\cdot\mathbf{v}}\right]\phi_{\text{total}}(\mathbf{k},\omega)$$
$$= eQ(\mathbf{k},\omega)\,\phi_{\text{total}}(\mathbf{k},\omega). \qquad (22.6)$$

This expression is identical to Eq. (22.1). Note that we have derived Eq. (22.6) from a collisionless Boltzmann equation in which the perturbing potential is the total electrostatic potential. As before, $\phi_{\text{total}} = \phi_{\text{ext}} + \phi_{\text{ind}}$, and ϕ_{ind} must be determined in a self-consistent way from Poisson's equation and Eq. (22.6). This combination of a collisionless Boltzmann equation with a self-consistently determined electrostatic potential is generally called the Boltzmann–Vlasov equation. It is the classical analog of the Hartree approximation and is valid under the same circumstances, namely when $V/K \ll 1$. More generally, if magnetic fields and transverse currents are present, Eq. (22.3) must be generalized as follows:

$$\partial f/\partial t + \mathbf{v}\cdot\nabla\mathbf{f} + (e/m)\left(\mathbf{E} + \frac{\mathbf{v}\times\mathbf{H}}{c}\right)\cdot\nabla_v f = 0, \qquad (22.7)$$

where $\mathbf{E}$ and $\mathbf{H}$ are the *total* electric and magnetic fields in the plasma. These fields must be calculated in a self-consistent way from the Maxwell equations. We will return to this point in Chapter VII.

Equation (22.2) can easily be integrated if $\omega = 0$. The result is

$$\epsilon(\mathbf{k},0) = 1 - (4\pi e^2/k^2)Q(\mathbf{k},0) = 1 + 1/\lambda_D^2 k^2, \qquad (22.8)$$

where λ_D (the Debye length) is defined by Eq. (5.5). From Eq. (15.1) we infer that the total potential about a charge e in a classical plasma has the form

$$\phi_{\text{total}}(r) = (e/r)\exp(-r/\lambda_D).$$

The screening is exponential, in contrast to the algebraic screening found in degenerate plasmas. However, plasmas having other sorts of classical velocity distributions give different types of screening. In particular, it has recently been shown (*11*) that *anisotropic*, classical plasmas can have an algebraic (rather than exponential) fall-off of the screening cloud. It may be, therefore, that exponential screening is a rather special case even in classical plasmas. This point deserves further study.

23. Plasma Waves in Maxwellian Plasmas

We next consider the response of Maxwellian plasmas to finite-frequency perturbations. The analysis of this case is slightly more complicated than that of the Fermi–Dirac distribution because the integrals which determine $Q(\mathbf{k}, \omega)$ yield transcendental functions. In general, these functions must be calculated numerically. However, in the limit $\omega/kv_{\text{th}} \gg 1$, Q can be evaluated by expanding the denominator of Eq. (22.2) in powers of $(\mathbf{k} \cdot \mathbf{v}/\omega)$. This procedure is valid over the important part $(v \lesssim v_{\text{th}})$ of the range of the velocity integration. In addition, there is a contribution to the integral from the pole at $\omega = \mathbf{k} \cdot \mathbf{v}$. If $\omega/kv_{\text{th}} \gg 1$, this term is quite small. Nevertheless, it must be retained since, when ω is real, it is the only imaginary contribution to $\epsilon(\mathbf{k}, \omega)$. A straightforward calculation yields the approximate formula

$$\epsilon(\mathbf{k}, \omega) \simeq \left\{1 - \frac{\omega_p^2}{\omega^2}\left[1 + \frac{(3k_BT)k^2}{m\omega^2}\right] - i\pi\left(\frac{4\pi ne^2}{mk^2}\right)\int \delta(\omega - \mathbf{k}\cdot\mathbf{v})\mathbf{k}\cdot\frac{\partial f_0}{\partial \mathbf{v}}\,d^3v\right\},$$

(23.1)

which is valid when $\omega/kv_{\text{th}} \gg 1$. In evaluating the last term which arises from the pole at $\omega = \mathbf{k} \cdot \mathbf{v}$ it has been assumed that the imaginary part of ω is small, as is actually the case for plasma waves whose phase velocities are large $(v_\phi \gg v_{\text{th}})$. The classical expression for $\epsilon(\mathbf{k}, \omega)$ always has a finite imaginary part. In this respect, there is a qualitative difference between Maxwellian and degenerate plasmas. For the latter, there is a strict upper bound (v_F) on the particle velocities and $\epsilon(\mathbf{k}, \omega)$ is purely real when $\omega/kv_F > 1$. On the other hand, in the Maxwellian case there are a few particles in the tail of the distribution whose velocities match that of the wave in question, even if $\omega/kv_{\text{th}} \gg 1$. As a consequence, plasma waves in Maxwellian plasmas are always Landau damped, though the effect is feeble when $\omega/kv_{\text{th}} \gg 1$.

The plasma frequency, as usual, is determined by the condition $\epsilon(\mathbf{k}, \omega) = 0$. From Eq. (23.1) one finds

$$\omega^2(\mathbf{k}) \simeq \omega_p^2\left\{1 + \frac{(3k_BT)k^2}{m\omega_p^2} + \left(\frac{i\pi\omega_p^2}{k^2}\right)\int \delta(\omega - \mathbf{k}\cdot\mathbf{v})\mathbf{k}\cdot\frac{\partial f_0}{\partial \mathbf{v}}\,d^3v\right\}$$

$$= \omega_p^2\left\{1 + \frac{k^2\langle v^2\rangle}{\omega_p^2} + i\pi\left(\frac{1}{k^2\lambda_D^2}\right)\left(\frac{\omega}{k}\right)\left(\frac{m}{2\pi k_BT}\right)^{1/2}\exp\left[-\frac{m\omega^2}{2k^2(k_BT)}\right]\right\}.$$

(23.2)

Note that in this formula the Landau damping, as previously indicated, is proportional to $\partial f_0/\partial v$ evaluated at the phase velocity of the wave.

Before leaving the subject of Landau damping, we should mention that this name is somewhat of misnomer—since the process is not truly one of damping at all. For instance, one can easily show that the Vlasov equation conserves the entropy,

$$S = \int f(\mathbf{v}) \ln[f(\mathbf{v})] \, d^3v \, d^3x, \tag{23.3}$$

of the electron system. One is forced to the conclusion that Landau damping is not dissipative. Instead, it is a process of phase mixing. Detailed calculations (12) show that the plasma wave, rather than being an eigenmode of the Vlasov equation, is a mixture of such eigenmodes. When the wave is excited, the various modes are in phase and there is an appreciable potential in the plasma. As time progresses, however, these phases change at different rates and a cancellation results, which is Landau "damping." When viewed in this light, the process is clearly not a dissipative one. In principle, it can be reversed. This has actually been done recently in plasma wave echo experiments in gas plasmas (13).

Finally, a word should be said about the evaluation of $\epsilon(\mathbf{k}, \omega)$ for arbitrary values of ω/k. If the velocity distribution is Maxwellian, the integral which determines Q can be written in the form

$$\begin{aligned}Q(\mathbf{k}, \omega) &= \left(\frac{n}{k_B T}\right) \int_{-\infty}^{\infty} \left(\frac{m}{2k_B T}\right)^{1/2} \exp\left(-\frac{mv^2}{2k_B T}\right) \frac{kv}{\omega - kv} \, dv \\ &= \left(\frac{n}{k_B T}\right) \pi^{-1/2} \int_{-\infty}^{\infty} e^{-x^2} \frac{x}{(\eta - x)} \, dx \\ &= \tfrac{1}{2}(n/k_B T) Z'(\eta), \end{aligned} \tag{23.4}$$

where

$$\eta = (m/2k_B T)^{1/2}(\omega/k) \tag{23.5}$$

and

$$Z(\eta) \equiv \pi^{-1/2} \int_{-\infty}^{\infty} \frac{e^{-x^2} \, dx}{(x - \eta)} \tag{23.6}$$

is the plasma dispersion function. $Z(\eta)$ has been tabulated, for complex values of η, by Fried and Conte (14). The dielectric constant is related to Z through the formula

$$\epsilon(\mathbf{k}, \omega) = 1 - \tfrac{1}{2}(1/\lambda_D^2 k^2) Z'(\eta). \tag{23.7}$$

REFERENCES

1. H. Ehrenreich and M. H. Cohen, *Phys. Rev.* **115**, 786 (1959).
2. J. Goldstone and K. Gottfried, *Nuovo Cimento* [10] **13**, 849 (1959).
3. D. Bohm and D. Pines, *Phys. Rev.* **92**, 609 (1953); M. Gell-Mann and K. A. Brueckner, *ibid.* **106**, 364 (1957); K. Sawada, K. A. Brueckner, N. Fukuda, and R. Brout, *ibid.*

108, 507 (1957); R. Brout, *ibid.* **108**, 515 (1957); J. Hubbard, *Proc. Roy. Soc. Ser A.* **243**, 336 (1957); P. Nozières and D. Pines, *Nuovo Cimento* [10] **9**, 470 (1958); D. F. DuBois, *Ann. Phys. (New York)* **7**, 174 (1959); **8**, 24 (1959).
4. J. Friedel, *Nuovo Cimento Suppl.* **7**, 287 (1958).
5. W. Kohn and S. H. Vosko, *Phys. Rev.* **119**, 912 (1960).
6. T. Rowland, *Phys. Rev.* **119**, 900 (1960).
7. J. Langer and S. J. Vosko, *Phys. Chem. Solids* **12**, 196 (1960).
8. L. D. Landau, *J. Phys. (USSR)* **10**, 25 (1946).
9. See T. H. Stix, "The Theory of Plasma Waves." McGraw-Hill, New York, 1962.
10. R. A. Ferrell, *Phys. Rev.* **107**, 450 (1957).
11. G. Joyce and D. Montgomery, *Phys. Fluids* **10**, 2017 (1967).
12. N. G. VanKampen, *Physica* **21**, 949 (1955); K. M. Case, *Ann. Phys. (New York)* **7**, 349 (1959).
13. R. W. Gould, T. M. O'Neil, and J. H. Malmberg, *Phys. Rev. Lett.* **19**, 219 (1967); R. W. Gould, *Phys. Letts. A* **25**, 559 (1967); J. H. Malmberg, C. B. Wharton, R. W. Gould, and T. M. O'Neil, *ibid.* **20**, 95 (1968).
14. B. D. Fried and S. D. Conte, "The Plasma Dispersion Function." Academic Press, New York, 1961.

IV. Comparison with Experiment — Degenerate Plasmas

24. RPA Formula for the Spectrum

In the preceding chapter, we have discussed the RPA approximation to the dielectric constant of an electron gas. In this chapter we use these results to interpret characteristic energy loss experiments which give direct information about the properties of plasmons in metals. Here, of course, one is dealing with the case of a degenerate plasma. Initially, we will assume that the plasma is translationally invariant, i.e., that band structure effects can be neglected. This is a good approximation in the simpler metals, such as Al, Na, K, etc. Later, we consider the modifications of the formalism that are required when the electrons move in a periodic crystal potential.

Our basic formula is Eq. (9.10) which relates the energy distribution of scattered electrons to the structure factor. $S(\mathbf{k}, \omega)$, in turn, is related to the dielectric functions via the fluctuation–dissipation theorem [Eq. (17.6)]. In RPA, $\epsilon(\mathbf{k}, \omega)$ is given by Eq. (19.5). Combining these equations yields the following approximate expression for $S(\mathbf{k}, \omega)$ in the low-temperature limit:

$$S^{\text{RPA}}(\mathbf{k}, \omega) = -(k^2/4\pi^2 e^2)\, \theta(\omega)\, \text{Im}[1/\epsilon(\mathbf{k}, \omega)]$$
$$= -(1/\pi)\, \theta(\omega) \frac{\text{Im}[Q(\mathbf{k}, \omega)]}{|\epsilon(\mathbf{k}, \omega)|^2}. \tag{24.1}$$

Equation (14.7) implies that

$$\theta(\omega)\, \text{Im}[Q(\mathbf{k}, \omega)] = -\pi \theta(\omega) \sum_{\mathbf{p}} \left\{ [n(\mathbf{p}) - n(\mathbf{p}+\mathbf{k})]\, \delta\!\left[\omega + \frac{p^2}{2m} - \frac{(\mathbf{p}+\mathbf{k})^2}{2m}\right] \right\}$$
$$= -\pi \sum_{\mathbf{p}} \left\{ n(\mathbf{p})[1 - n(\mathbf{p}+\mathbf{k})]\, \delta\!\left[\omega + \frac{p^2}{2m} - \frac{(\mathbf{p}+\mathbf{k})^2}{2m}\right] \right\}. \tag{24.2}$$

Upon comparing this result to Eq. (10.6), we see that $\theta(\omega)\,(\text{Im}\,Q)$ is proportional to the structure factor $S^0(\mathbf{k}, \omega)$ of the noninteracting electron gas. Thus, Eq. (24.1) can also be rewritten in the form

$$S^{\text{RPA}}(\mathbf{k}, \omega) = \frac{S^0(\mathbf{k}, \omega)}{|\epsilon(\mathbf{k}, \omega)|^2}. \tag{24.3}$$

Finally, by substituting this expression into Eq. (9.11), we obtain the following equation for the differential cross section:

$$d^2\sigma/d\Omega\, d\omega = (2me^2/k^2)^2 \frac{S^0(\mathbf{k},\omega)}{|\epsilon(\mathbf{k},\omega)|^2}.\qquad(24.4)$$

This formula will be used in analyzing the characteristic energy loss experiments.

25. Integrated Spectrum Characteristic Energy Loss Experiments

Broadly speaking, characteristic energy loss experiments fall into two categories: those in which one measures the energy distribution of all scattered electrons (irrespective of angle), and those in which the energy distribution is measured as a function of the scattering angle. The latter, though more difficult to perform, are also far more powerful since they determine $S(\mathbf{k}, \omega)$ as a function of both $\mathbf{k}$ and ω. Our most interesting conclusions will be drawn from experiments of this type. Historically, however, the first characteristic energy loss experiments were those in which the spectrum of all scattered electrons was measured (integrated-spectrum experiments). Since Eq. (24.4) heavily weights small k scatterings, such experiments are of principal value in determining the frequency ω_p of long-wavelength plasmons. Experiments of this kind have been performed in several ways (reflection, transmission, etc.) and in a wide variety of materials. We will not review the data in detail (1), but merely point out that in simple metals (particularly those without d bands) there is good agreement between measured and theoretical values of ω_p. Table IV-I shows typical results (1). In calculating the values of ω_p listed in the table, the electron density was determined from the number of valence electrons shown in the second column and the

Table IV-I

Element	Electron no.	$\hbar\omega_p$ (obs.) (eV)	$\hbar\omega_p$ (calc.) (eV)
Al	3	15.0	15.7
Be	2	18.9	18.8
Mg	2	10.5	10.9
Si	4	16.9	16.6
Ge	4	16.0	15.6
Sb	5	15.3	15.1
Na	1	5.7	5.9

free electron mass (rather than an effective mass) was used. The reader may be puzzled by the inclusion of two semiconductors (Si and Ge) in this tabulation. These elements do, in fact, show well-defined plasmon peaks in their characteristic energy loss spectra. The plasmon energy in these materials is so much greater than the band gap that the plasma can oscillate almost as a free electric gas, hardly sensing the presence of the gap. This argument, which is developed in detail in Section 31, also explains why the free electron mass rather than an effective mass is used in calculating ω_p.

26. Mean Free Path for Plasmon Emission

Another useful quantity that can be measured with integrated spectrum experiments (in transmission) is the mean free path for plasmon excitation. Experimentally, this parameter is determined by measuring the relative strengths of one, two, three, etc., plasmon lines in the characteristic energy loss spectrum (see Fig. I-2). The intensities are fitted to a Poisson distribution

$$I_n = (l/\lambda)^n \, e^{-l/\lambda}/n!, \tag{26.1}$$

in which I_n is the strength of the n-plasmon line, l the sample thickness, and λ the mean free path. The fit determines λ. To compare the experimental values with theory, one can derive an expression for the plasmon mean free path, using Eq. (21.5) to determine the plasmon contribution to the structure factor. The cross section (per electron) for plasmon scattering is

$$d\sigma/d\Omega \,|_{\text{plasmon}} = 2me^4/k^2 \omega_p. \tag{26.2}$$

To integrate this expression over angle, we note that k^2 is related to the scattering angle θ by the energy and momentum conservation laws:

$$k^2 = |\mathbf{k}_0 - \mathbf{k}_1|^2 = k_0^2 + k_1^2 - 2k_0 k_1 \cos\theta$$
$$k_0^2/2m \simeq (k_1^2/2m) + \omega_p. \tag{26.3}$$

As before, $\mathbf{k}_0$ and $\mathbf{k}_1$ are the initial and final momenta of the fast electron which excites the plasmon. For small scattering angles,

$$k^2 \simeq k_0^2(\vartheta^2 + \vartheta_E^2), \tag{26.4}$$

where

$$\vartheta_E = \omega_p/2E_0 \quad \text{and} \quad E_0 = k_0^2/2m. \tag{26.5}$$

Hence Eq. (26.2) can be rewritten in the form

$$d\sigma/d\Omega |_{\text{plasmon}} \simeq e^4/E_0\omega_p(\vartheta^2 + \vartheta_E^2), \qquad (26.6)$$

and integrated over angle to yield the result

$$1/\lambda_{\text{plasmon}} \simeq [2\pi n_0 e^4/E_0(\hbar\omega_p)] \ln(\vartheta_c/\vartheta_E), \qquad (26.7)$$

where $\vartheta_c = (k_c/k_0)$. This expression agrees reasonably well ($\pm 20\%$) with experiment. Its most important feature is the $(1/E_0)$ variation of the plasmon scattering rate. A typical experimental value (for Al) is $\lambda_{\text{plasmon}} = 600\text{Å}$ for $E_0 = 30$ keV.

27. Surface Plasmons

In many characteristic energy loss experiments a peak is observed at an energy below the volume plasmon loss. This feature is ascribed to *surface plasmon* excitation. The surface plasma wave propagates at the interface between the metal and the adjacent dielectric medium. Its frequency depends upon the geometry of the surface, as well as its composition. The properties of surface plasmons can be approximately calculated by treating the plasma as a dielectric medium with $\epsilon = (1 - \omega_p^2/\omega^2)$. For a plane surface, between the plasma and a medium of dielectric constant ϵ_0, solutions of Poisson's equation have the form

$$\phi \sim \exp[i(k_x x + k_y y)] \exp[-K|z|] \quad \text{where} \quad K^2 = k_x^2 + k_y^2. \quad (27.1)$$

The potential and the normal component of the displacement vector must be continuous at the surface. These boundary conditions lead to the equation

$$-K\epsilon_0 = K\{1 - \omega_p^2/\omega^2\}, \qquad (27.2)$$

which is satisfied if

$$\omega_{\text{surface}} = \omega_p/(1 + \epsilon_0)^{1/2}. \qquad (27.3)$$

Equation (27.3) gives the frequency of surface modes of moderate wave vector at a plane surface. For large k, this formula must be corrected for particle motion [as described by Eq. (21.2)]. At small wave vectors, the dispersion relation is modified by coupling of the surface modes to the electromagnetic field. This coupling, which does not occur for volume plasmons, takes place because the surface plasmon is a partially transverse wave. It has been investigated with optical experiments (2), particularly those involving rough or deliberately scribed surfaces.

While surface plasmons are often observed in characteristic energy loss experiments, they are not particularly useful for studying the bulk properties of plasmas in metals. Instead, they can be used as monitors of their surface condition. In particular, it is clear from Eq. (27.3) that the surface plasmon frequency can be used to measure the dielectric constant of an insulating layer on a metal. This fact has been used as the basis for experiments to determine the thickness of oxide layers on oxidizing metal surfaces (3).

28. Angular Variation of Characteristic Energy Loss Spectra

These experiments are the plasma analog of the inelastic neutron scattering experiments which have been so fruitful in revealing phonon dispersion relations in crystals. In the plasma case, of course, one measures the plasmon dispersion. Measurements of characteristic energy loss as a function of angle were first carried out by Watanabe (4), and have since been improved and extended by Kunz (5), Ninham, Powell, and Swanson (6), and Sueoka (7).

In principle, experiments of this type can be used to investigate both the collective ($k < \kappa_{FT}$) and single-particle ($k > \kappa_{FT}$) regimes in the electron gas. However, it is clear from Eq. (24.4) that measurements of the latter kind are more difficult than the former because the scattering cross section falls rapidly with increasing k. To date, detailed experiments have been performed only in the collective regime, though it would be interesting to extend them into the single-particle regime.

In the collective regime, the principal feature of the spectrum is a peak at $\omega = \omega(\mathbf{k})$ corresponding to plasmon excitation. Single-particle scattering is suppressed (as discussed in Chapter II) when $k \ll \kappa_{FT}$. The important features of the plasmon line are its position and width as a function of plasmon wave vector. Since the wave-vector transfer is related through Eq. (26.4) to the scattering angle, this information can be obtained by studying these quantities as a function of angle. Equations (21.2) and (26.4) indicate that, for small k, the plasma energy varies quadratically with angle ϑ. For experimental purposes, this variation is often characterized by a formula of the form:

$$\hbar\omega(\vartheta) = \hbar\omega_p + 2E_0\alpha\vartheta^2, \tag{28.1}$$

where α is a dimensionless parameter, independent of primary energy, that characterizes the plasmon dispersion. In RPA its value is

$$\alpha_{RPA} = \tfrac{3}{5}(E_F/\hbar\omega_p). \tag{28.2}$$

Experiments to measure α are performed at small scattering angles. As θ increases, the wave-vector transfer approaches the critical value ($k = k_c$) where Landau damping begins. In this region, the plasmon line rapidly broadens and disappears. A series of spectra showing the variation of the plasmon line with ϑ are illustrated in Fig. IV-1. These measurements were performed by Kunz (5), using 39-keV electrons on an Al film sufficiently thin that multiple plasmon excitation was infrequent. Such data can be used to determine both the plasmon frequency and

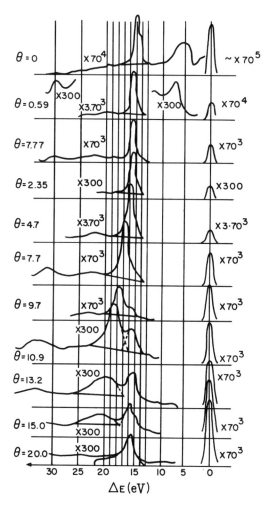

Fig. IV-1. Characteristic energy loss spectra for Al, at various scattering angles [after Kunz (5)]. Angles are measured in milliradians.

linewidth. Figures IV-2 and IV-3 show the variation of these quantities with scattering angle as determined from Kunz's data on Al. The abrupt broadening of the plasmon line due to Landau damping is clearly visible in Fig. IV-3.

The value of α for Al can be determined from Fig. IV-2. Values measured in this way for a series of elements are listed in Table IV-II (*1*). In this tabulation α_{RPA} is the value calculated from Eq. (28.2); α^* includes an exchange correction whose determination we will discuss presently. Notice that the values of α_{RPA} calculated from Eq. (28.2) are

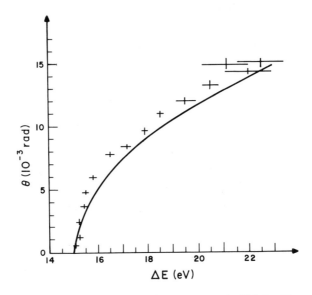

FIG. IV-2. Angular variation of plasmon energy in Al [after Kunz (*5*)].

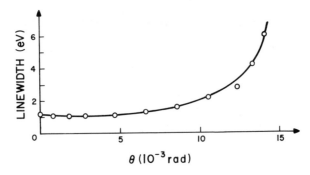

FIG. IV-3. Angular variation of plasmon line width in Al [after Kunz (*5*)].

29. EXCHANGE CORRECTION FOR RPA

TABLE IV-II

Element	α (obs.)	α (calc.) Eq. (28.2)	α* (calc.) with exchange
Be	0.42 ± 0.04	0.47	0.42
Al	0.35 ± 0.02	0.44	0.40
Mg	0.39 ± 0.01	0.39	0.37
Sb	0.37 ± 0.03	0.44	0.39
Na	0.29 ± 0.02	0.32	0.25

uniformly higher than those measured. In other words, at finite k the plasmon is somewhat "softer" than the RPA theory would predict. The discrepancy is believed to be due to exchange effects, which until now we have ignored. Because of the exclusion principle, parallel spin electrons are kept apart, thus reducing the direct Coulomb interaction and the restoring force of the plasma oscillation. However, since exchange forces are short range, they do not modify the frequency of the infinite-wavelength plasmon, which is detremined by the long-range Coulomb forces, but only affect the frequency of finite-wavelength plasma waves.

29. Exhange Corrections—The Generalized Random Phase Approximation

To estimate exchange corrections to RPA, we will study the Hartree–Fock equations for the perturbed electron gas. These equations are derived, via the SCF method, by using the *completely* factored approximation to the four-operator average indicated in Eq. (19.4). That is, we make the following replacements in Eq. (18.6):

$$\langle c_p^+ c_{p'-q}^+ c_{p'} c_{p+k-q} \rangle \cong [\langle c_p^+ c_{p+k-q} \rangle \langle c_{p'-q}^+ c_{p'} \rangle - \langle c_p^+ c_{p'} \rangle \langle c_{p'-q}^+ c_{p+k-q} \rangle] \quad (29.1)$$

$$\langle c_{p+q}^+ c_{p'-q}^+ c_{p'} c_{p+k} \rangle \cong [\langle c_{p+q}^+ c_{p+k} \rangle \langle c_{p'-q}^+ c_{p'} \rangle - \langle c_{p+q}^+ c_{p'} \rangle \langle c_{p'-q}^+ c_{p+k} \rangle].$$

Equation (18.6) then takes the form

$$i \frac{\partial}{\partial t} \langle c_p^+ c_{p+k} \rangle = \left[\frac{(\mathbf{p}+\mathbf{k})^2}{2m} - \frac{p^2}{2m} \right] \langle c_p^+ c_{p+k} \rangle$$

$$+ \sum_{p'',q}' \left\{ [\langle c_p^+ c_{p+k-q} \rangle - \langle c_{p+q}^+ c_{p+k} \rangle] \langle c_{p''-q}^+ c_{p''} \rangle \left(\frac{4\pi e^2}{q^2} \right) \right\}$$

$$- \sum_{p'',q}' \left\{ [\langle c_p^+ c_{p''} \rangle \langle c_{p''-q}^+ c_{p+k-q} \rangle - \langle c_{p+q}^+ c_{p''} \rangle \langle c_{p''-q}^+ c_{p+k} \rangle] \left(\frac{4\pi e^2}{q^2} \right) \right\}$$

$$+ e\phi_{\text{ext}}(\mathbf{k}, \omega) \, e^{-i\omega t} [\langle c_p^+ c_p \rangle - \langle c_{p+k}^+ c_{p+k} \rangle]. \quad (29.2)$$

After linearization [see Eq. (18.11)] and a certain amount of algebra, one obtains the following equation of motion:

$$i\frac{\partial}{\partial t}\langle c_p^+ c_{p+k}\rangle_1 = \left[\frac{(p+k)^2}{2m} - \frac{p^2}{2m}\right]\langle c_p^+ c_{p+k}\rangle_1$$

$$-\sum_{p'}\left\{n(p')\left[\frac{4\pi e^2}{|p'-p-k|^2} - \frac{4\pi e^2}{|p'-p|^2}\right]\right\}\langle c_p^+ c_{p+k}\rangle_1$$

$$-[n(p) - n(p+k)]\sum_q\left[\left(\frac{4\pi e^2}{q^2}\right)\langle c_{p+q}^+ c_{p+k+q}\rangle_1\right]$$

$$+[n(p) - n(p+k)]\,e\phi_{\text{total}}(k, \omega). \tag{29.3}$$

This equation is the Hartree–Fock analog of the Hartree equation (19.3). The approximation is also sometimes called the *generalized random phase approximation* (GRPA). It is important to realize that the GRPA is not a consistent approximation in the sense of including all corrections (to the next order in r_s) to the simple RPA. It includes only some of these terms. Despite this fact, GRPA has often been used to discuss the effects of exchange on the degenerate electron gas. Qualitatively speaking, its predictions are in agreement with experiment and what one expects on the basis of physical intuition.

It is interesting to observe that the Hartree–Fock equations can be derived from an *effective single-particle Hamiltonian* of the form

$$H_{\substack{\text{single}\\\text{particle}}} = \sum_{p'}\left[\frac{p'^2}{2m}(c_{p'}^+ c_{p'})\right]$$

$$+ \sum_{p',p'',q}[c_{p'+q}^+ c_{p'}\langle c_{p''-q}^+ c_{p''}\rangle(4\pi e^2/q^2)]$$

$$- \sum_{p',p'',q}[c_{p'+q}^+ c_{p''}\langle c_{p''-q}^+ c_{p'}\rangle(4\pi e^2/q^2)]$$

$$+ e\phi_{\text{ext}}(k, \omega)\,e^{-i\omega t}\sum_{p'}(c_{p'+k}^+ c_{p'}). \tag{29.4}$$

A straightforward calculation shows that

$$i(\partial/\partial t)\langle c_p^+ c_{p+k}\rangle = \left\langle\left[c_p^+ c_{p+k}, H_{\substack{\text{single}\\\text{particle}}}\right]\right\rangle \tag{29.5}$$

yields Eq. (29.3). The meaning of the four terms in the effective Hamiltonian is also fairly clear—the first is the kinetic energy, the second the Hartree field, the third the (Fock) exchange field, and the last a coupling to an external electrostatic perturbation. This method of deriving the Hartree–Fock operations closely parallels what we will later use to discuss the Landau Fermi-liquid equations. In particular, the key assumption of the Landau theory is that low-lying excitations (in wave vector and frequency) of a Fermi system can be described by a *single-particle Hamiltonian*, in which the single-particle energy is a *functional* of the electron distribution in other states. The third term of Eq. (29.4) has precisely such a form. This statement becomes particularly clear if it is rewritten, with a slight change of dummy variables, as

$$\sum_{\mathbf{p}',\mathbf{p}'',\mathbf{Q}} \left[c^+_{\mathbf{p}''+\mathbf{Q}/2} c_{\mathbf{p}''-\mathbf{Q}/2} \langle c^+_{\mathbf{p}'-\mathbf{Q}/2} c_{\mathbf{p}'+\mathbf{Q}/2} \rangle \frac{4\pi e^2}{|\mathbf{p}' - \mathbf{p}''|^2} \right]$$

$$= \sum_{\mathbf{p}',\mathbf{p}''} (2\pi)^{-3} \int d^3r \left[n(\mathbf{p}'', \mathbf{r}) \langle n(\mathbf{p}', \mathbf{r}) \rangle \frac{4\pi e^2}{|\mathbf{p}' - \mathbf{p}''|^2} \right]. \quad (29.6)$$

Here

$$n(\mathbf{p}, \mathbf{r}) \equiv \int e^{i\mathbf{Q}\cdot\mathbf{r}} c^+_{\mathbf{p}+\mathbf{Q}/2} c_{\mathbf{p}-\mathbf{Q}/2} d^3Q \quad (29.7)$$

is the operator which comes as close as is possible, within quantum mechanics, to the classical distribution function of $\mathbf{r}$ and $\mathbf{p}$ (8). The *form* of the interaction in Eq. (29.6) is the same as that which later appears in the Landau Fermi-liquid theory. In the Fermi-liquid theory, however, the Coulomb interaction ($4\pi e^2/|\mathbf{p}' - \mathbf{p}''|^2$) is replaced by a more complicated, phenomenological interaction function, $f(\mathbf{p}', \mathbf{p}'')$. Nevertheless, it is clear that the Hartree–Fock theory is formally similar to the Fermi-liquid theory. Readers interested in pursuing this analogy further are urged to compare Eqs. (29.6) and (52.13).

Equation (29.3) is a difficult integral equation. To study it we will make the rather drastic approximation of replacing the Coulomb potential by a delta function in the exchange terms. This is certainly a bad approximation to the bare Coulomb force. However, it is known that higher-order terms in the electron–electron interaction produce a screening of the exchange potential (as occurs in lower order, to the direct Coulomb interaction). Such screening is predicted by detailed many-body calculations (9). It is also needed to prevent unphysical divergences (10) in exchange corrections to the specific heat of an electron gas. For these reasons, we may expect the long-range exchange interaction in Eq. (29.3) to be replaced by an effective short-range potential. Such a change is physically reasonable, but it should be appreciated that the resulting equations are essentially phenomenological ones, with no real basis in a systematic perturbation theory expansion.

Once the exchange interaction is replaced by a screened potential, the delta function approximation becomes considerably better. Equation (29.3) then takes the form

$$i\frac{\partial}{\partial t}\langle c_p^+ c_{p+k}\rangle_1 = \left[\frac{(p+k)^2}{2m} - \frac{p^2}{2m}\right]\langle c_p^+ c_{p+k}\rangle_1$$
$$- [n(p) - n(p+k)]J\sum_q [\langle c_{p+q}^+ c_{p+k+q}\rangle_1]$$
$$+ [n(p) - n(p+k)]\,e\phi_{\text{total}}(k,\omega), \qquad (29.8)$$

where J is the strength of the delta-function exchange interaction. The solution of Eq. (29.8) leads to the following expression for the induced charge density

$$\sum_p \langle c_p^+ c_{p+k}\rangle_1 = e\phi_{\text{total}}(k,\omega)\,Q(k,\omega)/[1 + \tfrac{1}{2}JQ(k,\omega)]. \qquad (29.9)$$

Finally, with the aid of Eq. (15.2) we conclude that in Hartree–Fock (with delta-function exchange interaction)

$$\epsilon(q,\omega) = 1 - \frac{(4\pi e^2/k^2)\,Q(k,\omega)}{[1 + \tfrac{1}{2}JQ(k,\omega)]}$$
$$= \frac{1 - (4\pi e^2/k^2 - \tfrac{1}{2}J)Q(k,\omega)}{[1 + \tfrac{1}{2}JQ(k,\omega)]}. \qquad (29.10)$$

In the range $\omega/k \gg v_F$, the zeros of this equation occur at the points

$$\omega^2 \simeq \omega_p^2 + \left(\tfrac{3}{5}v_F^2 - \frac{n_0 J}{2m}\right)k^2$$
$$= \omega_p^2 + \tfrac{3}{5}v_F^2\left(1 - \tfrac{5}{12}\frac{n_0 J}{E_F}\right)k^2. \qquad (29.11)$$

Thus, in our simple model, exchange forces have the effect of reducing plasmon dispersion—a conclusion which is borne out by a variety of more detailed perturbation calculations based on Eq. (29.3). These calculations lead to an expression (*11*) for the plasma frequency of the form

$$\omega^2(k) = \omega_p^2 + \tfrac{3}{5}(1 - ar_s)\,k^2 v_F^2, \qquad (29.12)$$

where the numerical coefficient a is about 0.06. As can be seen from Table IV-2 (p. 71), Eq. (29.12) corrects the plasma dispersion relation in the proper direction and by about the right amount. The agreement between theory and experiment, for the simpler metals, is surprisingly good considering that r_s lies in the range 2–4.

Another possible approach to the exchange problem is to use the simple delta-function approximation [Eq. (29.11)], but determine the unknown parameter J from an independent experiment. In the delta-function approximation, the exchange enhancement of the Pauli spin susceptibility (over that of a noninteracting electron gas) is given by the formula

$$\frac{\chi(\text{spin enhanced})}{\chi(\text{Pauli})} = \frac{1}{(1 - 3Jn_0/4E_F)}. \qquad (29.13)$$

For a metal such as Na, this ratio is about 1.25. We conclude that

$$3Jn_0/4E_F \quad (\text{for Na}) \simeq 0.25;$$

whence, from Eq. (29.11) the exchange correction to the plasmon dispersion is estimated to be about 10%. This result is in reasonably good agreement with experiment.

30. The Cutoff Angle and Decay of Long-Wavelength Plasmons

Before discussing plasmas in periodic structures, there are two other features of the plasmon in translationally invariant systems that should be mentioned. The first is the cutoff angle, $\theta_c = k_c/k_0$, beyond which the plasma peaks broaden and disappear in the characteristic energy loss experiments. The cutoff angle θ_c has been carefully measured in Al. The observed value of the ratio $k_c/k_F = 0.83$ is in good agreement with the theory [see Eq. (21.7)].

A more interesting question is that of the damping of plasmons with $k < k_c$. In RPA such excitations are undamped because energy–momentum conservation forbids the transfer of excitation from a collective wave to a single electron–hole pair. This point is illustrated by Fig. IV-4 which is an often-used diagram indicating, in an (ω, k) plot, the collective mode and the allowed region for single-particle excitations of a degenerate gas. It is clear that, for small k, collective and single-particle modes have entirely different frequencies and cannot decay into one another. However, the conservation laws do not forbid the decay of low-k plasmons into *two* single-particle excitations. Such processes take place via electron–electron correlation, and are forbidden in RPA. There have been several detailed perturbation theory calculations (6, 12) of this effect, based on an expansion in terms of the coupling constant r_s. For $k < k_c$, the plasmon damping rate is given by the equation

$$\Gamma(k)/\omega_p = B(k/k_F)^2 = B'\theta^2, \qquad (30.1)$$

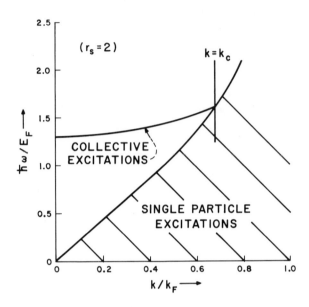

FIG. IV-4. ω–k relationships for excitations of a degenerate electron gas.

where $\Gamma(k)$ is the half-width of the plasmon line. The general form of this expression is easy to understand. In the long-wavelength limit ($k \to 0$) the current density operator commutes with the Coulomb interaction. Hence, following the discussion of Sections 6 and 7, the $k = 0$ plasmon is not damped by electron–electron interactions. Such interactions can only damp plasma waves of finite wave vector, i.e., in order k^2. A lengthy calculation, however, is required to determine the coefficient B in this expansion. Unfortunately, the values of B obtained by various authors differ by large factors (12). Moreover, the latest and most detailed calculations of B give a result which is almost an order of magnitude smaller than that of the most careful measurements (in Al) (6). It is conceivable that band structure effects are responsible for this discrepancy, but such an explanation seems unlikely in view of the fact that Al is, in other respects, a nearly free-electron-like metal. Another possibility is a failure of the r_s expansion used to derive Eq. (30.1) ($r_s = 1.8$ in Al). This question ought to be resolved since the two-electron plasmon decay is one of the few observable processes which is strictly forbidden in RPA.

31. Effects of Band Structure

Until now we have been discussing the properties of a translationally invariant electron gas, and using such a plasma as a model for the behavior of electrons in simple metals and semiconductors. In reality, of course, electrons in solids move through a periodic crystal potential and the unperturbed wave functions, rather than plane waves, are Bloch waves. An important question to answer is how our picture of the plasma is altered by this periodicity.

To discuss this question we will again use RPA. In the periodic problem, this approximation is obtained by replacing the function $\alpha(\mathbf{k} + \mathbf{K}, \mathbf{k} + \mathbf{K}'; \omega)$ of Section 15 by the corresponding function for a gas of *noninteracting* Bloch electrons, $\alpha^{(0)}(\mathbf{k} + \mathbf{K}, \mathbf{k} + \mathbf{K}'; \omega)$. A straightforward calculation yields the result

$$\alpha^{(0)}(\mathbf{k} + \mathbf{K}, \mathbf{k} + \mathbf{K}'; \omega)$$
$$\equiv \iint \alpha^{(0)}(\mathbf{r}, \mathbf{r}'; \omega) \, e^{-i(\mathbf{k}+\mathbf{K})\cdot\mathbf{r}} \, e^{i(\mathbf{k}+\mathbf{K}')\cdot\mathbf{r}'} \, d^3r \, d^3r' = e^2 \sum_{\mathbf{p}; ll'}$$
$$\left\{ \frac{[n(l, \mathbf{p}) - n(l', \mathbf{p} + \mathbf{k})]\langle l, \mathbf{p} \mid e^{-i(\mathbf{k}+\mathbf{K})\cdot\mathbf{r}} \mid l', \mathbf{p} + \mathbf{k}\rangle\langle l', \mathbf{p} + \mathbf{k} \mid e^{i(\mathbf{k}+\mathbf{K}')\cdot\mathbf{r}'} \mid l, \mathbf{p}\rangle}{[\omega + E(l, \mathbf{p}) - E(l', \mathbf{p} + \mathbf{k}) + i\delta]} \right\}$$
(31.1)

Here the indices l and $\mathbf{p}$ label the energy band and crystal momentum, respectively, of the Bloch states. $n(l, \mathbf{p})$ and $E(l, \mathbf{p})$ are the occupation number and energy of these states. To determine the RPA dielectric function one should, in principle, invert the matrix [see Eq. (15.0)]

$$1 - \frac{4\pi}{|\mathbf{k} + \mathbf{K}|^2} \alpha^{(0)}(\mathbf{k} + \mathbf{K}, \mathbf{k} + \mathbf{K}'; \omega). \tag{31.2}$$

This problem has been discussed by Adler (*13*) and Wiser (*14*), but is very complicated. For the present purpose, we are interested in the function $\alpha^{(0)}$ evaluated at a frequency $\omega \cong \omega_p$. In many materials ω_p is large compared to the important interband energies $[E(l, \mathbf{p}) - E(l', \mathbf{p} + \mathbf{k})]$ involved in Eq. (31.1). This condition is satisfied in simple metals (Na, K, Be, Mg, Al, etc.) and some semiconductors (Si, Ge, etc.). It fails in materials having d bands near the Fermi surface, such as Cu, Ag, Au, and most transition metals—the succeeding remarks do not apply to those cases. But in the simpler materials, one can take advantage of the fact that ω_p is large compared to the important interband energies to make an expansion of Eq. (31.1) in powers of $(1/\omega_p)$. From this series we will see that the off-diagonal elements of $\alpha^{(0)}$ $(\mathbf{k}+\mathbf{K}, \mathbf{k}+\mathbf{K}';\omega)$ are often relatively small compared to the diagonal ones. As a consequence, the approximation [see Eq. (15.12)]

$$\epsilon_M(\mathbf{k}, \omega) \simeq 1 - (4\pi/k^2)\, \alpha^{(0)}(\mathbf{k}, \mathbf{k}; \omega) \tag{31.3}$$

is valid in such materials. This result implies that plasma frequencies are not changed by local field corrections in nearly-free-electron-like crystals—a physically satisfying conclusion. Equation (31.3) will also enable us to understand why the *free* electron mass appears in the formula for the plasma, and how insulating materials (such as Si or Ge) can support a plasma mode.

The first nonvanishing term in the expansion of $\alpha^{(0)}$ in powers of $1/\omega$ is

$$\alpha^{(0)}(\mathbf{k}+\mathbf{K}, \mathbf{k}+\mathbf{K}'; \omega)$$
$$\simeq (e^2/\omega^2) \sum_{\mathbf{p}; l, l'} \{[n(l, \mathbf{p}) - n(l', \mathbf{p}+\mathbf{k})][E(l', \mathbf{p}+\mathbf{k}) - E(l, \mathbf{p})]$$
$$\times \langle l, \mathbf{p} | e^{-i(\mathbf{k}+\mathbf{K})\cdot\mathbf{r}} | l', \mathbf{p}+\mathbf{k}\rangle\langle l', \mathbf{p}+\mathbf{k} | e^{i(\mathbf{k}+\mathbf{K}')\cdot\mathbf{r}} | l, \mathbf{p}\rangle\}. \tag{31.4}$$

With a slight change of variables, this expression can be rewritten in the form

$$\alpha^{(0)}(\mathbf{k}+\mathbf{K}, \mathbf{k}+\mathbf{K}'; \omega)$$
$$\simeq (e^2/\omega^2) \sum_{\mathbf{p}; l, l'} [n(l, \mathbf{p})\{[E(l', \mathbf{p}+\mathbf{k}) - E(l, \mathbf{p})]\langle l, \mathbf{p} | e^{-i(\mathbf{k}+\mathbf{K})\cdot\mathbf{r}} | l', \mathbf{p}+\mathbf{k}\rangle$$
$$\times \langle l', \mathbf{p}+\mathbf{k} | e^{i(\mathbf{k}+\mathbf{K}')\cdot\mathbf{r}} | l, \mathbf{p}\rangle + [E(l', \mathbf{p}-\mathbf{k}) - E(l, \mathbf{p})]$$
$$\times \langle l, \mathbf{p} | e^{i(\mathbf{k}+\mathbf{K}')\cdot\mathbf{r}} | l', \mathbf{p}-\mathbf{k}\rangle\langle l', \mathbf{p}-\mathbf{k} | e^{-i(\mathbf{k}+\mathbf{K})\cdot\mathbf{r}} | l, \mathbf{p}\rangle\}]. \tag{31.5}$$

Next, we may evaluate the l' sum in Eq. (31.5) by an extension of the f-sum rule. This identity is derived by considering the diagonal matrix elements of the double commutator,

$$[e^{-i(\mathbf{k}+\mathbf{K})\cdot\mathbf{r}}, [e^{i(\mathbf{k}+\mathbf{K}')\cdot\mathbf{r}}, H_0]], \tag{31.6}$$

where H_0 is the Hamiltonian of the noninteracting Bloch electrons. The result is

$$\langle l, \mathbf{p} | [e^{-i(\mathbf{k}+\mathbf{K})\cdot\mathbf{r}}, [e^{i(\mathbf{k}+\mathbf{K}')\cdot\mathbf{r}}, H_0]] | l, \mathbf{p}\rangle$$
$$= -[(\mathbf{k}+\mathbf{K})\cdot(\mathbf{k}+\mathbf{K}')/m]\langle l, \mathbf{p} | e^{i(\mathbf{K}'-\mathbf{K})\cdot\mathbf{r}} | l, \mathbf{p}\rangle$$
$$= \sum_{l'} \{\langle l, \mathbf{p} | e^{-i(\mathbf{k}+\mathbf{K})\cdot\mathbf{r}} | l', \mathbf{p}+\mathbf{k}\rangle[E(l, \mathbf{p}) - E(l', \mathbf{p}+\mathbf{k})]\langle l', \mathbf{p}+\mathbf{k} | e^{i(\mathbf{k}+\mathbf{K}')\cdot\mathbf{r}} | l, \mathbf{p}\rangle$$
$$+ \langle l, \mathbf{p} | e^{i(\mathbf{k}+\mathbf{K}')\cdot\mathbf{r}} | l', \mathbf{p}-\mathbf{k}\rangle[E(l, \mathbf{p}) - E(l', \mathbf{p}-\mathbf{k})]\langle l', \mathbf{p}-\mathbf{k} | e^{-i(\mathbf{k}+\mathbf{K})\cdot\mathbf{r}} | l, \mathbf{p}\rangle\}. \tag{31.7}$$

Hence,

$$\alpha^{(0)}(\mathbf{k}+\mathbf{K}, \mathbf{k}+\mathbf{K}'; \omega) \simeq \left\{\left[\frac{e^2(\mathbf{k}+\mathbf{K})\cdot(\mathbf{k}+\mathbf{K}')}{m\omega^2}\right]\right.$$
$$\left. \times \sum_{\mathbf{p},l} [n(l,\mathbf{p})\langle l,\mathbf{p}\mid e^{i(\mathbf{K}'-\mathbf{K})\cdot\mathbf{r}}\mid l,\mathbf{p}\rangle]\right\}. \quad (31.8)$$

In this formula the l sum ranges over valence electron states—which give a relatively large contribution to Eq. (31.1)—but *not* over deep core states, whose contribution to $\alpha^{(0)}$ is small. The quantity

$$n_{\mathbf{K}'-\mathbf{K}} \equiv \sum_{\mathbf{p},l} n(l,\mathbf{p})\langle l,\mathbf{p}\mid e^{i(\mathbf{K}'-\mathbf{K})\cdot\mathbf{r}}\mid l,\mathbf{p}\rangle, \quad (31.9)$$

which appears in the formula for $\alpha^{(0)}$, is the Fourier transform of the valence electron density in the crystal. In free-electron-like materials (where valence electron wave functions are constant over most of the Wigner–Seitz cell) the Fourier components $n_{\mathbf{K}-\mathbf{K}'}$ with $\mathbf{K}' - \mathbf{K} \neq 0$, are small compared to n_0. Hence, one may neglect off-diagonal terms in the matrix $\alpha^{(0)}(\mathbf{k}+\mathbf{K}, \mathbf{k}+\mathbf{K}'; \omega)$. For such materials,

$$\epsilon_M(\mathbf{k},\omega) \simeq 1 - (4\pi/k^2)\,\alpha^{(0)}(\mathbf{k},\mathbf{k};\omega)$$
$$\simeq 1 - 4\pi n_0 e^2/m\omega^2. \quad (31.10)$$

This formula, which is valid for $\omega \gtrsim \omega_p$, implies that in free-electron-like crystals, the plasmon frequency is determined by the valence electron density and free electron mass. As indicated in Section 25, this approximation works well in explaining the positions of plasmon peaks in characteristic energy loss experiments. Apparently, the valence electron charge densities in the simple semiconductors (Si, Ge, etc.) are also sufficiently uniform that these arguments apply.

On the other hand, in d-band metals, such as Cu, Ag, and Au, the expansion leading to Eq. (31.10) fails and plasma frequencies are usually quite different from their free-electron values. In most such cases, the plasmons are also strongly damped by interband transitions [$\omega_p = E(l, \mathbf{p}+\mathbf{k}) - E(l, \mathbf{p})$ in Eq. (31.1)]. Ag is a unique and interesting exception. In this metal the onset of d → s–p transitions lies at relatively high energies (4 eV). At energies below 4 eV, virtual, interband d → s – p transitions give rise to a large, purely real d-band dielectric constant, which reduces the plasma frequency of the s–p electrons *below* 4 eV— thus moving it out of the lossy region. A sharp plasmon peak is observed in Ag at 3.7 eV. The corresponding situation does not occur in either Cu or Au because, in these metals, the onset of interband transitions is at lower energies (2 eV). These questions are discussed in detail by Ehrenreich and Philipp (*15*).

32. Plasmons in Doped Semiconductors

It is also interesting to examine the expression for $\alpha^{(0)}(\mathbf{k}, \mathbf{k}; \omega)$ in the case of a doped semiconductor. As before, there is a high-frequency collective mode due to virtual interband transitions of valence electrons at frequencies $\omega = \omega_p \gg (E_{l'} - E_l)$. In addition, doping gives rise to a low-frequency root, representing an oscillation of the *conduction* electrons. To study this mode we consider the formula for $\alpha^{(0)}(\mathbf{k}, \mathbf{k}; \omega)$ in the limit where ω is small compared to all interband energies. Splitting the contributions to $\alpha^{(0)}$ into interband and intraband terms gives the result

$$\alpha^{(0)}(\mathbf{k}, \mathbf{k}; \omega) \cong e^2 \sum_{\substack{l,l',\mathbf{p} \\ l \neq l'}} \frac{2n(l,\mathbf{p}) |\langle l, \mathbf{p} | e^{-i\mathbf{k}\cdot\mathbf{r}} | l', \mathbf{p} + \mathbf{k}\rangle|^2}{E(l,\mathbf{p}) - E(l',\mathbf{p}+\mathbf{k})}$$
$$+ e^2 \sum_{\mathbf{p},l} \frac{[n(l,\mathbf{p}) - n(l,\mathbf{p}+\mathbf{k})] |\langle l, \mathbf{p} | e^{-i\mathbf{k}\cdot\mathbf{r}} | l, \mathbf{p} + \mathbf{k}\rangle|^2}{\omega + E(l,\mathbf{p}) - E(l,'\mathbf{p}+\mathbf{k})}. \quad (32.1)$$

The first term in this formula is a contribution to the static dielectric constant of the semiconductor by virtual *interband* transitions; the second is plasmalike term involving *intraband* transitions of conduction electrons. The latter is finite only if some band is partially full, i.e., if the semiconductor is doped. To estimate this term we note that intraband matrix elements of $e^{i\mathbf{k}\cdot\mathbf{r}}$ are unity (to order k^2). Thus, the last part of Eq. (32.1) takes the form

$$\sum_{l,\mathbf{p}} \frac{n(l,\mathbf{p}) - n(l,\mathbf{p}+\mathbf{k})}{\omega + E(l,\mathbf{p}) - E(l,\mathbf{p}+\mathbf{k})}. \quad (32.2)$$

This result is the analog, for a conduction band plasma, of Eq. (14.7) which gives the susceptibility (in RPA) of a free electron gas. In a semiconductor such as n-InSb, which has a single, isotropic conduction band minimum, the energies which appear in Eq. (32.2) take the form

$$E_{\text{cond}}(\mathbf{p}) = p^2/2m^*. \quad (32.3)$$

The dielectric function then becomes

$$\epsilon(\mathbf{k}, \omega) = \epsilon_0 - (4\pi e^2/k^2) Q^*(\mathbf{k}, \omega), \quad (32.4)$$

where ϵ_0 is the dielectric constant of the semiconductor lattice and $Q^*(\mathbf{k}, \omega)$ the function Q [see Eq. (14.7)] evaluated with $E = p^2/2m^*$. For long wavelengths, Eq. (32.4) has a root at the frequency

$$\omega^2 = 4\pi n_c e^2/\epsilon_0 m^*, \quad (32.5)$$

where n_c is the *conduction* electron density. Thus, in addition to the high-frequency, interband plasmon which involves all the valence electrons, a doped semiconductor will also support a low-frequency plasma oscillation involving only the conduction electrons. For an isotropic, single-band material, such as n-type InSb, its properties are similar to those of a plasmon in a free electron gas. There are other cases, however, in which the conduction band structure is more complicated than that of free electrons. Well-known examples are n-type Si, Ge, and PbTe. These materials have multivalleyed conduction bands and, when doped, contain *multicomponent* plasmas. We will see later that a multicomponent plasma can support a greater variety of collective modes than does a single-component one.

Finally, we return briefly to the question of local field corrections. In deriving Eq. (32.4), we have assumed $\omega \ll E_G$, so the argument previously used (Section 31) to justify the neglect of local field corrections is invalid. Indeed, they may well be important in the semiconductor case. However, detailed calculations (*13*) which include local field corrections show that they only affect the *interband* portion [the first term of Eq. (32.1)] of the dielectric function. Thus, if the experimental value of ϵ_0 is used in Eq. (32.4), local field effects are automatically included. This artifice enables one to avoid the troublesome problem of calculating local field corrections to the properties of conduction band plasmons in semiconductors.

REFERENCES

1. These data are taken from a useful review article by H. Raether, *in* "Springer Tracts in Modern Physics," Vol. 38, p. 84. Springer-Verlag, Berlin and New York, 1965. The original references are given in this paper.
2. D. Beaglehole, *Phys. Rev. Lett.* **22**, 708 (1969); Y. Y. Teng and R. A. Stern, *ibid.* **19**, 511 (1967); N. Marshall, B. Fischer, and H. J. Queisser, *ibid.* **27**, 95 (1971).
3. C. J. Powell and J. B. Swan, *Phys. Rev.* **115**, 869 (1959); **118**, 640 (1960).
4. H. Watanabe, *J. Phys. Soc. Jap.* **11**, 112 (1956).
5. C. Kunz, *Z. Phys.* **167**, 53 (1962); **196**, 311 (1966).
6. B. W. Ninham, C. J. Powell, and N. Swanson, *Phys. Rev.* **145**, 209 (1966).
7. O. Sueoka, *J. Phys. Soc. Jap.* **12**, 2203 and 2249 (1965).
8. E. Wigner, *Phys. Rev.* **40**, 749 (1932).
9. J. Hubbard, *Proc. Roy. Soc. Ser. A* **243**, 336 (1957).
10. J. Bardeen, *Phys. Rev.* **50**, 1098 (1936); M. Gell-Mann, *ibid.* **106**, 369 (1957).
11. H. Kanazawa, S. Misawa, and E. Fujita, *Progr. Theor. Phys.* **23**, 426 (1960).
12. D. F. DuBois and M. G. Kivelson, *Phys. Rev.* **186**, 409 (1969) and D. F. DuBois, *Ann. Phys. (New York)* **8**, 24 (1959). M. Hasegawa and M. Watabe, *J. Phys. Soc. Jap.* **27**, 1393 (1969).
13. S. L. Adler, *Phys. Rev.* **126**, 413 (1962).
14. N. Wiser, *Phys. Rev.* **129**, 62 (1963).
15. H. Ehrenreich and H. R. Philipp, *Phys. Rev.* **128**, 1622 (1962).

V. Light Scattering Experiments

33. INTRODUCTION

We have discussed the properties of the high-frequency collective modes (plasmons) that occur in electron gases in Chapter IV. These modes have been extensively studied, by characteristic energy loss experiments, in a wide variety of solids. This technique is the only one which gives detailed information concerning collective modes of energy a few electron volts, or higher. As such, it has played an important role in leading to the presently accepted picture of the interacting electron gas.

We have also seen that some crystals—most notably doped semiconductors—will support lower-frequency plasma waves. These waves involve *only* the conduction electrons, in contrast to the much higher-frequency interband plasmons. Their energies are small (typically 0.01 eV or less), and they cannot easily be seen with characteristic energy loss experiments. Inelastic light scattering, on the other hand, is an almost ideal technique for studying them. In the past few years it has become a powerful tool for investigating plasma dynamics in semiconductors. This advance has been made possible by the development of high-power, infrared lasers whose frequencies are well matched to the semiconductors of primary interest as solid state plasma hosts.

Before we discuss these experiments, it should be emphasized that the physics aimed at here is rather different from that studied via the electron scattering technique. A doped semiconductor usually contains a small number of free carriers compared to the total number of atoms. As a consequence, the conduction electron plasma in such a crystal has relatively little influence on the overall structure of the solid. One does not, primarily, study such plasmas as a means of learning about crystals but as objects of interest in their own right. They have unique and unusual properties. Among solid state plasmas, they are the only ones in which all frequencies of interest—the plasma frequency, the cyclotron frequency, the Fermi energy, the phonon frequencies, and even, in special materials, the energy gap—can be made comparable. For this reason, mode coupling and quantum-mechanical effects are best studied in such systems. Semiconductor plasmas are also the only solid state

33. INTRODUCTION

plasmas in which it is possible to produce sizable departures from thermal equilibrium. Several types of instabilities can be excited in them, and may ultimately be investigated with light scattering techniques.

Carrier densities in doped semiconductors generally lie between 10^{14} and 10^{20} cm^{-3}. The corresponding screening wave vectors range from about 10^4 to 10^7 cm^{-1}. These figures nicely bracket the wave vectors (inside a crystal of dielectric constant $\epsilon_0 \simeq 10$) of the two most powerful infrared lasers; Nd:YAG at 1.06μm (1.2 eV) and CO_2 at 10.6μm (0.12 eV). Thus, one can hope to use light scattering to study both single-particle and collective excitations of such plasmas.

Measurements of equilibrium velocity distributions have been performed, via light scattering, in n-GaAs under a variety of conditions, in which the distribution ranges from degenerate to Maxwellian. Figure V-1 shows results of Mooradian (1) in which the evolution of one

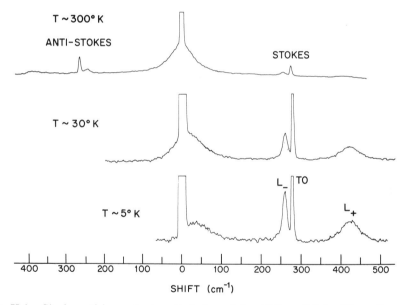

FIG. V-1. Single particle spectra in n-GaAs ($n = 1.4 \times 10^{18}$ cm^{-3}) [after Mooradian (1)]. As temperature increases, the portion of the spectrum near the laser line changes shape from a form typical of a Fermi distribution to that appropriate to a Maxwellian. Other features are due to phonon scattering.

distribution into the other can be followed in detail. The labeled features in this spectrum are all related to phonons and do not concern us here. The large sharp chopped-off feature in the center (zero shift) is stray laser light. The broad piece centered at zero and extending out about

$100\ cm^{-1}$ is the portion of the spectrum related to the free carriers. In these data the high-temperature runs have a gaussian single-particle spectrum, typical of a Maxwellian plasma, whereas the low-temperature traces have an entirely different shape, which can be shown to be characteristic of a Fermi–Dirac distribution. Such experiments have also been used (2) to determine nonequilibrium electron velocity distributions in GaAs. They provide the most direct measurement of a nonequilibrium velocity distribution yet performed in a solid. Such information is of basic interest and also is important in several technical applications, such as the design of Gunn oscillators.

Collective modes have been studied by light scattering in a number of semiconductors such as GaAs (3), InAs (4), InSb (5), and CdS (6). The first step in such an investigation is the identification, in the light scattering spectrum, of a feature which corresponds to Raman scattering from the mode in question. Subsequently, the behavior of this feature can be used to study damping, dispersion, mode coupling, and other effects. In this way, the plasma–phonon coupling has been investigated in detail. A number of magnetic field effects and field-induced mode couplings have also been studied—often under conditions in which the magnetic field drastically alters the properties of the waves which the plasma can support. To date, most of this work has been done in GaAs because it is particularly well matched to the Nd:YAG laser. However, the recent development (5) of the CO laser for scattering experiments suggests that similar work may soon be done in InSb and other small-gap semiconductors.

Proposals have also been made to use light scattering to observe more exotic (and hitherto unobservable) collective modes of the electron gas (7–9). One example is the acoustic plasma wave (the analog of the gas plasma ion-acoustic wave) which should occur in suitable multicomponent semiconductor plasmas, such as PbTe, or in a semimetal like Bi. Acoustic plasma waves have not, as yet, been observed in solids; but if they ever are, it is likely that it will be via light scattering. There is also a possibility that lasers can be used to excite collective modes in semiconductor plasmas to large amplitude. For some modes, the threshold power for stimulated light scattering is within range of present laser technology (8). If such experiments prove feasible, they might open the way to the study of nonlinear phenomena in solid state plasmas. The recent observation (10) of stimulated spin-flip scattering in n-InSb is a first step in this direction.

Early work on light scattering from semiconductor plasmas was interpreted in terms of the classical theory discussed in Chapter II. This theory works fairly well in describing collective mode scattering. However, there are serious discrepancies with regard to the single-particle portion of the spectrum (*1, 11*). For example, in the collective regime ($k \ll \kappa_{FT}$) the single-particle scattering is observed to be several orders of magnitude more intense than the classical theory would permit. It is now known that this effect is a specifically solid state phenomenon, having its origin in the fact that an electron in a crystal couples differently to a light wave than does a free, classical one. The modified coupling gives rise to several other novel processes, such as spin-flip Raman scattering (*12*) and multiphoton mixing (*13*) in materials like InSb, InAs, and PbTe. To understand these effects, it is necessary to develop a theory of light scattering which includes band structure. This development, and the comparison of the resultant theory with experiment, is one of the goals of Chapter V. The effects we are concerned with here are peculiar to mobile electrons in crystals and have no counterpart in the interaction of light with free electrons. On the other hand, there is a different class of phenomena—particularly those involving the effect of magnetic fields on the collective modes—for which the classical theory retains its validity. Ideally, the effects of band structure and magnetic field should be combined into one formula to describe all these experiments. Such formulas have been given (*9*), but they are exceedingly complicated and obscure much of the physics in a bewildering array of band indices and Landau-level quantum numbers. Where possible, therefore, we will separate the effects of band structure and magnetic field. Our program will be to discuss first those features of the scattering which can be understood in terms of the classical theory, or extensions of it. These include the simple plasmon scattering, magneto–plasma effects, and multicomponent plasmas. Subsequently, we will consider band structure effects. As noted above, they are responsible for the enhancement of the single-particle scattering. They also give rise to a variety of magneto–Raman processes, such as spin-flip scattering and Landau–Raman scattering.

34. Plasmon Light Scattering and Plasmon–Phonon Coupling

The first light scattering experiment involving plasma modes of a solid—and perhaps the first *direct* observation of the low-frequency plasmon in any semiconductor—was performed about five years ago by Mooradian and Wright (*3*). The plasma was created by n-type doping of GaAs; the light source was a Nd:YAG laser. This radiation has an energy well below the band gap of the GaAs crystal, yet sufficiently high that free carrier absorption is unimportant. It is also far above the plasma frequency. This relationship of the various frequencies ($\omega_p \ll \omega_0 < E_G$) is the optimal one from the experimental point of view. It makes the Nd:YAG laser an ideal tool for studying plasmas in semiconductors whose band gaps lie in the range 1.5–2.5 eV. GaAs, CdTe, InP, and GaP are examples of such materials.

Figure V-2 shows the spectrum of radiation scattered from n-GaAs, as measured by Mooradian and Wright (3). The two lower curves contain

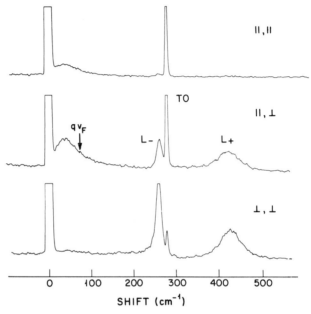

FIG. V-2. Spectrum of light scattered from heavily doped n-GaAs [after Mooradian and Wright (3)]. Polarized Raman spectra, GaAs, $n_c = 1.4 \times 10^{18}$ cm^{-3}, $T = 2°$K.

three prominent features, corresponding to Raman scattering from the longitudinal optical phonon ($\Delta\omega = \omega_0 - \omega_1 = \omega_L$), the transverse optical phonon ($\Delta\omega = \omega_T$), and the plasmon ($\Delta\omega = \omega_p$). The latter was positively identified as plasmon–Raman scattering by its variation with carrier concentration. For pure plasmon scattering, one expects a frequency shift $\Delta\omega = \omega_0 - \omega_1 = (4\pi n_c e^2/\epsilon_\infty m^*)^{1/2}$. This is exactly what is observed so long as the plasma frequency is well above the longitudinal optical phonon frequency ω_L. However, as $\omega_p \to \omega_L$, the two modes interact and behave in a more complicated way. This coupling is an interesting and important phenomenon that has been studied in great detail, particularly in GaAs. To see how the modes behave in the region $\omega_p \simeq \omega_L$, we consider the total dielectric constant of the crystal, including both plasma and phonon terms (14). It is sufficient to consider the behavior of this function in the limit $k \to 0$ since, when $\omega_p \simeq \omega_L$, the screening wave vector of the electron gas is large compared to that of

34. LIGHT SCATTERING AND COUPLING

the probing light. Under these conditions, and for the case of a cubic crystal such as GaAs which has only two atoms per unit cell, the dielectric function becomes

$$\epsilon(\omega) = \epsilon_\infty + \frac{(\epsilon_0 - \epsilon_\infty)\omega_T^2}{(\omega_T^2 - \omega^2)} - \left(\frac{4\pi n_c e^2}{m^* \omega^2}\right). \quad (34.1)$$

In this formula ϵ_∞ and ϵ_0 are the high- and low-frequency dielectric constants of the undoped crustal, and ω_T is the frequency of the zone-center, transverse optic mode. The simplicity of this formula results from the fact that we are considering a cubic (optically isotropic) crystal with only two atoms per unit cell. In more complicated and anisotropic materials, $\epsilon(\omega)$ is a tensor and several lattice resonances are involved. It is also important to realize that, in the $k = 0$ limit, the longitudinal and transverse dielectric functions of a cubic crystal are equal. Thus, Eq. (34.1) can equally well be used to describe the longitudinal and transverse response. In the undoped case ($n_c = 0$), the resonance at $\omega = \omega_T$ gives rise to the well-known reststrahl behavior of an ionic crystal. The zero of $\epsilon(\omega)$ then occurs at the point

$$\omega^2 = (\epsilon_0/\epsilon_\infty)\omega_T^2 = \omega_L^2, \quad (34.2)$$

The latter half of this equation is the Lyddane–Sachs–Teller relation.

Now consider the roots of the complete dielectric function [Eq. (34.1)]. This equation can be rewritten in the form

$$\epsilon(\omega) = \frac{[\omega^4 - (\omega_L^2 + \bar{\omega}_p^2)\omega^2 + \omega_T^2 \bar{\omega}_p^2]}{\omega^2(\omega^2 - \omega_T^2)} \epsilon_\infty = 0, \quad (34.3)$$

where the "high frequency" plasmon is defined by

$$\bar{\omega}_p^2 = 4\pi n e^2/\epsilon_\infty m^*. \quad (34.4)$$

The solutions are

$$\omega_\pm^2 = \tfrac{1}{2}(\omega_L^2 + \bar{\omega}_p^2) \pm \tfrac{1}{2}[(\omega_L^2 + \bar{\omega}_p^2)^2 - 4\bar{\omega}_p^2 \omega_T^2]^{1/2}. \quad (34.5)$$

Figure V-3 shows a comparison of the predictions of Eq. (34.5) with data of Mooradian and Wright. The agreement is excellent. For small n_c, the lower root ω_- (or L_-) of Eq. (34.5) starts out as a plasmon of frequency $(4\pi n_c e^2/\epsilon_0 m^*)^{1/2}$, then at higher carrier concentration becomes phononlike. In the limit of large n_c, it is the longitudinal, optical phonon mode of the crystal. The frequency is ω_T because the large electron density completely screens the ionic fields which, in a pure crystal, raise ω_L^2 compared to ω_T^2. Conversely, the root ω_+ (or L_+) starts out as the LO phonon and turns into a plasmon (now at frequency $\bar{\omega}_p$) as n_c increases.

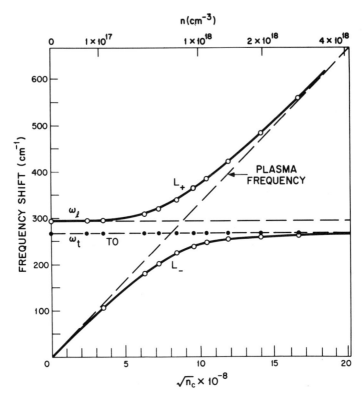

FIG. V-3. Comparison of theory and experiment for coupled plasmon–phonon modes in n-GaAs (300°K) [after Mooradian and Wright (3)].

As Fig. V-3 indicates, the variation of $\omega_\pm$ with carrier concentration conforms exactly to the predictions of Eq. (34.5). It is less easy to account for the strengths of these lines because, in addition to the electron scattering, the phonons in GaAs are also Raman active. The modes ω_+ and ω_- are mixtures of phonon and plasmon, so both contribute to the scattering. For the same reason, the polarization dependence of the Raman scattering is more complicated than that predicted by the classical formula (Eq. 11.9). A good deal of work has been done on these problems (15), and they are now fairly well understood, but we do not have the space to discuss them here.

To date, the most detailed work on plasmon light scattering has been done in n-GaAs. However, the effect has also been seen in several other semiconductors, such as CdTe, InP, InAs, InSb, and CdS. Plasmon–phonon coupling is observed in all of these materials, except InAs. In the cases where it is seen, the data fit Eq. (34.5) quite well.

The early experiments on plasmon light scattering in semiconductors were performed on fairly heavily doped materials, i.e., in the collective regime ($k \ll k_{FT}$). As techniques improved, the measurements have been extended to lightly doped samples. The portion of the spectrum near the laser line—single-particle scattering—has also been studied. As we have indicated, this work produced a major surprise. It was found that in the collective regime the single-particle scattering is very strong—orders of magnitude more intense than the classical theory would allow. This enhancement, which is now known to be a band structure effect, greatly facilitates the observation of the single particle scattering. So long as $k \ll \kappa_D$, the collective and single-particle portions of the spectrum are well separated from one another in frequency. A typical experimental curve in this regime is illustrated in Fig. V-4 (the lowest

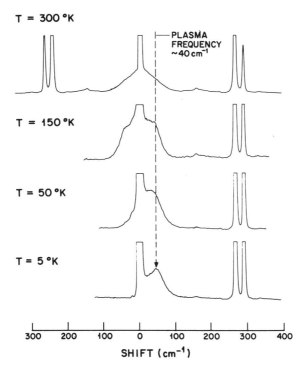

FIG. V-4. Scattering spectra illustrating Landau damping of the plasma mode in n-GaAs, $n_c = 1 \times 10^{16}$ cm^{-3} [courtesy Mooradian].

trace). As the temperature increases, the central peak broadens, while the position of the plasma line is unchanged. Gradually, the single-particle spectrum overlaps the plasma line. This occurs when $k \simeq \kappa_D$. The plasma wave then has a phase velocity about equal to the particle thermal velocity, and begins to be heavily Landau damped. As a consequence, the plasma line broadens and fades away as $k \rightarrow \kappa_D$. The damping and disappearance of the plasmon line are evident. It is fair to say that these experiments demonstrate the Landau damping of plasma waves in semiconductors.

35. The Magnetoplasma and Light Scattering Experiments Involving Magnetized Plasmas

So far in our discussions, we have only briefly considered the effects of magnetic fields on solid state plasmas. Our main concern has been plasma waves in metals. These are essentially unaffected by fields because the cyclotron frequency is invariably much lower than the plasma frequency ($\omega_c \ll \omega_p$). The one important effect of the field in metals is to create "windows" in the low-frequency portion of the spectrum ($\omega \leqslant \omega_c \ll \omega_p$). These are the propagating, low-frequency modes discussed in Chapters VI–X. On the other hand, in moderately dense semiconductor plasmas it is often possible to satisfy the condition $\omega_c \simeq \omega_p$ with attainable magnetic fields. For instance, in an n-type GaAs sample containing 10^{17} electrons cm^{-3}, $\omega_c = \omega_p$ when $H_0 \simeq 80$ kG. Such media are termed *magnetoplasmas*. Their properties differ in a number of respects from those of ordinary, unmagnetized plasmas. The doped semiconductors—n-GaAs, n-InSb, PbTe, Bi—are ideal systems in which to study these phenomena.

Magnetic fields influence the properties of solid state plasmas in two important ways. The first is classical; the second, quantum mechanical. The classical effect is the Lorentz force, which produces currents that are *not* parallel to the applied electric field. In this case the conductivity becomes a tensor—often a highly anisotropic one. The quantum-mechanical effect is the quantization of electron motion in the plane normal to the magnetic field. This produces a discrete set of energy levels (Landau levels) which, in the case of a parabolic energy–momentum relation, are those of a harmonic oscillator. A harmonic oscillator, however, is the one system for which classical and quantum-mechanical descriptions are most nearly equivalent. It is not surprising, therefore, that the properties of "classical" plasmas (those with parabolic energy–momentum relations) are not greatly modified by quantization. In other cases, quantization gives rise to interesting new effects. We will see some examples of these later.

35. THE MAGNETOPLASMA

For the time being, however, let us consider magnetoplasmas in which quantum-mechanical effects are unimportant. Our immediate aim will be to study how the properties of "classical" plasmas are modified by strong fields. Many semiconductor plasmas approximate such behavior, including most of those that have been studied with light scattering. We particularly wish to consider how the magnetoplasma scatters light, but in so doing will also develop expressions for the collective mode frequencies of such a medium.

For a classical ($E = p^2/2m^*$), single-component plasma, the light scattering spectrum is determined by the density–density correlation function via Eq. (11.9). The major problem in applying this formula is that of evaluating the structure factor $S(\mathbf{k}, \omega)$ of the magnetized electron gas. As before, we will use RPA to determine this function. The important new feature of the analysis is that the magnetic field induces transverse currents, and therefore transverse electromagnetic fields, in the plasma.

To proceed we imagine, as in the analysis preceding Eq. (13.16), that a weak, external, electrostatic potential is applied to the plasma. The induced charge density determines the response function $F(\mathbf{k}, \omega)$ through Eq. (13.15); and the structure factor can be calculated from the fluctuation–dissipation theorem [Eq. (17.6)]. Both of these relations are valid in the presence of a static magnetic field. In the spirit of the RPA, we calculate the current induced in the plasma by the *total* electric field, using the conductivity tensor of a magnetized, but *noninteracting*, electron gas. Thus

$$\mathbf{j}_{\text{ind}}(\mathbf{k}, \omega) = \boldsymbol{\sigma}(\mathbf{k}, \omega) \cdot \mathbf{E}_{\text{total}}(\mathbf{k}, \omega), \qquad (35.1)$$

where

$$\mathbf{E}_{\text{total}} = -\nabla \phi_{\text{ext}} + \mathbf{E}_{\text{induced}}, \qquad (35.2)$$

and E_{induced} will be determined in a self-consistent way from Maxwell's equations. The relevant equation is

$$(k^2 - \omega^2/c^2) \mathbf{E}_{\text{ind}} - \mathbf{k}(\mathbf{k} \cdot \mathbf{E}_{\text{ind}})$$
$$= -(4\pi/c^2) \partial \mathbf{j}_{\text{ind}}/\partial t = (4\pi i \omega/c^2) \boldsymbol{\sigma} \cdot \mathbf{E}_{\text{total}}. \qquad (35.3)$$

With the aid of Eq. (35.2), this equation can be rewritten in the form

$$\left[k^2 - \mathbf{k}\,\mathbf{k} - \frac{\omega^2}{c^2} \left(1 - \frac{4\pi \boldsymbol{\sigma}}{i\omega} \right) \right] \cdot \mathbf{E}_{\text{ind}} = \frac{4\pi i \omega}{c^2} \boldsymbol{\sigma} \cdot \mathbf{E}_{\text{ext}}. \qquad (35.4)$$

We define the tensor

$$\mathbf{D} \equiv k^2 - \mathbf{k}\,\mathbf{k} - \omega^2/c^2 \left(1 - \frac{4\pi \boldsymbol{\sigma}}{i\omega} \right) = k^2 - \mathbf{k}\,\mathbf{k} - (\omega^2/c^2)\,\boldsymbol{\varepsilon}(\mathbf{k}, \omega). \qquad (35.5)$$

In terms of **D**, the solution to Eq. (35.4) is

$$\mathbf{E}_{\text{ind}} = \mathbf{D}^{-1} \cdot (4\pi i \omega/c^2) \cdot \boldsymbol{\sigma} \cdot \mathbf{E}_{\text{ext}}. \tag{35.6}$$

The quantity we wish to calculate is the induced electron density, determined by Poisson's equation

$$\nabla \cdot \mathbf{E}_{\text{ind}} = 4\pi e n_{\text{ind}}. \tag{35.7}$$

Hence,

$$n_{\text{ind}} = \frac{i\omega}{e^2 c^2} (\mathbf{k} \cdot \mathbf{D}^{-1} \cdot \boldsymbol{\sigma} \cdot \mathbf{k}) \, e\phi_{\text{ext}}$$

$$= -\frac{1}{4\pi e^2} \left[k^2 + \frac{\omega^2}{c^2} (\mathbf{k} \cdot \mathbf{D}^{-1} \cdot \mathbf{k}) \right] e\phi_{\text{ext}}. \tag{35.8}$$

From Eq. (13.15) we then conclude that the density–density response function is

$$F(\mathbf{k}, \omega) = -(4\pi)^{-1}[k^2 + (\omega^2/c^2)(\mathbf{k} \cdot \mathbf{D}^{-1} \cdot \mathbf{k})], \tag{35.9}$$

and the structure factor

$$S(\mathbf{k}, \omega) = \frac{-1}{\pi e^2 (1 - e^{-\beta \omega})} \operatorname{Im}[F(\mathbf{k}, \omega)]$$

$$= \frac{1}{4\pi^2 e^2 (1 - e^{-\beta \omega})} \left(\frac{\omega^2}{c^2}\right) \operatorname{Im}[\mathbf{k} \cdot \mathbf{D}^{-1} \cdot \mathbf{k}]. \tag{35.10}$$

This is the basic formula which describes the scattering of light from a *single*-component magnetoplasma. It is the generalization—to include transverse currents and fields induced by the magnetic field—of Eq. (17.6).

To show that in the absence of a field Eq. (35.10) is equivalent to Eq. (17.6), we recall that under these circumstances $\epsilon(\mathbf{k}, \omega)$ is diagonal. **D** then takes the form

$$\mathbf{D} = \begin{pmatrix} (k^2 - \omega^2 \epsilon/c^2) & 0 & 0 \\ 0 & (k^2 - \omega^2 \epsilon/c^2) & 0 \\ 0 & 0 & -\omega^2 \epsilon/c^2 \end{pmatrix}, \tag{35.11}$$

where we have chosen the z axis in the **k** direction. Hence

$$\mathbf{k} \cdot \mathbf{D}^{-1} \cdot \mathbf{k} = k^2/D_{zz} = -c^2 k^2/\omega^2 \epsilon(\mathbf{k}, \omega). \tag{35.12}$$

The equivalence of Eqs. (35.10) and (17.6) follows directly from this result.

35. THE MAGNETOPLASMA

The significance of the tensor **D** can be seen from Eq. (35.4). If there is no driving field ($\mathbf{E}_{\text{ext}} = 0$), these equations will have a solution if, and only if, the determinant of their coefficients vanishes, i.e., if

$$\det(\mathbf{D}) = 0. \tag{35.13}$$

This is the secular equation which determines the dispersion relation of the modes of a magnetoplasma. In the unmagnetized case it becomes

$$\epsilon(\mathbf{k}, \omega)[k^2 - (\omega^2/c^2)\,\epsilon(\mathbf{k}, \omega)]^2 = 0. \tag{35.14}$$

The first root of this equation gives the longitudinal mode frequencies (plasma waves); the second determines the frequencies of the two transverse (electromagnetic) waves in the plasma. In a magnetoplasma, the situation is complicated by the fact that **D** has off-diagonal components. These couple longitudinal and transverse waves so that all modes (except in special cases) have both longitudinal and transverse currents.

With the aid of the well-known formula for the inverse of a matrix, we may write

$$\mathbf{k} \cdot \mathbf{D}^{-1} \cdot \mathbf{k} = (\mathbf{k} \cdot \mathbf{G} \cdot \mathbf{k})/\det(\mathbf{D}), \tag{35.15}$$

where

$$G_{ij} = \text{cofactor}(D_{ij}). \tag{35.16}$$

Thus

$$S(\mathbf{k}, \omega) = \left(\frac{\omega^2}{4\pi e^2}\right)\left(\frac{1}{1 - e^{-\beta\omega}}\right) \text{Im}[\mathbf{k} \cdot \mathbf{G} \cdot \mathbf{k}/\det(\mathbf{D})]. \tag{35.17}$$

This expression has singularities at points where $\det(\mathbf{D}) = 0$, corresponding to light scattering from the various possible modes of the plasma.

Equation (35.17) is a rigorous, but quite complicated, result. Fortunately, under the usual experimental conditions it can be greatly simplified. To see why this is so, we return to the expression for **D**, and consider the size of the various terms in it. In a typical experiment, light is scattered through a fairly large angle, and the frequency shift ω is small compared to the optical frequency ω_0. For example, in the GaAs experiments described previously $\omega \simeq \omega_p \leqslant 0.01\omega_0$. Under such conditions

$$k \simeq (2\omega_0/c)\sin(\theta/2) \gg \omega/c. \tag{35.18}$$

This inequality suggests that one should seek an expansion of Eq. (35.17) in powers of $1/n$, where

$$n = ck/\omega \gg 1. \tag{35.19}$$

$n = ck/\omega$ is the ratio of the velocity of light to the phase velocity ω/k of the wave being excited in the plasma. It can also be thought of as the effective refractive index of the medium for this wave. Consider first

$$D_{ij} = (\omega^2/c^2)[n^2(\delta_{ij} - \alpha_i\alpha_j) - \epsilon_{ij}], \quad (35.20)$$

where $\boldsymbol{\alpha}$ is the unit vector; $\boldsymbol{\alpha} \equiv \mathbf{k}/|\mathbf{k}|$. A straightforward expansion shows that $\det(\mathbf{D})$ is a polynominal of second degree in n^2:

$$\det(\mathbf{D}) = (\omega^2/c^2)^3(An^4 + Bn^2 + C) \quad (35.21)$$

with

$$A = (\boldsymbol{\alpha} \cdot \boldsymbol{\epsilon} \cdot \boldsymbol{\alpha}), \quad B = (\boldsymbol{\alpha} \cdot \boldsymbol{\epsilon} \cdot \boldsymbol{\epsilon} \cdot \boldsymbol{\alpha}) - (\boldsymbol{\alpha} \cdot \boldsymbol{\epsilon} \cdot \boldsymbol{\alpha})\,\mathrm{tr}(\boldsymbol{\epsilon}), \quad C = \det(\boldsymbol{\epsilon}). \quad (35.22)$$

If $n \gg 1$, we may approximate

$$\det(\mathbf{D}) \simeq (\omega^2/c^2)^3(\boldsymbol{\alpha} \cdot \boldsymbol{\epsilon} \cdot \boldsymbol{\alpha})n^4. \quad (35.23)$$

In a similar way it can be shown that, to lowest order in $1/n^2$,

$$G_{ij} \simeq k^2 k_i k_j. \quad (35.24)$$

Finally,

$$S(\mathbf{k}, \omega) \simeq \left(\frac{k^2}{4\pi e^2}\right)\left(\frac{1}{1 - e^{-\beta\omega}}\right) \mathrm{Im}[k^2/\mathbf{k} \cdot \boldsymbol{\epsilon} \cdot \mathbf{k}]. \quad (35.25)$$

This is the formula which has most commonly been used to discuss magnetoplasma scattering experiments. The approximation made in deriving it is generally called the *quasi-static* approximation because it neglects the transverse (truly electromagnetic) portion of the field. One can easily show that Eq. (35.25) follows from the truncated set of Maxwell equations;

$$-\nabla^2 \phi_{\mathrm{ind}} = 4\pi \rho_{\mathrm{ind}}, \quad \omega \rho_{\mathrm{ind}} - \mathbf{k} \cdot \mathbf{j}_{\mathrm{ind}} = 0, \quad \mathbf{j}_{\mathrm{ind}} = -\boldsymbol{\sigma} \cdot (\nabla \phi_{\mathrm{ext}} + \nabla \phi_{\mathrm{ind}}). \quad (35.26)$$

Further insight can be obtained by considering the solutions of Eq. (35.13). These are plotted, as a function of $|\mathbf{k}|$, for a typical magnetoplasma ($\omega_c = \omega_p; \mathbf{k} \perp \mathbf{H}_0$) in Fig. V-5. The right-hand portion of the graph, where the "electromagnetic" and "plasma" waves are well separated from one another, is that appropriate to most scattering experiments, and is the region in which the quasi-static approximation applies. In this range, the dispersion relation of the lower, plasmalike mode is determined by the equation

$$\mathbf{k} \cdot \boldsymbol{\epsilon} \cdot \mathbf{k} = 0. \quad (35.27)$$

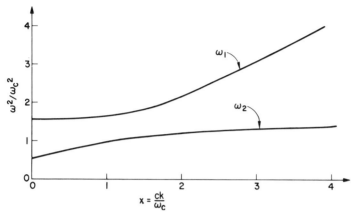

FIG. V-5. Dispersion of photon–plasmon modes when coupled by a magnetic field, $\omega_c = \omega_p$.

For smaller k, the two sorts of modes mix, and the quasi-static approximation is no longer valid. This interaction is similar to the mixing of phonon and light wave which creates the polariton. In our case, the *plasmon* and light wave couple to generate a mixed excitation that might be termed a plasma-polariton.

Returning to the quasi-static approximation, we wish to evaluate Eq. (35.25). For this purpose, we need an expression for the dielectric tensor of an electron gas in the presence of a dc magnetic field. Fortunately, in most magnetoplasma scattering experiments, k is small compared to the screening wave vector (κ_D or κ_{FT}) of the plasma. One may then use the local ($k = 0$) expressions for the conductivity in evaluating ϵ.

The results are (see Chapter VII)

$$\epsilon = \begin{pmatrix} \epsilon_\perp & \epsilon_\times & 0 \\ -\epsilon_\times & \epsilon_\perp & 0 \\ 0 & 0 & \epsilon_z \end{pmatrix}, \tag{35.28}$$

where

$$\epsilon_\perp = 1 - \omega_p^2/(\omega^2 - \omega_c^2) \tag{35.29}$$

$$\epsilon_\times = i\omega_c \frac{\omega_p^2}{\omega(\omega^2 - \omega_c^2)} \tag{35.30}$$

$$\epsilon_z = 1 - \omega_p^2/\omega^2. \tag{35.31}$$

$\omega_c = eH_0/mc$ is the cyclotron frequency, and the magnetic field $\mathbf{H}_0$ is assumed to be in the z direction. To keep these formulas simple, collision effects have been neglected ($\omega\tau \to \infty$).

The simplest geometry in which to study magnetoplasma effects is that in which $\mathbf{k} \perp \mathbf{H}_0$; i.e., the scattering takes place in the plane perpendicular to the field. One then finds

$$k^2/(\mathbf{k} \cdot \boldsymbol{\varepsilon} \cdot \mathbf{k}) = (\omega^2 - \omega_c^2)/(\omega^2 - \omega_c^2 - \omega_p^2). \tag{35.32}$$

This expression has a pole at the point $\omega = (\omega_p^2 + \omega_c^2)^{1/2}$ which is the frequency of a magnetoplasma wave propagating across the field. If $\omega_c \sim \omega_p$, the character of this wave is considerably different from that of an ordinary plasma wave. It is no longer purely longitudinal, but contains transverse currents comparable in magnitude to the longitudinal ones. These couple with the field to generate an additional restoring force and, thereby, account for the higher frequency of the wave. The frequency of this mode, $\omega = (\omega_p^2 + \omega_c^2)^{1/2}$, is known in gas plasma physics as the *upper hybrid frequency*. The pole at the hybrid frequency gives a contribution

$$S(\mathbf{k}, \omega)|_{\text{hybrid}} = \left\{ \frac{k^2}{8\pi e^2(1 - e^{-\beta\omega})} \frac{\omega_p^2}{(\omega_p^2 + \omega_c^2)^{1/2}} \delta[\omega - (\omega_p^2 + \omega_c^2)^{1/2}] \right\} \tag{35.33}$$

to the structure factor. On comparing this result to Eq. (21.5), we notice that the scattering strength of the magnetoplasma wave is somewhat lower [by a factor $\omega_p/(\omega_p^2 + \omega_c^2)^{1/2}$] than that of the plasma wave in an unmagnetized plasma.

The condition $\omega_c \sim \omega_p$ can be achieved, with reasonable doping density and reasonable magnetic field, in several of the common semiconductors with light effective mass, such as n-type GaAs, InAs, and InSb. To date, magnetoplasma waves have mainly been studied (*16*) by light scattering in GaAs. This is a favored material for such investigations because it is transparent to the 1.06μm Nd:YAG laser radiation. To probe the narrower gap materials, one needs an intense light source in the farther infrared portion of the spectrum. The CO_2 laser has been used in this way to study the magnetoplasma in n-type InAs (*4*). However, it is a less effective tool than the Nd:YAG laser because infrared detectors are slower, and very much less sensitive, at 10.6 than at 1.06μm.

In GaAs, magnetoplasma wave scattering has been observed in two geometries: $\mathbf{k} \perp \mathbf{H}_0$ (discussed above) and $\mathbf{k}$ at 45° to $\mathbf{H}_0$. The former is the simpler from the theoretical point of view, since it can be shown that there is no Landau damping of plasma waves that propagate directly across the magnetic field. Correspondingly, the dielectric tensor $\boldsymbol{\epsilon}(\mathbf{k}, \omega)$ has a relatively simple analytic form (as a function of frequency) in this geometry, whereas it is far more complicated in most other cases. Despite these facts, there is also some interest in the oblique (45°)

geometry, since the magnetoplasma then supports two collective modes for a given **k**, whose frequencies are

$$\omega^2 = \tfrac{1}{2}[(\omega_p^2 + \omega_c^2) \pm (\omega_p^4 + \omega_c^4)^{1/2}]. \tag{35.34}$$

This expression follows directly from the equation **kεk** = 0. It points up an interesting feature of Raman (and other) scattering experiments— that these are experiments in which the wave vector is fixed by the scattering geometry [$k = 2k_0 \sin(\theta/2)$] and the scattering system picks out the frequencies of the modes with this k. By contrast, in most other measurements one probes a system with a given frequency, and the system determines the concomitant wave vector.

In most of the field range over which it is observed, the magnetoplasma wave scattering from n-GaAs has the field and carrier concentration variation and intensity predicted by Eq. (35.33) or (35.34). The line width is also in agreement with that calculated from mobility data. In this sense, there is good agreement between the classical theory and experiment. However, a sizable discrepancy occurs, in the $\mathbf{k} \perp \mathbf{H}_0$ geometry, at the field where $\omega_p = \sqrt{3}\,\omega_c$. In this range the measured frequency of the magnetoplasma wave falls below that predicted by Eq. (35.33). At the same time, the line width becomes anomalously large. This deviation from the classical theory is an interesting effect which had previously been predicted (*17*). It is a mode coupling phenomenon, brought about by the fact that the wave vector k, though small, is not zero. To see how it comes about we must investigate, at least to lowest order in k^2, the k variation of the dielectric tensor. For $\epsilon_{xx}(\mathbf{k}, \omega)$ one finds for example (see Chapter VIII), an expression of the form

$$\epsilon_{xx}(k, \omega) = 1 - \omega_p^2 \left[\frac{1-\lambda}{\omega^2 - \omega_c^2} + \frac{\lambda}{\omega^2 - 4\omega_c^2} + \cdots\right], \tag{35.35}$$

where $\lambda = k^2 v_{\text{th}}^2/\omega_c^2$ is a measure of the nonlocality. Equation (35.35) applies to a Maxwellian plasma, as was used in the magnetoplasma light scattering experiments, and for simplicity has been written without collision terms ($\omega\tau \to \infty$). The important point is that the second term in the brackets in this equation gives rise to a new root of the dispersion relation **kεk** = $\epsilon_{xx}(\mathbf{k}, \omega) = 0$, with frequency near $2\omega_c$. This solution is a *Bernstein mode* (see Chapter VIII). It is degenerate with the hybrid mode when

$$(\omega_p^2 + \omega_c^2)^{1/2} = 2\omega_c, \quad \text{i.e., when} \quad \omega_p = 3^{1/2}\omega_c. \tag{35.36}$$

At this point there is mode mixing, which is best studied by considering the complete solutions of the dispersion relation, $\epsilon_{xx}(\mathbf{k}, \omega) = 0$. They are

$$\omega_{1,2}^2 = \tfrac{1}{2}\{(5\omega_c^2 + \omega_p^2) \pm [(3\omega_c^2 - \omega_p^2)^2 + 12\lambda\omega_p^2\omega_c^2]^{1/2}\}. \quad (35.37)$$

A graph of these solutions, as a function of ω_c, is shown in Fig. V-6 for a particular value of k. Except near the point $\omega_p = 3^{1/2}\omega_c$, the two

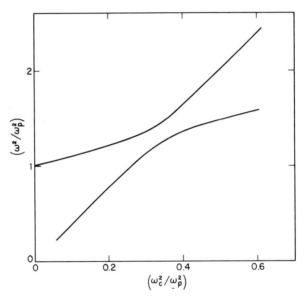

FIG. V-6. Interaction of the lowest order Bernstein mode with the hybrid plasma mode, as the field $\mathbf{H}_0$ is varied. The splitting at $\omega_p^2 = 3\omega_c^2$ is linear in k.

modes are quite distinct from one another. It can also be shown that, in the regions away from the crossing, the Bernstein ($2\omega_c$) mode essentially does not scatter light because there is only a small electron density fluctuation associated with it. This is not true, however, near the crossing. At that point there is a sharing of the oscillator strength for light scattering. The mixing produces a considerable distortion and broadening of the magnetoplasma light scattering peak. Detailed calculations (*18*) show that the mode-coupling theory we have outlined explains all the peculiarities of the light scattering data in GaAs. This experiment, therefore, can be considered as a direct observation of the Bernstein mode. It is the first such in a solid state plasma, and the only observation of these modes by light scattering.

36. Multicomponent Plasmas

So far in this chapter, we have discussed plasma waves and light scattering for single-component plasmas only. These are typified by most metal plasmas and the simpler semiconductor plasmas, such as those in n-GaAs, InSb, InAs, CdS, etc. However, in many other common semiconductors the conduction (or valence) band is multivalleyed; i.e., there can be carriers present with several different values of effective mass. Such a situation occurs in n-type Si, Ge, PbTe, PbSe, etc., and in p-type PbTe, PbSe. The semimetal Bi also contains a multicomponent plasma. These are examples of multicomponent plasmas that exist under equilibrium conditions. It is also possible to generate *nonequilibrium*, multicomponent plasmas by injecting carriers into a crystal. For example, electron bombardment or optical excitation (*19*) can produce fairly large electron–hole densities in intrinsic samples of semiconductors such as Ge, GaAs, InSb, etc. These plasmas are quite interesting media, though they have not yet been studied by light scattering. They deserve further investigation.

Most of the formalism of the preceding sections can be used in the discussion of multicomponent plasmas. Our main aim will be to calculate the response functions of such a plasma. As before, we imagine the plasma, to be perturbed by a weak electrostatic potential, and seek its linear response. The perturbation is of the form

$$\sum_i \left[e_i \int n_i(\mathbf{r}) \phi_{\text{ext}}(\mathbf{r}, t) \, d^3r \right], \tag{36.1}$$

where $n_i(\mathbf{r})$ is the particle density operator of the ith species, and e_i its charge. Using the methods of the preceding chapters, it is now easy to show that, in a translationally invariant plasma, the induced charge density of species i (at wave vector $\mathbf{k}$ and frequency ω) is given by

$$\langle \rho_i(\mathbf{k}, \omega) \rangle = \sum_j [F_{ij}(\mathbf{k}, \omega) \phi_{\text{ext}}(\mathbf{k}, \omega)], \tag{36.2}$$

where

$$F_{ij} = -i \int_0^\infty e^{i(\omega + i\delta)t} \langle [\rho_i(\mathbf{k}, t), \rho_j(-\mathbf{k}, 0)] \rangle \, dt. \tag{36.3}$$

These equations are natural generalizations of Eqs. (13.15) and (13.16). It is also convenient, as before, to introduce functions $\alpha_i(\mathbf{k}, \omega)$ which relate the induced charge of the ith species to the *total* electrostatic potential in the plasma:

$$\langle \rho_i(\mathbf{k}, \omega) \rangle = \alpha_i(\mathbf{k}, \omega) \phi_{\text{total}}(\mathbf{k}, \omega). \tag{36.4}$$

By using Poisson's equation to determine the induced potential,

$$k^2 \phi_{\text{ind}}(\mathbf{k}, \omega) = 4\pi \sum_i [\langle \rho_i(\mathbf{k}, \omega) \rangle], \tag{36.5}$$

one can show that the functions $F_{ij}(\mathbf{k}, \omega)$ and $\alpha_i(\mathbf{k}, \omega)$ are related by the formula

$$\alpha_i \left[1 + (4\pi/k^2) \sum_{jk} (F_{jk}) \right] = \sum_j [F_{ij}]. \tag{36.6}$$

Finally, the plasma dielectric constant is given by

$$\epsilon(k, \omega) \equiv \phi_{\text{ext}}/\phi_{\text{total}} = 1 - \phi_{\text{ind}}/\phi_{\text{total}}$$

$$= 1 - \sum_i [(4\pi/k^2) \alpha_i(\mathbf{k}, \omega)]$$

$$= \frac{1}{1 + (4\pi/k^2) \sum_{ij} (F_{ij})}. \tag{36.7}$$

Note that the last two lines of Eq. (36.7) are compatible with Eq. (36.6). The combination $\sum_{ij} (F_{ij})$ which appears in the denominator of Eq. (36.7) is equal to [from Eq. (36.3)]

$$\sum_{ij} (F_{ij}) = -i \int_0^\infty e^{i(\omega + i\delta)t} \langle [\rho(\mathbf{k}, t) \rho(-\mathbf{k}, 0)] \rangle \, dt \tag{36.8}$$

where

$$\rho(\mathbf{r}) = \sum_i \rho_i(\mathbf{r}) \tag{36.9}$$

is the total charge density operator for the plasma. Using the fluctuation–dissipation theorem, we may now use this result to determine the structure factor for *charge density* fluctuations

$$S(\mathbf{k}, \omega) \equiv (2\pi)^{-1} \int_{-\infty}^\infty e^{i\omega t} \langle \rho(\mathbf{k}, t), \rho(-\mathbf{k}, 0) \rangle \, dt$$

$$= \left(\frac{k^2}{4\pi^2 e^2} \right) \left(\frac{1}{e^{-\beta \omega} - 1} \right) \text{Im} \left(\frac{1}{\epsilon(\mathbf{k}, \omega)} \right). \tag{36.10}$$

Since an electron beam couples to the plasma charge density, characteristic energy loss experiments measure $S(\mathbf{k}, \omega)$ and, therefore, $\text{Im}[1/\epsilon]$. As we have seen this is also true, in a single-component plasma, of light scattering experiments. The situation is different, however, in a multicomponent plasma. Light couples differently [as $(e_i^2/m_i c^2)^2$] to each type of particle. The scattering, as we shall see presently, measures a somewhat more complicated correlation function than $S(\mathbf{k}, \omega)$.

36. MULTICOMPONENT PLASMAS

For the time being, however, let us consider the normal modes of the plasma. These, as before, are determined by the zeros of the dielectric function, which we again calculate in RPA:

$$\epsilon(k, \omega) = 1 - (4\pi/k^2) \sum_i e_i^2 Q_i(k, \omega). \tag{36.11}$$

Here $Q_i(k, \omega)$ is determined from Eq. (14.7) for each species. Detailed evaluation of Eq. (36.11) requires, of course, some specification of the various carriers in the plasma (charge, effective mass tensor, etc.). In most solid state cases, unfortunately, multicomponent plasmas are made up of carriers having highly anisotropic, tensor effective masses, (e.g., n-Si, n-Ge, Bi, PbTe). This fact complicates the analysis and obscures much of the physics. We will be content to treat simple model plasmas for which the analysis is fairly straightforward and the physics easily visualized. More complicated cases are no harder to treat, in principle, but fairly extensive algebra may be required.

Consider, then, the simplest possible multicomponent plasma consisting of electrons and holes with isotropic masses, m_1 and m_2. We will discuss cases in which these components are either degenerate or Maxwellian. An approximation to the isotropic two-component Maxwellian plasma can be achieved fairly easily by warming an intrinsic piece of InSb or InAs. The isotropic degenerate case is more difficult to realize and, to the author's knowledge, does not occur under equilibrium conditions in any solid. However, a sufficiently large density of electrons and holes can be injected into InSb to produce a nonequilibrium degenerate plasma. We will first discuss the degenerate case, since the analysis of its properties is somewhat simpler than for the corresponding Maxwellian distribution. Later, in the light of this analysis, we will consider the two-component Maxwellian plasma.

For degenerate distributions, the integrals which determine the functions $Q_i(k, \omega)$ can easily be evaluated in the small k limit ($k/k_F \ll 1$). One finds

$$Q_i(k, \omega) \simeq \frac{m_i k_{Fi}}{2\pi^2} \left[\left(\frac{\omega}{k v_{Fi}}\right) \ln\left(\frac{\omega + k v_{Fi}}{\omega - k v_{Fi}}\right) - 2 \right]. \tag{36.12}$$

Here k_{Fi} and v_{Fi} are the Fermi wave vector and velocity of the ith carrier type. In Eq. (36.12), ω has a small positive imaginary part, and the logarithm is complex valued for $-v_F \leqslant \omega/k \leqslant v_F$. For a two-component plasma, there are two terms (that we denote as Q_1 and Q_2) which contribute to $\epsilon(k, \omega)$. Then, $\epsilon(k, \omega)$ is a function of three variables having the dimensions of velocity, (ω/k), v_{F1}, and v_{F2}. For simplicity, we will assume that the two species (electrons and holes) have the same density, so that $k_{F1} = k_{F2} \equiv k_F$. Then

$$v_{F1} = k_F/m_1; \qquad v_{F2} = k_F/m_2. \tag{36.13}$$

The most interesting case is that in which the masses of the two components are quite different from one another. Thus, in our subsequent work, we will often have in mind the situation in which $m_1 \ll m_2$, so that $v_{F1} \gg v_{F2}$. In InSb the electron-hole mass ratio is about 1/30, which fairly well satisfies this condition.

Let us now study the behavior of the dielectric function,

$$\epsilon = 1 - (4\pi e^2/k^2)(Q_1 + Q_2),$$

in various limiting cases. The simplest is that in which $\omega/k \gg v_{F1}$ and v_{F2}. It immediately follows from our preceding analysis for the single-component plasma that

$$\epsilon \simeq 1 - \omega_{p1}^2/\omega^2 - \omega_{p2}^2/\omega^2 \tag{36.14}$$

where

$$\omega_{pi}^2 = 4\pi n_i e_i^2/m_i \epsilon_0 \tag{36.15}$$

is the plasma frequency for the ith species. The condition $\epsilon = 0$ gives the frequency of the high-frequency collective mode:

$$\omega^2 = \omega_{p1}^2 + \omega_{p2}^2. \tag{36.16}$$

This result can easily be generalized to any multicomponent plasma. One can also show that in the oscillation described by Eq. (36.16) the charge densities of the two species add to one another. As a consequence, the oscillator has a greater restoring force, and higher frequency, than either of the separate plasmas. In analogy with phonon terminology in a crystal containing two atoms per unit cell, such a mode might be called an *optic plasma wave*. This nomenclature immediately leads one to ask the question, "Can a two-component plasma also support a low-frequency, acoustic plasma wave, in which the charge densities of the two species oscillate out of phase with one another"? We will see that this is, indeed, possible, though the mode is only weakly damped in a degenerate plasma when the mass ratio m_1/m_2 is small compared to unity.

To study this acoustic plasma wave, we assume that $v_{F1} \gg v_{F2}$ ($m_1/m_2 \ll 1$) and look for a mode whose phase velocity lies *between* the two Fermi velocities:

$$v_{F2} \ll \omega/k \ll v_{F1}. \tag{36.17}$$

36. MULTICOMPONENT PLASMAS

In analyzing the dielectric function we will therefore use a *low-frequency* approximation for Q_2. The result, which follows immediately from Eq. (36.12), is

$$\epsilon \simeq 1 - \frac{\omega_{p2}^2}{\omega^2} + \frac{\kappa_{FT1}^2}{k^2}\left[1 + \frac{i\pi}{2}\left(\frac{\omega}{kv_{F1}}\right)\right]. \quad (36.18)$$

It is important to notice that ϵ is complex, indicating that the acoustic plasma wave is Landau damped. This damping is due to the faster species in the plasma (species 1 in our notation). There is no Landau damping by species 2 in the degenerate case since $\omega/k > v_{F2}$. The dispersion relation for the acoustic plasma wave is determined by the condition $\epsilon(\mathbf{k}, \omega) = 0$. In the long-wavelength limit ($k/\kappa_{FT1} \ll 1$), the unity term in Eq. (36.18) may be dropped, and the dispersion relation becomes

$$\left(\frac{\omega}{k}\right)^2 = \left(\frac{\omega_{p2}^2}{\kappa_{FT1}^2}\right)\left[\frac{1}{1 + \frac{1}{2}i\pi(\omega/kv_{F1})}\right]. \quad (36.19)$$

Assuming, for the moment, that the imaginary term is relatively small, we see that the phase velocity of the wave is

$$\omega/k \simeq \omega_{p2}/\kappa_{FT1} = (4\pi ne^2/m_2)^{1/2}(m_1 v_{F1}^2/12\pi ne^2)^{1/2}$$
$$= (\tfrac{1}{3}v_{F1}v_{F2})^{1/2}. \quad (36.20)$$

Thus, in a degenerate gas, the acoustic plasma wave travels at a speed that is essentially the geometric mean of the two Fermi velocities. In the limit $v_{F1} \gg v_{F2}$, the damping term in the denominator of Eq. (36.19) is relatively small and the complete dispersion relation becomes

$$\omega/k \simeq (\tfrac{1}{3}v_{F1}v_{F2})^{1/2}[1 - \tfrac{1}{4}i\pi(v_{F2}/3v_{F1})^{1/2}]. \quad (36.21)$$

The condition $v_{F1} \gg v_{F2}$ is necessary to ensure that the Landau damping of the acoustic plasma wave (by the faster species) be relatively small. This is also a necessary condition for the existence of a well-defined acoustic wave; if $v_{F1} \cong v_{F2}$, the wave damps within one or two periods.

There are several solids which contain multicomponent, degenerate plasmas which might support an acoustic plasma wave. Two well-known examples are the semimetal Bi and the semiconductor PbTe. In Bi, the Fermi surface consists of small, highly anisotropic electron and hole pockets—three of the former and one of the latter. As viewed along the trigonal axis, the electron ellipsoids are equivalent and have a relatively small effective mass ($m_e^* \simeq 0.01 m_0$ in this direction); the corresponding hole mass is about 100 times greater. As a consequence, the Fermi velocities of electrons and holes parallel to the trigonal axis are rather different from one another, as required for a weakly damped acoustic wave. Quite similar statements apply to the doped semiconductor PbTe, which has multivalleyed conduction and valence bands consisting of quite prolate, $\langle 111 \rangle$-oriented ellipsoids. In this plasma, the acoustic wave would be made up of two different types of electrons oscillating out of phase with each other. Electron and hole collision times can be quite long in both Bi and PbTe; so from this point of view, these materials are good candidates in which to search for acoustic waves.

Some years ago, McWhorter and May (20) used microwaves in an attempt to excite acoustic plasmons in Bi slabs. They looked for energy transmission through the slabs, but saw no effects that could be attributed to acoustic wave propagation. This failure is not, perhaps, surprising. The acoustic wave is longitudinal and contains little charge, so would not be expected to couple directly to a transverse electromagnetic wave. The lack of charge also makes the coupling to an electron beam relatively weak. Thus, at present, the best hope for observing acoustic plasma waves in solids, as we shall see, is via light scattering. This is an interesting problem, to which we will return presently.

Now, however, let us briefly consider the properties of acoustic waves in other types of plasmas. The simplest extension is to the case of a plasma whose lighter species is degenerate (species 1), but whose heavier constituent is Maxwellian (species 2). Here, again, we use low- and high-frequency approximations to evaluate Q_1 and Q_2. In the high-frequency limit, however, the form of Q_2 is the same for a Maxwellian as for a Fermi–Dirac distribution, namely, $4\pi Q_2/k^2 \simeq \omega_{p2}^2/\omega^2$. Thus, in this case, the form of the dispersion relation is precisely the same as that for an entirely degenerate two-component plasma, and can be written in the form

$$v_\phi \equiv \omega/k = [\tfrac{1}{3}(m_1/m_2)]^{1/2} v_{F1} \{1 - \tfrac{1}{2} i\pi [\tfrac{1}{3}(m_1/m_2)]^{1/2}\}. \qquad (36.22)$$

Two-component, degenerate-Maxwellian plasmas can occur (with large mass ratio) in warm intrinsic crystals of semiconductors such as InSb or InAs. However, a considerably more interesting example of a two-component solid state plasma, again with large mass ratio, is an alkali metal. The ion cores in these crystals are far apart (well out of contact) so only electrostatic forces act between the various constituents of the solid. These crystals are also quite soft and shear easily, so the forces which hold the ions in the crystal lattice must be relatively weak. Under these circumstances, it is reasonable to approximate the properties of such a solid by that of a two-component plasma consisting of ions and electrons. This is the famous "jellium" model (21). It will not, of course, support transverse waves (as a true crystal does), but gives a good qualitative picture of the longitudinal phonons. The phase velocity calculated from Eq. (36.22) (using from m_1 the electron mass and m_2 the ion mass) is in surprisingly good agreement with measured longitudinal sound velocities in alkali metals.

Finally, we briefly consider the properties of acoustic waves in two-component Maxwellian plasmas. In such media, it is possible (and often happens in gas plasmas) for the electron temperature to be considerably greater than that of the positive charges. We will preserve this freedom in our analysis. Indeed, it turns out that a well-defined acoustic wave only exists in the limit $T_1 \gg T_2$.

For Maxwellian distributions the susceptibilities are given by

$$Q_i = \frac{1}{k_B T_i} \int \frac{(\mathbf{k} \cdot \mathbf{v}) f_i(v) \, d^3v}{(\omega - \mathbf{k} \cdot \mathbf{v} + i\delta)}. \tag{36.23}$$

Once again we evaluate Q_1 in the low-frequency limit, and Q_2 in the high-frequency approximation. For Q_1, we find

$$Q_1 \simeq -\frac{n_0}{k_B T_1} - \frac{i\pi n_0}{k_B T_1} \int (\mathbf{k} \cdot \mathbf{v}) \, \delta(\omega - \mathbf{k} \cdot \mathbf{v}) f_1(v) \, d^3v$$

$$= -\frac{n_0}{k_B T_1} \left[1 + i\pi \left(\frac{\omega}{k} \right) \left(\frac{m_1}{2\pi k_B T_1} \right) \exp\left(-\frac{m_1 \omega^2}{2k^2 k_B T_1} \right) \right]. \tag{36.24}$$

In evaluating Q_2, we will ignore Landau damping due to this species, assuming that the phase velocity of the wave is considerably greater than the thermal speed of species 2. This approximation is only valid if $T_1 \gg T_2$. If $T_1 \simeq T_2$, both species cause appreciable Landau damping. The analysis is then considerably more complicated and, in any case, the acoustic wave is a poorly defined excitation. For $T_1 \gg T_2$,

$$\frac{4\pi e^2 Q_2}{k^2} \simeq \frac{\omega_{p2}^2}{\omega^2} \left[1 + \frac{3 k_B T_2 k^2}{m_2 \omega^2} \right]. \tag{36.25}$$

In the long-wavelength limit ($k/\kappa_D \ll 1$), the dispersion relation becomes

$$\left(\frac{\omega}{k} \right)^2 \simeq \left(\frac{k_B T_1 + 3 k_B T_2}{m_2} \right) \left\{ \frac{1}{1 + i[\frac{1}{2}\pi(m_1/m_2)(1 + 3T_2/T_1)]^{1/2}} \right\}. \tag{36.26}$$

Note that the phase velocity is approximately

$$\omega/k \simeq (T_1/T_2)^{1/2} (k_B T_2/m_2)^{1/2}, \tag{36.27}$$

which is a factor $(T_1/T_2)^{1/2}$ times the thermal velocity of the slower particle (species 2). Landau damping due to these particles (the imaginary part of Q_2) can only be ignored if the phase velocity is appreciably bigger than $(k_B T_2/m_2)^{1/2}$; i.e., in the limit $T_1 \gg T_2$. However, if $T_1 \gg T_2$, the acoustic plasmon *is* a well-defined and relatively long-lived excitation. It has been extensively studied in gas plasma physics.

Another interesting situation to consider is that in which one constituent of the plasma (say species 1) drifts with respect to the other, but still has a Maxwellian distribution. It is then easy to show that the expression for Q_1 is modified to

$$Q_1 = -\frac{n_0}{k_B T}\left[1 + i\pi\left(\frac{\omega - \mathbf{k}\cdot\mathbf{v}_D}{k}\right)\left(\frac{m_1}{2\pi k_B T_1}\right)^{1/2}\exp\left(-\frac{m_1(\omega - \mathbf{k}\cdot\mathbf{v}_D)^2}{2k^2 k_B T_1}\right)\right] \quad (36.28)$$

where $\mathbf{v}_D$ is the drift velocity. The important feature of Eq. (36.28) is the fact that the drift velocity term tends to reduce its imaginary part. When $(\mathbf{k}\cdot\mathbf{v}_D)/k$ exceeds the phase velocity ω/k, the imaginary part of Q_1 becomes negative, and the acoustic wave is unstable. This is the famous *two-stream instability*, whose properties have been thoroughly studied by gas plasma physicists (22). The treatment we have given here is a much simplified one, only applicable to situations ($T_1 \gg T_2$) in which a well-defined acoustic wave exists. Nevertheless, this limiting case illustrates the essential physics of the instability, which is a decrease and reversal of Landau damping as the drift velocity of the electrons approaches and then exceeds the speed of the acoustic plasma wave. The mechanism is the same as that of a traveling wave tube, in which a fast electron beam travels through a slow wave structure at a speed greater than the phase velocity of electromagnetic waves on the structure. Notice, also, that this explanation of the two-stream instability implies that the phase velocity of the acoustic wave is to be measured with respect to the slow (ion) plasma species, i.e., that the wave propagates in the ion system. This is indeed the case, as can be seen by rewriting the dispersion relation in the form

$$\omega^2 \simeq \frac{\omega_{p2}^2}{\kappa_{D1}^2/k^2} \simeq \frac{\omega_{p2}^2}{\epsilon_1(k)} \quad (36.29)$$

where $\epsilon_1(k) = (\kappa_{D1}^2/k^2)$ is the static dielectric constant of species 1. We see, therefore, that the acoustic wave is essentially *a plasma oscillation of the heavier species, screened by the static dielectric constant of the lighter carriers.* It is in this sense that the acoustic wave may be said to propagate in the heavier portion of the plasma.

The two-stream instability is well known in gas plasma physics, but to date has not been observed in solids. The difficulty, in all cases, is that collisional damping is greater than the growth rate of the instability. This is true in both Maxwellian and degenerate solid state plasmas. A further problem, in degenerate cases, is that enormous current densities are required to match the drift velocity to the phase velocity of the waves. The best possibility is probably a plasma generated by injection or avalanche breakdown in InSb but even here, as was shown some years ago by Pines and Schrieffer (23), the mobilities are rather small to permit instability to occur.

37. Light Scattering by Multicomponent Plasmas

As indicated earlier, the problem of light scattering from a multicomponent plasma is more complicated than that of scattering from a single-component one. The reason is easy to see from the Hamiltonian which couples two types of particles to the electromagnetic field:

$$H = \sum_i \frac{(\mathbf{p}_{1i} - (e_1/c)\,\mathbf{A}_{1i})^2}{2m_1} + \sum_j \frac{(\mathbf{p}_{2j} - (e_2/c)\,\mathbf{A}_{2j})^2}{2m_2} + \text{(coulomb terms)}. \quad (37.1)$$

As usual, light couples to particles in the plasma via $\mathbf{p} \cdot \mathbf{A}$ and A^2 terms and, as before, the $\mathbf{p} \cdot \mathbf{A}$ terms make a small contribution to the scattering. The majority of the scattering comes from the terms

$$\frac{e_1^2}{2m_1 c^2} \sum_i (A_i^2) + \frac{e_2^2}{2m_2 c^2} \sum_j (A_j^2) \quad (37.2)$$

which couple the electromagnetic field to particle densities of species 1 and 2. Note, however, that the coupling strengths differ by a large factor if $m_1 \ll m_2$. In this limit, the scattering is essentially caused by *density fluctuations of particle* 1, i.e., the scattering cross section is

$$\frac{d^2\sigma}{d\omega\, d\Omega} = \left(\frac{e_1^2}{m_1 c^2}\right)^2 \left(\frac{\omega_1}{\omega_0}\right) (\boldsymbol{\varepsilon}_0 \cdot \boldsymbol{\varepsilon}_1)^2 (2\pi)^{-1} \int_0^\infty e^{i\omega t} \langle n_1(\mathbf{k},t)\, n_1(-\mathbf{k},0)\rangle\, dt, \quad (37.3)$$

where n_1 is the particle density operator of species 1. The correlation function which appears in Eq. (37.3) is considerably different from the density correlation function which determines the dielectric constant. To evaluate

$$S_1(\mathbf{k}, \omega) \equiv (2\pi)^{-1} \int_0^\infty e^{i\omega t} \langle n_1(\mathbf{k}, t)\, n_1(-\mathbf{k}, 0)\rangle\, dt, \quad (37.4)$$

we again use the fluctuation–dissipation theorem which relates S_1 to the corresponding response (or Green's) function, namely,

$$S_1(\mathbf{k}, \omega) = \frac{1}{\pi e^2(e^{-\beta\omega} - 1)} \operatorname{Im}[F_{11}(\mathbf{k}, \omega)], \qquad (37.5)$$

where F_{11} is defined by Eq. (36.3). Physically, F_{11} represents the response of the plasma to a fictitious external potential which couples *only* to particles of species 1 via an interaction of the form

$$H_1 = e_1 \int n_1(\mathbf{r}) \phi_{\text{ext}}(\mathbf{r}, t) \, d^3r. \qquad (37.6)$$

The induced density of species 1 is then given by

$$e_1 \langle n_1(\mathbf{k}, \omega) \rangle = F_{11}(\mathbf{k}, \omega) \phi_{\text{ext}}(\mathbf{k}, \omega). \qquad (37.7)$$

Of course, this induced particle density (and also that of species 2) will generate a charge in the plasma that in turn, via Poisson's equation, will produce an induced potential. As previously, it is easiest in calculating these induced particle densities to consider the response of the plasma to the *total* potential seen by each species. Thus, we write

$$\langle \rho_1(\mathbf{k}, \omega) \rangle = \alpha_1(\phi_{\text{ext}} + \phi_{\text{ind}}), \qquad \langle \rho_2(\mathbf{k}, \omega) \rangle = \alpha_2 \phi_{\text{ind}} \qquad (37.8)$$

and, finally, Poisson's equation

$$k^2 \phi_{\text{ind}} = 4\pi [\langle \rho_1 \rangle + \langle \rho_2 \rangle]. \qquad (37.9)$$

It is important to notice that in Eqs. (37.8) *only species* 1 responds to the fictitious external potential, whereas all particles respond to ϕ_{ind}. This is as it should be since all particles are coupled to one another by Coulomb forces, and this interaction is exactly what is described by ϕ_{ind}.

We may now eliminate $\langle \rho_2(\mathbf{k}, \omega) \rangle$ and $\phi_{\text{ind}}(\mathbf{k}, \omega)$ from Eqs. (37.7)–(37.8) to obtain an expression for $\langle \rho_1(\mathbf{k}, \omega) \rangle$ in terms of ϕ_{ext}, i.e., an expression for F_{11}. The result is

$$\langle \rho_1(\mathbf{k}, \omega) \rangle = \frac{\alpha_1[1 - (4\pi\alpha_2/k^2)] \phi_{\text{ext}}}{[1 - (4\pi\alpha_1/k^2) - (4\pi\alpha_2/k^2)]} \qquad (37.10)$$

whence, via Eqs. (37.5) and (37.7),

$$S_1(\mathbf{k}, \omega) = \frac{1}{\pi e^2(e^{-\beta\omega} - 1)} \operatorname{Im}\left[\frac{\alpha_1(1 - (4\pi\alpha_2/k^2)}{(1 - (4\pi\alpha_1/k^2) - (4\pi\alpha_2/k^2)}\right]$$

$$= \frac{1}{\pi e^2(e^{-\beta\omega} - 1)} \left\{ \frac{|1 - (4\pi\alpha_2/k^2|^2}{|\epsilon(\mathbf{k}, \omega)|^2} \operatorname{Im}[\alpha_1] + \left(\frac{4\pi}{k^2}\right)^2 \frac{|\alpha_1|^2}{|\epsilon(\mathbf{k}, \omega)|^2} \operatorname{Im}[\alpha_2] \right\}.$$
$$(37.11)$$

$S_1(\mathbf{k}, \omega)$ has been evaluated in great detail for the two-component Maxwellian plasma, using RPA to calculate the susceptibilities ($\alpha_i \cong e_i^2 Q_i$). The spectra have entirely different forms in the single-particle and collective regimes. In the single-particle regime ($k \gg \kappa_D$)

$$S_1(\mathbf{k}, \omega) = \frac{\text{Im}(Q_1)}{\pi(e^{-\beta\omega} - 1)} \tag{37.12}$$

and essentially measures the velocity distribution of the lighter species in the plasma. In particular, the width of the spectrum is about $\Delta\omega \sim k(k_B T/m_1)^{1/2}$. This limit is the same as that for a single-component Maxwellian plasma made up of species 1. In the collective regime, on the other hand, the spectrum of light scattered from a two-component plasma is quite different from that produced by a single-component plasma. Besides the weak plasmon satellite [of strength $(e^2/m_1 c^2)^2(k^2/\kappa_D^2)$] there is now a strong central line, whose intensity (per particle) is about $\tfrac{1}{2}(e^2/m_1 c^2)^2$, and whose width is determined by the thermal velocity of the *heavier* species in the plasma [$\Delta\omega \simeq (k_B T/m_2)^{1/2} k$]. This central line is an entirely new feature, which has no counterpart in the spectrum of light scattered from a single-component plasma. Scattering in the central line is caused by density fluctuations among species 2, which are nearly perfectly screened by species 1. Such fluctuations are essentially uncharged, and occur with their statistical probability. Hence, this scattering is not reduced in intensity by the factor $(k/\kappa_D)^2$, as is the plasmon scattering. The particles which actually cause the scattering are, of course, those of species 1, which are screening the density fluctuation in species 2.

Similar calculations, for degenerate plasmas, have been carried out by one of the authors (7). As in the classical case, he finds that in the collective limit ($k \ll \kappa_{FT}$) the spectrum of light scattered from a two-component plasma is quite different from that produced by a single-component plasma. Figure V-7 shows typical spectra obtained from this calculation. There is again a strong central line of width $\Delta\omega \simeq v_{F2} k$. In addition, if $m_1 \ll m_2$, there is a well-defined acoustic plasma wave resonance in Eq. (37.10), whose form is

$$S_1(\mathbf{k}, \omega)\big|_{\text{acoustic}} \simeq \frac{(3n_c/4)(k_B T/E_{F1})(1/kv_{F1})}{[1 - (\omega/kv_\phi)^2]^2 + (\pi\omega/2kv_{F1})^2}, \tag{37.13}$$

where $v_\phi = (\tfrac{1}{3} v_{F1} v_{F2})^{1/2}$ is the phase velocity of the acoustic plasma wave. The integrated cross section, per particle, in the acoustic plasmon line is

$$d\sigma/d\Omega\big|_{\text{acoustic}} \simeq (e^2/m_1 c^2)^2 (3k_B T/4E_{F1}). \tag{37.14}$$

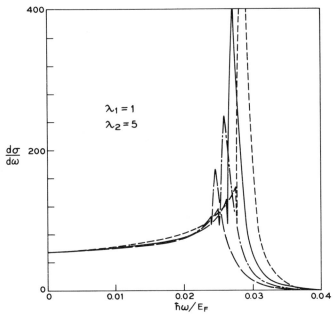

FIG. V-7. Light scattering spectra for a degenerate, two-component plasma. The differential scattering cross section per unit volume per unit solid angle per particle normalized to the Thomson cross section r_0^2 for a degenerate two-component plasma. The Fermi energy is fixed at 30 mV and the mass ratio $m_1/m_2 \equiv \lambda_2/\lambda_1 = 5$ is that appropriate to n-type Si. The different curves correspond to various drift velocities of the two species. —— $v_D = 0$; ——— $v_D = 0.1$; —— $v_D = 0.2$; --- $v_D = 0.3$.

This formula is valid in the limit $k \ll \kappa_{FT2}$. Note that the cross section per particle is of order $(e^2/m_1 c^2)^2$ and *does not* contain the reduction factor (k^2/κ_{FT1}) that appears in the optic plasmon scattering cross section. This factor is absent because the acoustic plasma is a nearly neutral excitation. As a consequence, fluctuations in the amplitudes of such waves are not suppressed by Coulomb interactions. In materials with small effective masses, the cross section for acoustic plasmon scattering can become quite sizable. It now appears that light scattering experiments, in materials such as PbTe or Bi, may be the best way of detecting acoustic plasma waves in solids.

Finally, it should be mentioned that there is a possibility of achieving *stimulated* Raman scattering from the acoustic plasmon. McWhorter (8) has calculated thresholds for this process in n-PbTe (with a 10.6μm CO_2 laser pump) and Bi (with a 28μm H_2O laser pump) and finds values of 100 kW cm^{-2} and 300 W cm^{-2}, respectively. Both of these figures are within reach of present laser technology. In stimulated plasmon scattering

38. Band Structure Effects in Light Scattering

So far in our discussion of light scattering from plasmas in semiconductors, we have used a simple classical Hamiltonian in describing the electron's motion. Band structure effects have only been included to the extent of replacing the free electron mass by an effective mass m^*. This approximation, as we will see presently, is valid when all frequencies are small compared to the band gap ($\omega_0 \ll E_G$). Such a condition is not satisfied in many experiments. Thus, we are led to consider the problem of light scattering at finite frequencies. We will soon see that this is a considerably more complicated, and richer, problem than the classical one.

In discussing the finite-frequency problem, one must use a complete set of band states to describe the scattering. We will begin by considering light scattering from a single electron in a semiconductor, i.e., a situation in which the valence bands of the crystal are completely full, and there is *one* electron in the conduction band. For the time being, we will also neglect electron–electron interactions and assume that the system is adequately described by a one-electron Hamiltonian. The Hamiltonian of this system is

$$H = (p^2/2m) + V(r) + (\hbar/4m^2c^2)[\nabla V \times \mathbf{p} \cdot \boldsymbol{\sigma}], \quad (38.1)$$

where $V(\mathbf{r})$ is the periodic crystal potential and the last term is the spin–orbit coupling. It is assumed that the eigenfunctions (Bloch waves) and eigenvalues of this Hamiltonian are known:

$$H\phi_{\mathbf{p},\mu} = E_\mu(p)\phi_{\mathbf{p},\mu}. \quad (38.2)$$

Here $\phi_{\mathbf{p},\mu}$ is a Bloch wave of crystal momentum $\mathbf{p}$ in band μ, and $E_\mu(\mathbf{p})$ is the corresponding energy. Now let us couple these Bloch electrons to the light field via the replacement $\mathbf{p} \to (\mathbf{p} - (e/c)\mathbf{A})$ in Eq. (38.1). As before, there are two important types of electron–photon interactions[†] the terms

$$-(e/mc)[\mathbf{p} + (\hbar/4mc^2)(\boldsymbol{\sigma} \times \nabla V)] \cdot \mathbf{A} \equiv -(e/c)(\mathbf{v} \cdot \mathbf{A}), \quad (38.3)$$

[†] Here we are neglecting magnetic dipole transitions caused by the $\mu g \boldsymbol{\sigma} \cdot \mathbf{H}$ term in the complete Hamiltonian.

where the velocity operator is defined by

$$\mathbf{v} = \mathbf{p}/m + (\hbar/4m^2c^2)(\boldsymbol{\sigma} \times \nabla V); \qquad (38.4)$$

and the terms $e^2A^2/2mc^2$. In the case of free electrons, the $\mathbf{v} \cdot \mathbf{A}$ term hardly contributes to the light scattering cross section—because there are two second-order matrix elements of the $\mathbf{v} \cdot \mathbf{A}$ interaction that almost exactly cancel.

In the Bloch electron problem, this cancellation persists as far as *intraband* matrix elements of the $\mathbf{v} \cdot \mathbf{A}$ interaction are concerned, but there are also *interband* matrix elements, which give an important contribution to the cross section. These are of the form (ignoring the small momentum of the light waves)

$$\left(\frac{e}{c}\right)^2 \frac{2\pi\hbar c^2}{(\omega_0\omega_1)^{1/2}} \sum_{\substack{\mu' \\ (\mu' \neq \mu)}} \Bigg\{ \frac{[\boldsymbol{\epsilon}_0 \cdot \langle \mu\mathbf{p} | \mathbf{v} | \mu'\mathbf{p}\rangle][\langle \mu'\mathbf{p} | \mathbf{v} | \mu\mathbf{p}\rangle \cdot \boldsymbol{\epsilon}_1]}{E_\mu(\mathbf{p}) - E_{\mu'}(\mathbf{p}) - \omega_0}$$

$$\frac{[\boldsymbol{\epsilon}_1 \cdot \langle \mu\mathbf{p} | \mathbf{v} | \mu'\mathbf{p}\rangle][\langle \mu'\mathbf{p} | \mathbf{v} | \mu\mathbf{p}\rangle \cdot \boldsymbol{\epsilon}_0]}{E_\mu(\mathbf{p}) - E_{\mu'}(\mathbf{p}) + \omega_1} \Bigg\}, \qquad (38.5)$$

where the (single) electron is initially in the state $(\mu, \mathbf{p})$. In this expression the sum is carried out over all bands (filled as well as empty). This is the usual situation in perturbation theory—that the exclusion principle plays no role in virtual, intermediate states.

To get the total matrix element for light scattering, one must add to Eq. (38.5) the contribution, in first order, of the A^2 term. The result is

$$M = \left(\frac{e}{c}\right)^2 \frac{2\pi\hbar c^2}{(\omega_0\omega_1)^{1/2}} \Bigg\{ \left(\frac{\boldsymbol{\epsilon}_0 \cdot \boldsymbol{\epsilon}_1}{m}\right) + \sum_{\substack{\mu' \\ (\mu' \neq \mu)}} \bigg(\frac{[\boldsymbol{\epsilon}_0 \cdot \langle \mu\mathbf{p} | \mathbf{v} | \mu'\mathbf{p}\rangle][\langle \mu'\mathbf{p} | \mathbf{v} | \mu\mathbf{p}\rangle \cdot \boldsymbol{\epsilon}_1]}{E_\mu(\mathbf{p}) - E_{\mu'}(\mathbf{p}) - \omega_0}$$

$$+ \frac{[\boldsymbol{\epsilon}_1 \cdot \langle \mu\mathbf{p} | \mathbf{v} | \mu'\mathbf{p}\rangle][\langle \mu'\mathbf{p} | \mathbf{v} | \mu\mathbf{p}\rangle \cdot \boldsymbol{\epsilon}_0]}{E_\mu(\mathbf{p}) - E_{\mu'}(\mathbf{p}) + \omega_1} \bigg) \Bigg\}. \qquad (38.6)$$

In the zero frequency limit $(\omega_0, \omega_1 \to 0)$ the expression in brackets is the familiar sum rule:

$$\frac{\partial^2 E_\mu(p)}{\partial p_i \, \partial p_j} \equiv \frac{\delta_{ij}}{m} + \sum_{\substack{\mu' \\ (\mu' \neq \mu)}} \Bigg\{ \frac{\langle \mu\mathbf{p} | v_i | \mu'\mathbf{p}\rangle\langle \mu'\mathbf{p} | v_j | \mu\mathbf{p}\rangle}{E_\mu(\mathbf{p}) - E_{\mu'}(\mathbf{p})}$$

$$+ \frac{\langle \mu\mathbf{p} | v_j | \mu'\mathbf{p}\rangle\langle \mu'\mathbf{p} | v_i | \mu\mathbf{p}\rangle}{E_\mu(\mathbf{p}) - E_{\mu'}(\mathbf{p})} \Bigg\}. \qquad (38.7)$$

Thus, as $\omega_0 \to 0$, the matrix element becomes

$$M = \left(\frac{e}{c}\right)^2 \frac{2\pi\hbar c^2}{(\omega_0\omega_1)^{1/2}} \left[\epsilon_0 \cdot \frac{\partial^2 E_\mu(\mathbf{p})}{\partial \mathbf{p}\, \partial \mathbf{p}} \cdot \epsilon_1\right] \tag{38.8}$$

which, for a single, isotropic conduction band yields the effective mass formula

$$M = \left(\frac{e^2}{m^*c^2}\right) \frac{2\pi\hbar c^2}{(\omega_0\omega_1)^{1/2}} (\epsilon_0 \cdot \epsilon_1). \tag{38.9}$$

This result is the justification of the use of the effective mass approximation when $\omega_0 \ll E_G$.

It is clear from Eq. (38.6) that, at finite frequencies, the light scattering cross section must differ from the effective mass value $(e^2/m^*c^2)^2$. To make estimates of this effect, it is convenient to symmetrize Eq. (38.6) as follows, assuming $\omega_1 \simeq \omega_0$:

$$M = \left(\frac{e}{c}\right)^2 \frac{2\pi\hbar c^2}{(\omega_0\omega_1)^{1/2}} \Bigg[\left(\frac{\epsilon_0 \cdot \epsilon_1}{m}\right)$$
$$+ \sum_{\substack{\mu' \\ (\mu'=\mu)}} \Bigg\{\frac{[\epsilon_0 \cdot \mathbf{v}_{\mu\mu'}(\mathbf{p})\mathbf{v}_{\mu'\mu}(\mathbf{p}) \cdot \epsilon_1 + \epsilon_1 \cdot \mathbf{v}_{\mu\mu'}(\mathbf{p})\mathbf{v}_{\mu'\mu}(\mathbf{p}) \cdot \epsilon_0][E_\mu(\mathbf{p}) - E_{\mu'}(\mathbf{p})]}{[E_\mu(\mathbf{p}) - E_{\mu'}(\mathbf{p})]^2 - \omega_0^2}$$
$$+ \frac{\omega_0[\epsilon_0 \cdot \mathbf{v}_{\mu\mu'}(\mathbf{p})\,\mathbf{v}_{\mu'\mu}(\mathbf{p}) \cdot \epsilon_1 - \epsilon_1 \cdot \mathbf{v}_{\mu\mu'}(\mathbf{p})\,\mathbf{v}_{\mu'\mu}(\mathbf{p}) \cdot \epsilon_0]}{[E_\mu(\mathbf{p}) - E_{\mu'}(\mathbf{p})]^2 - \omega_0^2}\Bigg\}\Bigg] \tag{38.10}$$

where $\mathbf{v}_{\mu\mu'} = \langle \mu\mathbf{p}|\mathbf{v}|\mu'\mathbf{p}\rangle$. The first sum in Eq. (38.10) is closely related to that which appears in Eq. (38.7) and determines the effective mass, whereas the second is related to the magnetic moment. To evaluate these sums in detail requires, of course, considerable knowledge of the band structure and interband matrix elements. There are several important cases, however, in which the properties of the conduction band of a semiconductor are largely determined by its interaction with a single, lower-lying valence band (examples are InSb, InAs, PbTe). Within this simple two-band model there is only a single term in each of the sums appearing in Eq. (38.10). For this two-band case, one finds

$$M \simeq \left(\frac{e}{c}\right)^2 \frac{2\pi\hbar c^2}{(\omega_0\omega_1)^{1/2}} \Bigg\{\frac{(\epsilon_0 \cdot \epsilon_1)}{m^*}\left[\left(\frac{E_G^2}{E_G^2 - \omega_0^2}\right)\left(\frac{1}{m^*} - \frac{1}{m}\right) + \frac{1}{m}\right]$$
$$+ \left(1 - \frac{m_s}{m}\right)\left(\frac{\omega_0 E_G}{E_G^2 - \omega_0^2}\right)\frac{\boldsymbol{\sigma}\cdot(\epsilon_0 \times \epsilon_1)}{m_s}\Bigg\}, \tag{38.11}$$

where m_s is the spin mass (the g value of the electron is $|g| = 2m/m_s$).

Equation (38.11) has two interesting features. We see that the matrix element is enhanced by a factor $E_G^2/(E_G^2 - \omega_0^2)$ as the light frequency approaches the gap. As a consequence, all cross sections contain a factor $E_G^2/[E_G^2 - \omega_0^2]$ which diverges as $\omega_0 \to E_G$. The other point to notice is that the second term of Eq. (38.11) is a spin-dependent. Its occurrence suggests the possibility of spin-flip light scattering. This process does, indeed, occur—the spin flip cross section is given (within the two-band model) by the formula

$$\sigma(\text{spin-flip}) = \left\{\left(\frac{e^2}{m_s c^2}\right)^2 \left(\frac{\omega_0 E_G}{E_G^2 - \omega_0^2}\right)^2 |\langle \alpha' | \, \sigma \cdot \epsilon_0 \times \epsilon_1 | \alpha \rangle|^2 \left(1 - \frac{m_s}{m}\right)^2\right\},$$
(38.12)

where the labels α, α' refer to the spin state of the electron.

Spin-flip scattering, first predicted by Yafet (*12*), is best observed as a Raman process by applying a dc magnetic field to the sample. One then observes Raman scattered light at frequencies $\omega_0 \pm \omega_s$, where ω_s is the spin resonance frequency. This process has been studied in n-InSb (*13*), n-InAs (*4*), and n-PbTe (*24*), using a 10.6μm CO_2 laser as the source. The measured cross sections agree within experimental accuracy with those predicted by Eq. (38.12). Because of the large g values of electrons in these materials ($g = -50$ in InSb) appreciable frequency shifts are attainable. Recently, stimulated spin-flip scattering has been observed in InSb (*10*). This experiment provides an intense, extremely sharp (<0.05 cm^{-1} half-width), magnetically tunable, source of infrared radiation. Such a device may be useful for high resolution and time-resolved spectroscopy in the 5–15μm wavelength range.

The achievement of CW stimulated spin-flip scattering (*24a*), with a CO laser pump, is particularly important from the technological point of view. The low thresholds found in these experiments are a dramatic proof of the existence of the resonance enhancement predicted by Eq. (38.11). These measurements imply that the spin-flip cross section at the CO laser frequency (which is very near E_G in InSb) is more than 100 times larger than that measured at 10.6μm!

Another interesting case is that of n-PbTe (*24*), where the conduction band minima are a set of ellipsoids of revolution, oriented along (111) directions. For each ellipsoid, there is a g tensor with values, measured by light scattering, of $g_\parallel = 58$, $g_\perp = 15$. Here, as in InSb, the Raman lines are quite sharp and stimulated scattering may again be a possibility.

We have seen [Eq. (38.12)] that in a magnetic field an electron in a semiconductor can give rise to spin-flip Raman scattering. In the light of this result, it is natural to inquire whether such electrons can also cause Raman scattering involving a change of *orbital* state (Landau level) in the magnetic field (25). Also, does the orbital quantization affect the elastic (Thomson) scattering cross section (26)?

Let us first consider the effect of orbital quantization on the elastic scattering, using the effective mass approximation to describe the electron motion. The Hamiltonian is then

$$H = [\pi - (e/c)\mathbf{A}]^2/2m^* \tag{38.13}$$

where $\pi = \mathbf{p} - (e/c)\mathbf{A}_0$, $\mathbf{A}_0$ is the vector potential of the dc magnetic field (which we assume to be in the z direction) and $\mathbf{A}$ the vector potential of the electromagnetic field. The eigenfunctions of the unperturbed Hamiltonian, $\pi^2/2m^*$, are the familiar Landau levels, whose energies are quantized in units of the cyclotron energy, $\omega_c = eH_0/m^*c$. To calculate the scattering of light from such an electron we again consider the various electron–photon coupling terms which are now of the form $(e/m^*c)(\pi \cdot \mathbf{A})$ and $(e^2/2m^*c^2)A^2$. Earlier we saw that for a completely free electron the first-order electron–photon coupling terms $\pi \cdot \mathbf{A}$ do not contribute, and the whole cross section arises from the second-order A^2 term. This is no longer true for an electron in a magnetic field. In the magnetic case, the $\pi \cdot \mathbf{A}$ interaction has matrix elements which change the Landau level quantum number ($\Delta n = \pm 1$ in dipole approximation). These terms do not cancel, as in the free electron case, but give an important contribution to the matrix element. The calculation is straightforward so we will not present any of the details, but merely give the complete formula for the matrix element [including $\pi \cdot \mathbf{A}$ and A^2 contributions]. It is

$$M = \left(\frac{e^2}{m^*c^2}\right)\left[\left(\frac{\omega_0}{\omega_0 - \omega_c}\right)\epsilon_1^+\epsilon_0^- + \left(\frac{\omega_0}{\omega_0 + \omega_c}\right)\epsilon_1^-\epsilon_0^+ + \epsilon_{0z}\epsilon_{1z}\right], \tag{38.14}$$

where ω_0 is the light frequency;

$$\epsilon_0^\pm = 2^{-1/2}(\epsilon_{0x} \pm i\epsilon_{0y}) \quad \text{and} \quad \epsilon_1^\pm = 2^{-1/2}(\epsilon_{1x} \pm i\epsilon_{1y})$$

are polarization vectors for incident and scattered photons. Equation (38.14) indicates that the elastic scattering cross section should be strongly enhanced when $\omega_0 \simeq \omega_c$ (26). This result, which is akin to resonance fluorescence, is hardly surprising. The scattered light tends to be circularly (or elliptically) polarized under such circumstances.

Next, we consider the possibility of Raman scattering involving a change of orbital (Landau level) quantum number. In the dipole approximation such inelastic scatterings can only be caused by the $\pi \cdot \mathbf{A}$ terms of Eq. (38.13). A straightforward calculation shows that when all terms are summed the resultant matrix element vanishes. In retrospect, this result is hardly surprising. The electron described by Eq. (38.13) behaves as a simple harmonic oscillator in the magnetic field. The incoming photon is another harmonic oscillator. When two harmonic oscillators are coupled (via the $\pi \cdot \mathbf{A}$ interaction), one does not expect mixed frequencies (such as $\omega_0 \pm \omega_c$ or $\omega_0 \pm 2\omega_c$) in the response. But such a mixed frequency current is exactly what is required to generate Raman scattered radiation at frequencies $\omega_0 \pm \omega_c$ or $\omega_0 \pm 2\omega_c$. This physical argument tells us why there is no Landau–Raman scattering from a classical electron (one with Hamiltonian $p^2/2m^*$) and indicates that one can never expect Raman scattering from a purely harmonic system.

The argument given above depends crucially upon the fact that the electron's Hamiltonian is of the form $p^2/2m^*$. In many semiconductors (particularly narrow gap materials such as InSb or InAs) there are important corrections to such a formula which are manifestations of band nonparabolicity. In InSb, for example, the Hamiltonian can be written more accurately as

$$H = [(E_G/2)^2 + E_G(p^2/2m^*)]^{1/2} - \tfrac{1}{2} E_G$$
$$\simeq (p^2/2m^*) - (1/E_G)(p^2/2m^*)^2. \qquad (38.15)$$

Equation (38.15) has the same form as the relativistic energy–momentum relation and indicates that, as the electron climbs in the energy band, its mass becomes larger. This is a well-known effect which has often been observed in InSb and other narrow-gap semiconductors. In a magnetic field, the Landau levels of such an electron are not equally spaced in energy, but crowd together as the electron energy increases. Such an electron no longer constitutes a perfectly harmonic system, and can give rise to orbital magnetic Raman scattering. The cross sections can be estimated (25) from the Hamiltonian of Eq. (38.15). It turns out that two such Raman processes ($\Delta n = \pm 1, \pm 2$) are allowed. Both have been observed with CO_2 laser scattering experiments in InSb (13) and InAs(4). The calculated cross section for the $\omega_0 \pm 2\omega_c$ process, for example, is of the order of magnitude

$$d\sigma/d\Omega \simeq (e^2/m^*c^2)^2(\hbar\omega_c/E_G)^2. \qquad (38.16)$$

Note that the cross section for this $2\omega_c$ process is lower, by the factor $(\hbar\omega_c/E_G)^2$, than that for elastic scattering. The reduction factor is a measure of the anharmonicity that is needed to produce Landau–Raman scattering. In InSb at 30 kG the calculated cross section is about 10^{-22} cm²/steradian. This value agrees fairly well with low-field experiments. However, the magnetic field variation of the cross section is quite different from that predicted by Eq. (38.16). Rather than rising quadratically with H, the cross section falls abruptly to zero near 40 kG. So far this effect has not been explained.

To conclude this rather lengthy chapter, we return to the question of the screening of single-particle scattering. In the classical plasma (see Chapter II) the screening is predicted to be almost complete when $k \ll \kappa_{FT}$. On the other hand, Mooradian (*11*) found that in n-GaAs the scattering remains strong even when $k \ll \kappa_{FT}$. The discrepancy between theory and experiment is an enormous one—more than two orders of magnitude.

The polarization of the scattered radiation provides a clue to the resolution of this difficulty. Mooradian observed that in the more heavily doped GaAs samples, the simple-particle scattering is polarized *perpendicular* to that of the incident laser beam. Such behavior is precisely the reverse of that expected for density fluctuation scattering [see Eq. (11.9)], but is in agreement with the selection rules for spin-flip scattering [Eq. (38.12)]. This observation lead Hamilton and McWhorter (*28*) to suggest that *spin-density fluctuations* are responsible for the enhanced single-particle scattering when $k \ll \kappa_{FT}$. Detailed calculations confirm this view (*28*). The essential point is that spin-density flutuations, which have no charge, do not induce a potential in the plasma and therefore are *unscreened*. The calculation of spin-density fluctuation scattering is performed as if one were dealing with a noninteracting electron gas. Results of such calculations are in quantitative agreement with Mooradian's experiments.

REFERENCES

1. A. Mooradian, *Proc. Int. Conf. Light Scattering Spectra Solids*, p. 285. Springer-Verlag, Berlin and New York, 1969.
2. A. Mooradian and A. G. Foyt, *Bull. Amer. Phys. Soc.* **15**, 303 (1970); A. Mooradian and A. L. McWhorter, *Proc. 10th Int. Conf. Physics Semiconductors*, p. 380. U.S. Atomic Energy Commission, 1970.
3. A. Mooradian and G. B. Wright, *Phys. Rev. Lett.* **16**, 999 (1966); A. Mooradian and A. L. McWhorter, *Proc. Int. Conf. Light Scattering Spectra Solids*, p. 297. Springer-Verlag, Berlin and New York, 1969.
4. C. K. N. Patel and R. E. Slusher, *Phys. Rev.* **167**, 413 (1968).
5. F. A. Blum and A. Mooradian, *Proc. 10th Int. Conf. Phys. Semiconductors*, p. 755. U.S. Atomic Energy Commission, 1970.
6. T. C. Damen, R. C. C. Leite, J. F. Scott, and J. Shah, *Bull. Amer. Phys. Soc.* **15**, 42 (1970).
7. P. M. Platzman, *Phys. Rev.* **139**, A379 (1965).
8. A. L. McWhorter, *in* "Physics of Quantum Electronics" (P. L. Kelley, B. Lax, and P. E. Tannenwald, eds.), p. 111. McGraw-Hill, New York, 1966.

9. P. A. Wolff, *Phys. Rev. B* **1**, 164 (1970); F. A. Blum, *ibid.* *B* **1**, 1125 (1970); F. A. Blum and R. H. Davies, *ibid. B* **3**, 3270 (1971).
10. C. K. N. Patel and E. D. Shaw, *Phys. Rev. Lett.* **24**, 451 (1970).
11. A. Mooradian, *Phys. Rev. Lett.* **20**, 1102 (1968).
12. Y. Yafet, *Phys. Rev.* **152**, 858 (1966); R. E. Slusher, C. K. N. Patel, and P. A. Fleury, *Phys. Rev. Lett.* **18**, 77 (1967).
13. C. K. N. Patel, R. E. Slusher, and P. A. Fleury, *Phys. Rev. Lett.* **17**, 1011 (1966); P. A. Wolff and Gary A. Pearson, *ibid.* **17**, 1015 (1966).
14. B. B. Varga, *Phys. Rev. A* **137**, 1896 (1965).
15. A. Mooradian and A. L. McWhorter, *Phys. Rev. Lett.* **19**, 849 (1967); see also Mooradian and Foyt (2).
16. C. K. N. Patel and R. E. Slusher, *Phys. Rev. Lett.* **21**, 1563 (1968).
17. P. M. Platzman, N. Tzoar, and P. A. Wolff, *Phys. Rev.* **174**, 489 (1968).
18. N. Tzoar and E.-N. Foo, *Phys. Rev.* **180**, 535 (1969).
19. Most experiments of this type have been performed to achieve laser action. See, for example, C. Benoit a la Guillanme and J. M. Debever, *in* "Physics of Quantum Electronics" (P. L. Kelley, B. Lax, and P. Tannenwald, eds.), p. 397. McGraw-Hill, New York, 1966. R. J. Phelan, Jr., and R. H. Rediker, *Appl. Phys. Lett.* **6**, 70 (1965). Generation of plasmas by two-photon absorption is discussed by R. E. Slusher, W. Giriat, and S. R. J. Brueck, *Phys. Rev.* **183**, 758 (1969).
20. A. L. McWhorter and W. G. May, *IBM J. Res. Develop.* **8**, 285 (1964).
21. D. Pines and P. Nozières, "The Theory of Quantum Liquids," Vol. 1. Benjamin, New York, 1966.
22. See, for example, B. D. Fried and R. W. Gould, *Phys. Fluids* **4**, 139 (1961).
23. D. Pines and J. R. Schrieffer, *Phys. Rev.* **124**, 1387 (1961).
24. C. K. N. Patel and R. E. Slusher, *Phys. Rev.* **177**, 1200 (1969).
24a. A. Mooradian, S. R. J. Brueck, and F. A. Blum, *Appl. Phys. Lett.* **17**, 481 (1970).
25. P. A. Wolff, *Phys. Rev. Lett.* **16**, 225 (1966).
26. B. Lax, *Proc. 9th Int. Conf. Phys. Semiconductors*, p. 253. "Nauka," Leningrad, 1968.
27. G. B. Wright, P. L. Kelley, and S. H. Groves, *Proc. Int. Conf. Light Scattering Spectra Solids* (1969).
28. D. C. Hamilton and A. L. McWhorter, *Proc. Int. Conf. Light Scattering Spectra Solids*, p. 309. Springer-Verlag, Berlin and New York, 1969.

VI. Waves in Metals

39. Introductory Remarks

The first half of this book has dealt almost exclusively with a rather general but simple microscopic description of solid state plasmas and the application of these ideas to the longitudinal excitations that can propagate in them. While the behavior of metallic plasmas in a magnetic field and the possibility of transverse wave propagation, were briefly alluded to, there was little discussion of such phenomena. In the second half of this book, we will go on to a detailed discussion of low-frequency ($\omega \ll \omega_\text{p}$) electromagnetic (EM) waves in metallic plasmas (1).

The phenomenon of EM propagation in metals is in itself an interesting and important problem. An understanding of the general features of these waves leads one to a study of many aspects of the nature of cooperative motions of electrons in solids. It is important to think of these propagating disturbances as spectroscopic probes of the bulk metallic medium. We have already stressed the fact that metals are poor plasmas ($r_\text{s} > 1$). Their behavior is never quantitatively described by the simple RPA or mean field approximation presented in Chapter III. The metallic plasma and its collective motion is complicated by the existence of explicit short-range electron interaction effects. Since the transport properties of such strongly interacting systems cannot be calculated from first principles, an experimental spectroscopic probe of their collective behavior is extremely valuable. The transverse EM waves provide spectroscopic information about this strongly coupled system, information which cannot be obtained by any other means.

Since electronic and magnetic currents set up by the wave are an integral part of the phenomenon, it is quite clear that the band structure, i.e., the Fermi-surface topology of the metal will play an important role in determining the characteristics of the waves. In most metals the complexities introduced by the Fermi-surface topology usually preclude any possibility of gaining deep insight into the wave phenomenon itself. The alkali metals Na, K, and Rb are unique and important exceptions. In these metals there are practically no band structure or Fermi-surface effects. They are the closest real-world approximation to an isotropic interacting electron gas. It is the relative absence of band structure effects and the presence of electron–electron interaction effects that makes the alkalis intriguing. They are the ideal materials for the confrontation between experiment and the many-body theory of normal metals.

The discussion of the next few chapters will center around an analysis of the propagation characteristics of relatively low-frequency ($\omega \ll E_F \sim \omega_p$) EM waves in simple metals of which the alkalis are the outstanding example. More complicated metals, such as the noble metals, are of course important and we do not skip over a study of their EM properties for lack of interest. The alkalis are simply the testing ground for our ideas. They are the simplest examples of metallic plasmas which support EM waves. Once we understand the phenomenon in these simple metals, we may ultimately be able to extend our ideas to more complex systems.

Before presenting a detailed analysis of waves in alkali metals, it is useful to list the three distinct types of wave propagation that are known and to discuss the utility of these waves as probes of the electron gas. In all three cases, the waves are associated with a single-particle resonance in the metal (2). There are two such resonances—cyclotron resonance (3, 4) (see Section 7) and spin resonance (5).

As an electron moves in a uniform magnetic field, its transverse velocity precesses with a characteristic frequency $\omega_c = eH_0/m^*c$. The quantity m^* is the effective mass of the particle. It is not generally equal to the free electron mass m_0 even in the absence of band structure. (This point will be discussed later.) However, in the alkalis, unlike the semiconductors, the mass m^* is quite close to its free electron value. For example, $m^*/m_0 = 1.21$ for Na. If a suitable electric field, at a frequency ω, is applied to such an electron, then resonance occurs whenever

$$\omega = n\omega_c, \quad n = 1, 2, 3\ldots . \tag{39.1}$$

Equation (39.1) simply states that the applied rf field has a fixed phase with respect to the electrons orbital motion. This condition in turn implies that there is a net energy flow from the field into the electrons orbital motion. There are two quite distinct sets of waves whose properties depend on the existence of cyclotron resonance.

The first and most thoroughly studied of these waves is the "helicon," which was mentioned in Section 7. In solids, the possibility of propagating these waves was pointed out by Aigrain (6), and by Konstantinov and Perel (7). Since then many groups have investigated their behavior in a wide range of materials (8, 9). These disturbances are essentially transverse circularly polarized waves that propagate in uncompensated metals and doped semiconductors. They may propagate either parallel to, or at a small angle relative to, the static magnetic field. In principle, as we shall see, their frequency (ω_H) may approach the cyclotron frequency from the low side. However, we will also be able to show that the existence of high-frequency helicons $\omega_c \cong \omega_H$ in metals requires very large magnetic fields, so that in practice, with fields less than 100 kG, $\omega_H/\omega_c \ll 1$. In this regime these waves are similar to the disturbances known as "whistlers" in ionospheric physics (10).

Helicons are simple excitations. They can propagate in a gas of noninteracting electrons. To a very good approximation, their propagation characteristics only depend on the density of electrons. If one analyzes their propagation characteristics very carefully, one finds effects which can be used to measure the Fermi momentum p_F which, for a spherical Fermi surface, is equivalent to measuring the electron density n_0. We will see, quite generally, that their propagation characteristics are nearly unaffected by electron–electron interaction effects. Hence, as probes of the many-body aspects of the metallic medium, the helicons are relatively useless. They are, however, interesting from several other points of view.

Because of their low frequency, we will see that helicons propagate at velocities which are comparable to sound velocities. In this regime they can be made to interact with sound waves. This phenomenon has been observed by several workers (11) We will discuss the helicon sound-wave interaction. Helicons also show interesting observable damping effects as one tilts the magnetic field away from the propagation direction (12, 13).

We will discuss helicons from the point of view of simply understanding the wide range of phenomenon associated with them rather than thinking of them as spectroscopic probes of the many-body system. The techniques we employ will lay the groundwork for a discussion of the other two, more sophisticated, types of wave propagation.

The second fairly well-understood class of waves in metals was discovered in 1965 (14). Like the helicons, these disturbances have propagation characteristics that depend strongly on the existence of cyclotron resonance. They propagate most readily perpendicular to the static magnetic field in the vicinity of cyclotron resonance and its harmonics. Because they occur for $\omega \cong n\omega_c$, $n = 1, 2...$, they will be referred to as cyclotron waves (CW) as contrasted with the helicon mode of propagation where $\omega \ll \omega_c$. The CW propagate with velocities which are comparable to particle velocities, i.e., much larger than the sound velocity. Thus, unlike the helicons, they do not interact with phonons.

We will show that it is possible to understand the existence and general features of the CW in terms of an independent-particle picture of the electron gas. In this model the dispersion of the CW depends on the Fermi momentum p_F and the mass m^* of the particles. The CW are a more sophisticated phenomenon than the helicons. Even in the noninteracting picture, their propagation characteristics are dependent on the one dynamical variable in the theory, the mass m^*.

Recently, it has become increasingly clear that the propagation characteristics of the CW depend, in a nontrivial way, on the Coulomb interactions among the electrons (15). At present the primary interest in the CW stems from the fact that they may be used as probes of many body effects in the metal. However, we should emphasize that the interaction effects that modify CW propagation are essentially corrections on a noninteracting picture which describes these waves fairly well.

There is another wavelike mode tied to cyclotron resonance which has been observed (16, 17). This wave is interesting because it does *not* exist in the absence of Coulomb interactions among the electrons. It propagates parallel to a magnetic field at a frequency close to cyclotron resonance. Under ordinary conditions this wave is strongly attenuated so that its wavelike properties, i.e., its dispersion, cannot be measured directly. A small anomaly in the surface impedance of a slab of metal has been attributed to its presence. Because of the rather fragmentary evidence available, we will only discuss this wave briefly in Section 57.

The third major class of waves known to exist in metals are of a completely different character. These modes are the paramagnetic spin waves (PSW) which were first discovered by Schultz and Dunifer (18) and analyzed by Silin (19) and the authors (20). They depend critically on the single-particle resonance resulting from the electrons spin degree of freedom. They propagate at frequencies close to the electron spin resonance frequency, i.e., $\omega \simeq \omega_s \equiv g(eH_0/2m_0c)$. (For the alkali metals, g is very near the free electron value, $g = 2.0023$.)

The spin waves are, in a real sense, the most sophisticated of the three types of waves we will discuss. They do not exist in a noninteracting gas of particles, but owe their existence to the exchange interactions (see Section 29) among electrons. The fact that they have been observed experimentally is a clear indication that a noninteracting picture of a metal is an incomplete one. The spin waves represent the first real qualitative break with the simple self-consistent-field or noninteracting, picture of a metal. The dispersion characteristics of PSW, give explicit information concerning the nature of the "exchange" interactions among electrons. The "exchange" effects are (in the absence of spin-orbit coupling) complementary to the direct Coulomb effects observed in an analysis of the CW experiments. It is perfectly conceivable, therefore, that the experiments of CW and PSW will provide us with an almost complete picture of the low-lying excited states of an interacting electron gas.

40. General Formulation of the Wave Propagation Problem

In order to describe wave propagation in an infinite metallic medium we need a set of equations which characterize the macroscopic electromagnetic field in the medium. These equations are Maxwell's equations augmented by the constitutive relations for current and magnetization:

$$\nabla \cdot \mathbf{B} = 0 \tag{40.1}$$

$$\nabla \cdot \mathbf{E} = 4\pi\rho \tag{40.2}$$

$$\nabla \times \mathbf{E} = -(1/c)\, \partial \mathbf{B}/\partial t \tag{40.3}$$

$$\nabla \times \mathbf{H} = (1/c)\, \partial \mathbf{E}/\partial t + 4\pi \mathbf{j}/c \tag{40.4}$$

with

$$\mathbf{B} = \mathbf{H} + 4\pi \mathbf{M}. \tag{40.5}$$

The field quantities $\mathbf{E}$ and $\mathbf{B}$ are the total macroscopic (see Sections 16 and 22) self-consistent fields present in the medium. The charge ρ, current density $\mathbf{j}$, and the magnetization $\mathbf{M}$ are the induced charge current and magnetization, respectively. For small enough fields, $\mathbf{j}$ and $\mathbf{M}$ [see Eq. (16.9)] will be linearly related to the total $\mathbf{E}$ and $\mathbf{H}$ (as in Section 16) by the constitutive relations

$$j_\alpha(\mathbf{r}, t) = \int \sigma_{\alpha\beta}(\mathbf{r}, \mathbf{r}', t - t')\, E_\beta(\mathbf{r}', t')\, d^3r'\, dt' \tag{40.6}$$

$$M_\alpha(\mathbf{r}, t) = \int \chi_{\alpha\beta}(\mathbf{r}, \mathbf{r}', t - t)\, H_\beta(\mathbf{r}', t')\, d^3r'\, dt'. \tag{40.7}$$

For a spherical Fermi surface, $\boldsymbol{\sigma}$ and $\boldsymbol{\chi}$ are tensor in character because of the presence of the static magnetic field $\mathbf{H}_0$. In general, in a medium with an anisotropic Fermi surface, they will be tensors even in the absence of a magnetic field.

Maxwell's equations plus the two constitutive relations Eqs. (40.6) and (40.7) make up the complete set of relations governing the macroscopic fields in the medium. Clearly the physics of the medium, i.e., the dynamics of the electron gas, is contained in the two quantities $\sigma_{\alpha\beta}$ and $\chi_{\alpha\beta}$.

In an infinite translationally invariant medium $\sigma_{\alpha\beta}$ and $\chi_{\alpha\beta}$ are only functions of the difference of the coordinates $\mathbf{r}$ and $\mathbf{r}'$, i.e.,

$$\sigma_{\alpha\beta}(\mathbf{r}, \mathbf{r}', t - t') = \sigma_{\alpha\beta}(\mathbf{r} - \mathbf{r}', t - t') \tag{40.8}$$

and

$$\chi_{\alpha\beta}(\mathbf{r}, \mathbf{r}', t - t') = \chi_{\alpha\beta}(\mathbf{r} - \mathbf{r}', t - t') \tag{40.9}$$

so that the two Fourier transforms,

$$\sigma_{\alpha\beta}(\mathbf{k}, \omega) = \int d^3r\, dt\, e^{-i(\mathbf{k}\cdot\mathbf{r} - \omega t)}\, \sigma_{\alpha\beta}(\mathbf{r}, t) \tag{40.10}$$

$$\chi_{\alpha\beta}(\mathbf{k}, \omega) = \int d^3r\, dt\, e^{-i(\mathbf{k}\cdot\mathbf{r} - \omega t)}\, \chi_{\alpha\beta}(\mathbf{r}, t) \tag{40.11}$$

contain all of the information pertinent to our studies of wave propagation.

It is the behavior of $\sigma_{\alpha\beta}(\mathbf{k}, \omega)$ that will determine the properties of the waves which are related to the orbital or cyclotron motion of the carriers, the helicons and CW. In considering the properties of PSW we will be interested in the form of $\chi_{\alpha\beta}(\mathbf{k}, \omega)$. The effective decoupling between the two types of propagation occurs because of the fact that we will neglect spin-orbit coupling and because the magnetism of the electrons is very weak.

The electric (or orbital) and the magnetic (or spin) problems have a certain symmetry with respect to one another. However, there is an important quantitative difference between the two cases. This difference stems from the size of the two quantities χ and σ/ω. The magnitude of χ, roughly speaking, tells us how magnetic the material is. For the alkali metals, and most other nonmagnetic metals, the truth is that these materials are very nonmagnetic. In cgs units χ is dimensionless and is of the order of 10^{-6} cgs.[†] The smallness of χ results from the fact that only a tiny fraction (ω_s/E_F) of the electrons partake in the magnetism, the rest being frozen out by the Fermi statistics. Because of the smallness of χ, we need only consider linear terms in the parameter χ. Since the magnetization,

$$M_\alpha(\mathbf{k}, \omega) = \chi_{\alpha\beta}(\mathbf{k}, \omega) H_\beta(\mathbf{k}, \omega), \tag{40.12}$$

is small, i.e., linearly proportional to χ, the $H_\beta(\mathbf{k}, \omega)$ in Eq. (40.12) may be determined from a solution of the electromagnetic problem neglecting the magnetism of the medium. There is no self-consistency problem involved in calculating $\mathbf{M}$. We will make no distinction between $\mathbf{B}$ and $\mathbf{H}$ in Eq. (40.12).

[†] Near a resonance χ may get as large as 10^{-3} cgs.

When we consider the orbital motion, we are faced with a much tougher problem. Here the dimensionless quantity characterizing the size of the currents induced in the medium is σ/ω which in turn is related to the dielectric constant of the medium,

$$\epsilon_{\alpha\beta} = \delta_{\alpha\beta} - (4\pi/i\omega)\, \sigma_{\alpha\beta}(\mathbf{k}, \omega). \tag{40.13}$$

In good metals at microwave frequencies the magnitude of $\epsilon_{\alpha\beta}$ is typically of the order of 10^{10}. Physically the large size of σ/ω results from the large plasma frequency, or equivalently, from the high density of electrons, i.e.,

$$4\pi\sigma(\mathbf{k}, \omega)/i\omega \sim -\omega_p^2/\omega^2. \tag{40.14}$$

The sixteen orders of magnitude which separate χ and σ/ω make a qualitative difference in our analysis.

In studying the properties of the waves associated with cyclotron resonance, we will neglect the magnetism of the material and solve Maxwell's equations self-consistently to get the dispersion relation of the modes in the infinite metal. Since wave propagation is a bulk phenomenon, we will see that the infinite-medium dispersion relations will enable us to predict successfully the *position* of resonances observed in several experiments. To go beyond such a simple analysis is exceedingly difficult. In order to predict the line shapes, phases, and relative intensities of experimentally observed peaks, one must solve the skin-effect problem (*23*). Several attempts have been made to solve these exceedingly difficult boundary value problems by approximate schemes. We will discuss one of the simplest ones applied to the transmission problem in Section 49 (*24*). It is fair to say that such "approximate" formulas should be taken with a grain of salt. They only give us a crude idea of what the solution of the real boundary value problem might be. Pushed too far, these formulas can, in some cases, yield completely erroneous predictions.

In the magnetic problem we examine the behavior of $\chi(\mathbf{k}, \omega)$. Because of the smallness of $\chi(\mathbf{k}, \omega)$, and the narrowness in space of the skin region, we will be able to solve the problem of resonance transmission through a slab. We will be able to predict line shapes, phases, and relative intensities, as well as the positions of the resonances.

REFERENCES

1. E. A. Kaner and V. G. Skobov, "Electromagnetic Waves in Metals in a Magnetic Field." Taylor & Francis, London, 1968.
2. A. B. Pippard, *in* "Low Temperature Physics" (C. DeWitt, B. Drefus, and P. G. DeGennes, eds.), Section IV. Gordon & Breach, New York, 1961.

3. M. Ya Azbel and E. A. Kaner, *J. Phys. Chem. Solids* **6**, 113 (1958).
4. E. Fawcett, *Phys. Rev.* **103**, 1582 (1956).
5. G. Feher and A. F. Kip, *Phys. Rev.* **98**, 337 (1955).
6. P. Aigrain, *Proc. Int. Conf. Semicond. Phys.*, *1960*, p. 224.
7. O. V. Konstantinov and V. I. Perel, *Zh. Eksp. Teor. Fiz.* **45**, 1638 (1960) [Engl. transl., *Sov. Phys.—JETP* **18**, 1125 (1960)].
8. R. Bowers, C. Legendy, and F. Rose, *Phys. Rev. Lett.* **7**, 330 (1961).
9. S. J. Buchsbaum, *Proc. Symp. Plasma Effects Solids*, *1964*, p. 3 (1965).
10. T. Stix, "Theory of Plasma Waves." McGraw-Hill, New York, 1962.
11. C. C. Grimes and S. J. Buchsbaum, *Phys. Rev. Lett.* **12**, 357 (1964).
12. E. A. Kaner and V. I. Skobov, *Zh. Eksp. Teor. Fiz.* **45**, 610 (1963) [Engl. transl., *Sov. Phys.—JETP* **18**, 419 (1964)].
13. J. R. Houck and R. Bowers, *Phys. Rev.* **166**, 397 (1968).
14. W. M. Walsh and P. M. Platzman, *Phys. Rev. Lett.* **15**, 784 (1965).
15. P. M. Platzman, W. M. Walsh, and E. Ni-Foo, *Phys. Rev.* **172**, 689 (1968).
16. Y. C. Cheng, J. S. Clark, and N. D. Mermin, *Phys. Rev. Lett.* **20**, 1486 (1968).
17. G. A. Baraff, C. C. Grimes, and P. M. Platzman, *Phys. Rev. Lett.* **22**, 590 (1969).
18. S. Schultz and G. Dunifer, *Phys. Rev. Lett.* **18**, 283 (1967).
19. V. P. Silin, *Zh. Eksp. Teor. Fiz.* **33**, 1227 (1957) [Engl. transl., *Sov. Phys.—JETP* **6**, 1945 (1958)].
20. P. M. Platzman and P. A. Wolff, *Phys. Rev. Lett.* **18**, 280 (1967).
21. L. D. Landau, *Zh. Eksp. Teor. Fiz.* **30**, 1058 (1956) [Engl. transl., *Sov. Phys.—JETP* **3**, 920 (1956)].
22. V. P. Silin, *Zh. Eksp. Teor. Fiz.* **33**, 495 (1957) [Engl. transl., *Sov. Phys.—JETP* **6**, 945 (1958)].
23. G. E. H. Reuter and E. H. Sondheimer, *Proc. Roy. Soc. (London)* **195**, 136 (1948).
24. P. M. Platzman and S. J. Buchsbaum, *Phys. Rev.* **132**, 2 (1963).

VII. The Local Limit

41. The Dispersion Relation for Helicons

The first evidence for the existence of low-frequency wavelike excitations capable of relatively undamped propagation in metals was observed in sodium by Bowers, Legendy, and Rose (*1*). Their experimental configuration is shown in Fig. VII-1. The sample, a thin cylinder

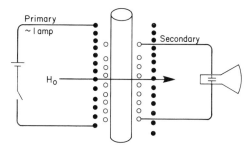

FIG. VII-1. A schematic representation of the experimental arrangement in the first solid state helicon experiment. The sample, a thin cylinder of Na, is shown at the center of the drawing.

of very pure Na, was placed in a large magnetic field ($H_0 \approx 10^4$ G) perpendicular to the axis of the cylinder. The switch to the primary coil was closed suddenly. The signal, the time-dependent rf fields produced by eddy currents induced in the specimen, was picked up in the secondary windings. A typical set of experimental traces is shown in Fig. VII-2. In zero magnetic field the upper trace shows a pure exponential decay. The time constant of the decay is a direct measure of the dc resistivity of the specimen. In the presence of a large magnetic field the ensuing traces exhibit a series of very well-defined oscillations ($\omega/2\pi \approx 30$ Hz!!) superimposed on the exponential decay. Roughly speaking, the frequency of these oscillations is proportional to H_0 and inversely proportional to the square of the radius of the cylinder. These oscillations are the signature of resonant standing helicon waves which have been excited in the specimen.

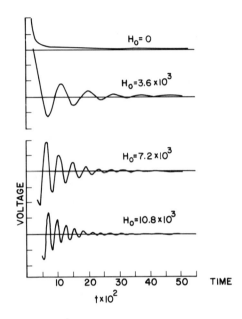

FIG. VII-2. The voltage in the secondary winding of the coil in the Bowers–Legendy–Rose experiment for different dc magnetic fields. Time is measured in seconds and the magnetic field in gauss.

To interpret these "extra" oscillations we look to a solution of Maxwell's equations. We will find that the infinite medium can support waves that propagate along the magnetic field. The oscillations in the resistance of the rod are due to standing wave excitation of these modes in this finite geometry situation. The situation is similar to the Fabry–Perot oscillations in the response to light of a dielectric slab. When one excites a particular standing wave mode in the sample, the external field feeds energy into the mode. The energy is dissipated in the sample and the impedance of the sample has an extremum. For a fixed H_0, the frequency of oscillation is determined solely by the dispersion relation. Thus we can hope to extract the dispersion relation from such data. This speculation is completely borne out by a detailed comparison of theory with experiment.

The theoretical analysis of the problem is simplified if we consider plane waves in the geometry defined by Fig. VII-3. Since many of the later helicon experiments (2)[†] were done in this type of geometry, our

---

[†] C. C. Grimes, unpublished data. (Dr. Grimes has kindly supplied us with several plots of unpublished data from his work on helicons.)

41. DISPERSION RELATION FOR HELICONS

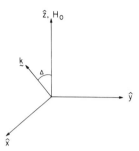

FIG. VII-3. The coordinate system for plane wave propagation in an infinite medium.

results will be directly applicable to such measurements. The cylindrical geometry is slightly more complicated. We will touch on it briefly following an analysis of the plane wave situation.

Neglecting the magnetic properties of the medium for the time being ($\chi = 0$), and assuming that all fields vary as

$$\mathbf{E} \equiv \mathbf{e} e^{i(\mathbf{k}\cdot\mathbf{r}-\omega t)} \tag{41.1}$$

the set of Maxwell equations, Eqs. (40.1)–(40.5) are all contained in the homogeneous wave equation,

$$\mathbf{k} \times (\mathbf{k} \times \mathbf{e}) + k_0^2 \boldsymbol{\varepsilon} \cdot \mathbf{e} = 0. \tag{41.2}$$

The quantity $\boldsymbol{\varepsilon}$ is defined by Eq. (40.13).

In order to "solve" Eq. (41.2) for the normal modes, we must have an expression for $\boldsymbol{\sigma}$. That is, we must calculate the current induced in the medium by a total electric field of the form given in Eq. (41.1). We shall not at this point make any assumptions about the dependence of $\boldsymbol{\sigma}$ on the dynamical variables of the system, the frequency ω, or the wave vector $\mathbf{k}$ of the field. Utilizing only the symmetry which follows from the special choice of coordinate axis given in Fig. VII-3 and from the axial nature of the magnetic field $\mathbf{H}_0$ we are immediately led to a considerable simplification of the wave equation, Eq. (41.2).

According to the so-called Onsager relations (3), the subscripts on the conductivity tensor are interchanged when the sign of an axial vector ($\mathbf{H}_0$) is reversed, but are not affected by a change in sign of a polar vector such as the propagation vector ($\mathbf{k}$), i.e.,

$$\sigma_{ij}(\mathbf{H}_0, \mathbf{k}, \omega) = \sigma_{ji}(-\mathbf{H}_0, \mathbf{k}, \omega) = \sigma_{ji}(-\mathbf{H}_0, -\mathbf{k}, \omega). \tag{41.3}$$

In Fig. VII-3 the propagation vector $\mathbf{k}$ is oriented so that the $\hat{z}$, $\hat{y}$, and $\hat{x}$ axis are along $\mathbf{H}_0$, $\mathbf{H}_0 \times \mathbf{k}$ and $(\mathbf{H}_0 \times \mathbf{k}) \times \mathbf{k}$, respectively. Thus, changing the sign of $\mathbf{H}_0$ and $\mathbf{k}$ changes the direction of the x and z axis but not that of y. If the medium is isotropic, the signs of σ_{xy} and σ_{yz} are changed, but σ_{xz} and the diagonal elements σ_{ii} are left unchanged by such a transformation. It then follows from Eqs. (41.3) that for the isotropic situation,

$$\sigma_{xy} = -\sigma_{yx}, \qquad \sigma_{yz} = -\sigma_{zy}, \qquad \sigma_{zx} = \sigma_{xz}. \qquad (41.4)$$

Equations (41.4) reduce the number of independent tensor components in $\boldsymbol{\sigma}$ to six.

In general (see the discussion preceding Eq. (35.13)) $\boldsymbol{\sigma}(\mathbf{k}, \omega)$ is a complicated transcendental function of $\mathbf{k}$ and the equations obtained by annulling the 3 × 3 determinant of Eq. (41.2), in order to obtain the dispersion relation, i.e., the relation between $\mathbf{k}$ and ω that allows a non-trivial solution of Eq. (41.2) to exist, is quite complicated. This is particularly true for an arbitrary angle of propagation. The number of possible modes at such angles is probably quite large. For special angles, such as $\mathbf{k} \parallel \mathbf{H}_0$ or $\mathbf{k} \perp \mathbf{H}_0$, the problem is greatly simplified. For the most part we will avoid the problem of an arbitrary angle of propagation and consider only the (almost) parallel and perpendicular cases. The reader who is interested in the general case is referred to an excellent review article by Kaner and Skobov (4). To date there has been no really new physics associated with arbitrary angle propagation. Thus, specializing as we do, we will still cover most of the *physically* interesting phenomena.

The dispersion relation for waves propagating parallel to a static magnetic field in the z direction is given by

$$\begin{vmatrix} \epsilon_{xx} - k^2/k_0^2 & \epsilon_{xy} & \epsilon_{xz} \\ -\epsilon_{xy} & \epsilon_{yy} - k^2/k_0^2 & \epsilon_{yz} \\ \epsilon_{xz} & -\epsilon_{yz} & \epsilon_{zz} \end{vmatrix} = 0. \qquad (41.5)$$

For this particular geometry, assuming only that the medium is rotational symmetric about the z axis it follows that,

$$\epsilon_{xx} = \epsilon_{yy} \qquad (41.6)$$

and

$$\epsilon_{xz} = \epsilon_{yz} = 0. \qquad (41.7)$$

Equation (41.6) is simply a statement of the fact that in this geometry the x and y axes are equivalent. Equation (41.7) is true because of the fact that there is no coupling between the z component of the electrons motion and its x–y motion.

Expanding the determinant given in Eq. (41.5) and utilizing Eqs. (41.6) and (41.7), we find that

$$(k^2/k_0^2 - \epsilon_+)(k^2/k_0^2 - \epsilon_-)\epsilon_{zz} = 0, \tag{41.8}$$

where

$$\epsilon_\pm = \epsilon_{xx} \pm i\epsilon_{xy}. \tag{41.9}$$

The modes associated with the two roots

$$k^2/k_0^2 = \epsilon_\pm \tag{41.10}$$

are purely transverse and circularly polarized while the root

$$\epsilon_{zz} = 0. \tag{41.11}$$

corresponds to a purely longitudinal mode—the plasmon.

42. The Dielectric Tensor

Equations (41.10) and (41.11) are as far as we can go without saying something about the dynamics of the electron gas, i.e., without computing $\boldsymbol{\varepsilon}(\mathbf{k}, \omega)$. If we neglect the $\mathbf{k}$ dependence of the dielectric constant, we may calculate the conductivity tensor for the noninteracting gas in an extremely simple manner. Let us assume, for now, that this neglect is justified. We will see later under what conditions this assumption is justified. Neglecting the $\mathbf{k}$ dependence of $\boldsymbol{\sigma}(\mathbf{k}, \omega)$ is equivalent to assuming that its spatial dependence is proportional $\delta(\mathbf{r} - \mathbf{r}')$ or equivalently, that the driving electric field is uniform. We may evaluate the current by calculating the motion of the electrons in uniform electric and magnetic fields.

We will soon see that the properties of helicons in the low-frequency regime $\omega/\omega_c \ll 1$ are accurately described within the framework of such a uniform field approximation. Of the three types of waves we will discuss, they are the only ones for which this is the case. CW do not exist in the uniform field limit. PSW exist only in the presence of both finite short-range interaction among the electrons and nonuniform field effects.

In the uniform field approximation

$$\mathbf{j}(\omega) = \boldsymbol{\sigma}(\omega) \cdot \mathbf{E}(\omega) \tag{42.1}$$

When the electrons are noninteracting and have no average drift motion,

$$\mathbf{j}(\omega) = n_0 e \mathbf{v}(\omega). \tag{42.2}$$

Here $\mathbf{v}(\omega)$ is determined from the equation of motion

$$m^*\dot{\mathbf{v}} = e[\mathbf{E} + (\mathbf{v}/c) \times \mathbf{H}_0] - m^*\mathbf{v}/\tau. \quad (42.3)$$

where $\mathbf{H}_0$ is the external field taken in the z direction and m^* is the effective mass of the electrons. The last term in in Eq. (42.3) is a phenomenological collision term which describes, in an approximate way, the dissipative effects of the medium.

Solving Eq. (42.3) and substituting the results into Eq. (42.2) we find that $\boldsymbol{\varepsilon}(\omega)$ has the components,

$$\epsilon_{xx} = \epsilon_{yy} = 1 - \frac{\omega_p^2}{\omega\tilde{\omega}} \frac{1}{[1 - (\omega_c/\tilde{\omega})^2]} \quad (42.4)$$

$$\epsilon_{xy} = \epsilon_{yx} = -\frac{i\omega_p^2}{\omega\tilde{\omega}} \frac{(\omega_c/\tilde{\omega})}{[1 - (\omega_c/\tilde{\omega})^2]} \quad (42.5)$$

and

$$\epsilon_{zz} = 1 - \omega_p^2/\omega\tilde{\omega} \quad (42.6)$$

with

$$\tilde{\omega} = \omega + i/\tau. \quad (42.7)$$

For the moment consider the limit of zero magnetic field. In this case the dielectric tensor becomes diagonal, i.e., $\epsilon_{xy} = 0$ and,

$$\epsilon_{xx} = \epsilon_{yy} = \epsilon_{zz} = 1 - \omega_p^2/\omega\tilde{\omega}. \quad (42.8)$$

The two transverse waves ($\mathbf{k} \cdot \mathbf{e} = 0$) satisfy identical dispersion relations

$$k^2/k_0^2 = 1 - \omega_p^2/\omega\tilde{\omega}. \quad (42.9)$$

The third purely longitudinal mode [see Eq. (41.10)] is found by setting,

$$\epsilon_{zz} = 1 - \omega_p^2/\omega\tilde{\omega} = 0. \quad (42.10)$$

In the limit $\omega\tau \gg 1$ this root is simply,

$$\omega = \omega_p \quad (42.11)$$

i.e., the plasmon discussed in Chapter III.

In the long-wavelength, uniform-field approximation the longitudinal mode has no dispersion, no k dependence, so that there must indeed be a physically accessible region where $\omega = \omega_p$ and the uniform approximation is the correct result. We know this to be the case from our earlier discussion in Chapter III of the RPA and the behavior of plasmons in the long-wavelength limit.

42. THE DIELECTRIC TENSOR

For the transverse mode, the situation is quite different. There is no propagation for frequencies $\omega < \omega_p$. The wave vector

$$k = (i\omega_p/c)(1 + i/\omega\tau)^{-1/2}. \tag{42.12}$$

In the so-called classical limit ($\omega\tau \ll 1$), the wave vector has equal real and imaginary parts, i.e.,

$$k = 2^{-1/2}(1 + i)(\omega_p/c)(\omega\tau)^{1/2}. \tag{42.13}$$

The field, whose spatial dependence is characterized by Eq. (42.13) is nonpropagating. However, since $\operatorname{Im} k \sim (\omega)^{1/2}$ it is possible, for sufficiently low frequencies, to have the field vary very slowly in space. This fact implies that there is a low-frequency regime in which the uniform field approximation is indeed a valid description.

On the other hand, at high frequencies in sufficiently pure metals ($\omega\tau \gg 1$) the assumption that $\boldsymbol{\sigma}$ is independent of $\mathbf{k}$ leads to a contradiction. In this limit

$$k \cong i\omega_p/c; \tag{42.14}$$

and the field falls off exponentially in a distance c/ω_p, which is typically 10^{-6} cm in a good metal. By contrast we expect that the conductivity tensor will vary on a distance scale described by the mean free path $l \equiv v_F\tau$ of the carriers. The length l is the distance over which current is easily transported by charged particles moving with some typical velocity v_F. In the course of this motion the carriers sample the electric and magnetic fields which are present. For pure metals, at low temperatures $l \cong 0.1$ cm and the conductivity varies slowly compared to the computed field variation! We are faced with a contradiction. It is necessary to consider the $\mathbf{k}$ dependence of $\boldsymbol{\sigma}(\mathbf{k}, \omega)$. In the absence of a field $\mathbf{H}_0$, the resulting problem is the celebrated anomalous-skin-effect problem. It has been treated in great detail by several authors (5). We will not discuss these analyses here since they do not involve an understanding of wave phenomenon. However, we will discuss in some detail the $\mathbf{k}$ dependence of the conductivity tensor in Chapter VIII and make brief contact with the anomalous-skin-effect calculations in Section 49 when we discuss the boundary value problem.

The main conclusion of the various anomalous-skin-effect calculations is that the field, instead of being damped in distance $\delta_{cl} = c/\omega_p$, penetrates a distance $\delta_A = \delta_{cl}^{2/3} l^{1/3}$. In very pure metals ($\tau \cong 10^{-9}$ sec), the anomalous-skin depth δ_A can be as long as several thousand Ångstroms. Nevertheless, our qualitative conclusion, i.e., the absence of nonpropagating transverse waves is still valid. Thus we conclude that in the absence of a field, damped modes, or equivalently skin-effect behavior, is the dominant characteristic of a metallic medium. The situation is quite different in the presence of a static magnetic field as we have indicated in Section 7.

Since we will concern ourselves with the behavior of metals in the microwave regime we have $\omega \ll \omega_p$. In this regime we can neglect the displacement current [the $\delta_{\alpha\beta}$ in Eq. (40.13)] relative to the conduction current. In this case the uniform-field approximation implies that the two transverse waves which are solutions of Eq. (41.10) satisfy the dispersion relation,

$$\frac{k^2}{k_0^2} = -\frac{\omega_p^2}{\omega(\omega \mp \omega_c + i/\tau)} \equiv \epsilon_{\pm}. \tag{42.15}$$

The dielectric constant ϵ_+ shows a resonance (infinity) at the electrons gyrofrequency. For the moment, let us consider the ϵ_+ mode in the limit $\tau \to \infty$. We will completely ignore the ϵ_- mode which is nonresonant and nonpropagating (Im $k \geqslant$ Re k). In the $\tau \to \infty$ limit Eq. (42.15) is

$$k_+^2/k_0^2 = -\omega_p^2/\omega(\omega - \omega_c). \tag{42.16}$$

The dielectric constant is large if $\omega \ll \omega_p$. For $\omega < \omega_c$, it is a large negative number and the wave vector k_+ is a pure imaginary number. No propagation is possible. As ω passes through ω_c, the dielectric constant changes sign, k_+ becomes real and propagation is possible. The cyclotron resonance frequency marks the boundary between the propagating and damped solution of our approximate set of equations.

An examination of the local dispersion relation Eq. (42.16) is physically useful since it clearly reveals the "origin" of propagation. In the limit $\omega_c \tau \to \infty$, the transverse components of the conductivity tensor undergo a phase reversal at the cyclotron frequency. For example, let us consider $\epsilon_{xy}(\omega)$. This component of the dielectric tensor describes the current response in the x direction due to a driving field in the y direction. This current may be in phase ($\omega_c/\omega < 1$) or out of phase ($\omega_c/\omega > 1$) with the driving electric field. The possibility of the phase reversal of the currents relative to the driving electric fields is the origin of propagation in metals. This is always the case.

The local dispersion relation [Eq. (42.16)] implies that helicons propagate at all frequencies such that $\omega \leqslant \omega_c$. However, it is clear that the local dispersion must fail if we allow ω to approach arbitrarily close to ω_c since we have assumed in our analysis that the electric field was uniform, i.e., $k \simeq 0$. The dispersion relation tells us that at ω_c, $k = \infty$, so that our locality assumption fails. In fact we will see that the local dispersion relation breaks down at much lower frequencies, and that the "correct" nonlocal dispersion relation predicts that helicons will not propagate in a metal unless $\omega \ll \omega_c$. However, when the frequency is sufficiently small, then the local dispersion relation is valid and the helicon wave has the dispersion relation given by

$$k_H^2/k_0^2 = \omega_p^2/\omega\omega_c . \tag{42.17}$$

In this regime the phase velocity of the helicon is

$$V_H = \omega/k_H = c(\omega\omega_c/\omega_p^2)^{1/2} \equiv \omega^{1/2}\left(\frac{H_0 c}{4\pi n_0 e}\right)^{1/2} \tag{42.18}$$

The principal features of Eq. (42.18) are that the phase velocity is independent of m^* and proportional to the square root of the magnetic field and the frequency. At typical metallic densities $\omega_p \simeq 10^{16}$ sec^{-1}; for an $\omega \simeq 10^7$ sec^{-1} and $\omega_c \simeq 10^{11}$ sec^{-1}, the phase velocity of a helicon is approximately 3×10^3 cm sec^{-1}. Helicons, in metals, are extremely slow waves. Their velocity is small relative to the velocity of light and small compared to typical particle velocities ($v_F \simeq 10^8$ cm sec^{-1}). This situation obtains because the effective dielectric constant for the medium is proportional to the square of the plasma frequency which is very large relative to the other characteristic frequencies in the problem. The only velocity that can be made comparable to helicon velocities in metals is the sound velocity. This equality can lead, as we shall see later, to some interesting mode-coupling effects in metals.

Two other properties of a helicon are worth pointing out at this time. The first obtains from the dispersion relation Eq. (41.10), i.e.,

$$E_x/E_y = i \tag{42.19}$$

The helicon wave is circularly polarized. In addition the third Maxwell equation, [Eq. (40.4)] implies that a helicon is almost pure magnetic field since

$$|H_x/E_x| = c/V_H \gg 1 . \tag{42.20}$$

The field configuration of a helicon is shown schematically in Fig. VII-4.

The Bowers–Legendy–Rose experiment (*1*) can now be qualitatively understood. The impulsive rf field sets up a standing helicon wave in the cylindrical specimen. The wave vector of the helicon is fixed by the geometry and the frequency is then determined by a dispersion relation similar to Eq. (42.17). The situation is similar to the modes of vibration of a drum head struck by an impulsive blow. In principle many spatial modes with different frequencies can be excited, apparently only one is in

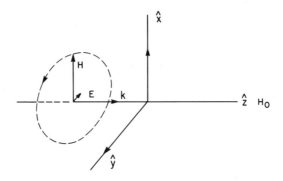

Fig. VII-4. The configuration of fields in a helicon wave.

the Bowers–Legendy–Rose experiment. For the fields used in the experiment ($H_0 \simeq 3$ kG), a frequency $\omega \simeq 2 \times 10^2$ sec^{-1} corresponds to a $k \simeq 10$ cm^{-1} in Eq. (42.17). This k corresponds, very roughly, to the small dimension of the cylinder. However, it is clear that a quantitative comparison of the theory of wave propagation with impulsive standing-wave experiments in finite cylinders requires a solution of Maxwell's equations for the dielectric cylinder. We will not undertake such a solution. It is much easier and more illuminating to compare our infinite plane wave solutions Eq. (42.17) with the results of *transmission* experiments.

The helicon transmission experiments are illustrative of all transmission experiments. Since we will have occasion to discuss the transmission experiment of Schultz and Dunifer (6) in connection with the CW and PSW, it is worthwhile to analyze this type of experiment in more detail than was done in Section 12. Typically, in a transmission experiment, a source of electromagnetic energy drives currents on one face of a slab of metal. These currents excite the modes (or perhaps single-particle excitations) of the slab. Energy is then carried by the mode (or equivalently by the electrons) to the other side of the slab, where it is reradiated into a receiver. Figure VII-5 is a schematic of such an experimental configuration. The transmitted $\mathbf{E}_T$ or $\mathbf{H}_T$ field contains both phase and power (intensity) information. Since there are no intrinsically phase-sensitive detectors, the phase information is determined by adding the transmitted field, whose phase is $\phi_T(H_0)$, to a large reference field, E_R, whose phase ϕ_R is fixed relative to the incident field. The reference field is indicated by a solid line going around the specimen in Fig. VII-6. The total field,

$$E = E_R \exp(i\phi_R) + E_T \exp[i\phi_T(H_0)] \qquad (42.21)$$

42. THE DIELECTRIC TENSOR

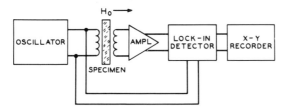

FIG. VII-5. A block diagram of the experimental arrangement in a typical helicon transmission experiment.

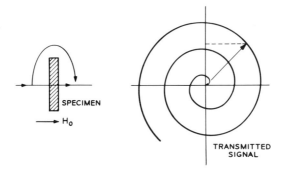

FIG. VII-6. A schematic of a phase sensitive transmission experiment.

is then fed into a square-law detector which is sensitive to E^2. To leading order in E_T

$$E^2 \simeq E_R^2 + 2E_T E_R \cos(\phi_R - \phi_T(H_0)). \tag{42.22}$$

The signal (term proportional to E_T) is simply the projection of E_T onto a fixed reference. The phase ϕ_R in Eq. (42.22) can be adjusted by external devices so that it is possible to pick either the real or imaginary parts of E_T.

The spiral on the right-hand side of Fig. VII-6 shows the phase and amplitude variations of the transmitted signal. As the magnetic field increases, the phase of the signal cartwheels around the origin. Since the phase of the field at the far surface of the specimen is proportional to kL, the product of the wave vector of the mode and the thickness of the specimen, i.e.,

$$\exp[i\phi_T(H_0)] \sim \exp[ik(H_0)L]. \tag{42.23}$$

each time the wave vector k goes through $2\pi/L$ the projection of the transmitted field on the fixed reference goes through a complete cycle. Measuring the period of the oscillation and knowing the thickness of the specimen enables one to extract the dispersion relation of the wave. Since there is damping associated with the wave, the successive maxima of the beat pattern will vary in magnitude as well. The intensity envelope of the transmitted beat pattern gives information about this damping, i.e., $\text{Im}[k(H_0)]$.

An oscillating gradually changing envelope is characteristic of the transmission patterns which are observed when *waves* are being excited. In Fig. VII-7 we show some data taken by Grimes.[†] The samples

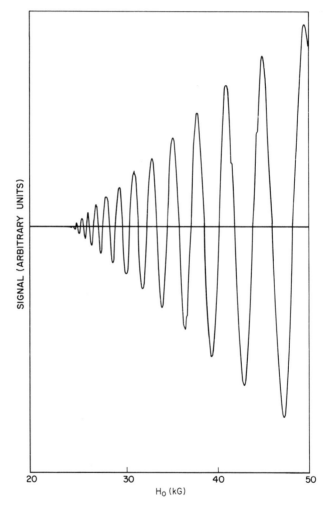

FIG. VII-7. A typical transmission pattern taken in a K sample in the helicon wave regime. $d = 0.464$ mm; $T = 4.2°$K; $\omega = 1.26 \times 10^8$ Hz.

[†] See footnote p. 128.

were thin slabs of potassium driven at a frequency $\omega = 1.26 \times 10^8$ sec^{-1}. The pattern is as described. The successive maxima are the phase oscillations of the electric field transmitted by the helicon waves launched at the front surface. It is important to realize that the beats are *not* standing wave Fabry–Perot resonances which would manifest themselves as oscillations in the power, but are simply magnetic-field dependent phase variations of the signal.

The kind of wave pattern displayed in Fig. VII-7 is to be contrasted with the spectrum observed in transmission near a single particle resonance like cyclotron resonance. Figure VII-8 shows such a trace

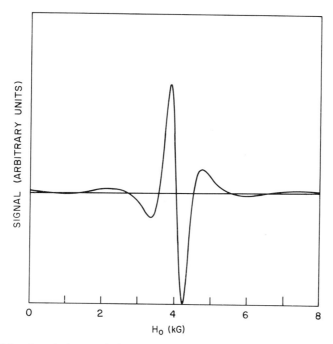

FIG. VII-8. A typical transmission pattern taken in Na. The single peak in this field normal geometry is at $\omega_c = \omega$. $d = 0.163$ mm; $T = 1.3°$K; $\omega = 5.8 \times 10^{10}$ Hz.

taken in Na by Schultz and Dunifer.[†] In this case the magnetic field is oriented *perpendicular* to the slab. There is a single, reasonably simple, peak with a fixed (usually π) phase shift across the resonance.

For the moment we will concentrate on the helicon data shown in Fig. VII-7. The transmitted field is zero until H_0 reaches a value

[†] Drs. Schultz and Dunifer have kindly supplied us with several pieces of unpublished data such as that shown in Fig. VII-8.

$H_D \simeq 24 \times 10^3$ G. The cyclotron resonance field, where $\omega = \omega_c$, (for this frequency) is at the surprisingly small field of 8 G. Our theory as presented does not explain the absence of propagation for fields such that $H_C < H_0 < H_D$. We have hinted that the discrepancies are due to nonlocal or finite k effects which have been omitted in obtaining Eq. (42.16). This point will be discussed in some detail shortly.

The positions of the maxima may be compared with the simple dispersion relation Eq. (42.17). The comparison is shown in Fig. VII-9.

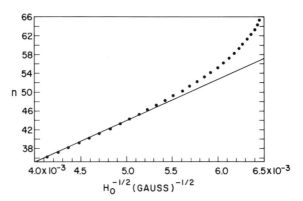

FIG. VII-9. Comparison of the calculated local dispersion relation with the helicon transmission data displayed in Fig. VII-7. $d = 0.464$ mm; $T = 4.2°$K; $\omega = 1.26 \times 10^8$ Hz.

The integer n, $k = n\pi/L$, is plotted versus $H_0^{-1/2}$. If the simple dispersion relation Eq. (42.17) were correct, the straight line shown in black would fit the data for all $H_0 > H_C$. As H_0 decreases and approaches H_D, the data points begin to deviate from the straight line. Since $H_0 \gg H_C$ throughout it is clear that the theory as presented does not correctly predict the dispersion relation. It does however, fit the data for fields such that $H_0 \gg H_D$.

Our treatment of helicon propagation was approximate. We have assumed that the electrons were noninteracting. Later calculations will show that this is a good approximation. In addition, we have completely neglected the $\mathbf{k}$ dependence of the dielectric constant $\epsilon_+(\mathbf{k}, \omega)$. For small magnetic fields, this approximation is a poor one. The failure of the local approximation is responsible for the discrepancies apparent in Fig. VII-9.

Without going into any great detail it is possible to see when we might expect nonlocal (finite $\mathbf{k}$) effects to become important. We resort to a crude argument which will be made quantitative in Chapter VIII. Suppose that all the electrons in the solid are moving with some velocity

$\mathbf{V}^D$, in the absence of any driving fields. Loosely speaking, one could take this motion into account by transforming to a frame of reference moving with this velocity. The result would be that the effective microwave frequency the electrons see is Doppler shifted by the amount $\mathbf{k} \cdot \mathbf{V}^D$. In this uniform streaming model Eq. (42.16) becomes

$$\frac{k^2}{k_0^2} = -\frac{\omega_p^2}{\omega(\omega - \omega_c + \mathbf{k} \cdot \mathbf{V}^D + i/\tau)}. \qquad (42.24)$$

A more complete treatment, requires an average over electron velocities, but qualitatively Eq. (42.24) is quite accurate. This formula tells us two things. First we see that, even at low frequencies, nonlocal effects become important when $kV_D/\omega_c \to 1$. For a metal, V^D is of the order of the Fermi velocity v_F. By substituting the expression for k given by Eq. (42.16) into Eq. (42.24) we then see that the inequality

$$(v_F/c)(\omega_p/\omega_c)(\omega/\omega_c)^{1/2} < 1 \qquad (42.25)$$

defines the limits of the region of validity of the uniform field approximation which leads to the dispersion relation Eq. (42.16). For magnetic fields of the order of 10^4 G, $\omega_p/\omega_c \simeq 10^4$. Since $v_F/c \simeq 10^{-3}$ for a metal, ω/ω_c must be less than 10^{-2} in order that Eq. (42.25) be satisfied.

For $kv_F/\omega_c > 1$ it is clear from Eq. (42.24) that there will be a singularity in the dielectric constant for some group of electrons, i.e., those with the correct projection of $\mathbf{V}^D$ on $\mathbf{k}$. The imaginary part of ϵ_+ for these electrons becomes very large. This phenomenon, so-called cyclotron damping, strongly suppresses helicon propagation whenever $kv_F/\omega_c > 1$. It is the origin of the long cut-off region between H_c and H_D as shown in Fig. VII-7.

The singularity in Eq. (42.24) and the associated damping of the waves have quite a simple physical interpretation. When $kv_z = \omega_c - \omega$, the electrons experience (in their rest frame) an electric field, rotating at a frequency precisely equal to ω_c. Since the electrons are themselves rotating in the dc field at this rate, they are in effect subjected to a static electric field as they ride synchronously with the wave. This static electric field accelerates the electrons in a plane perpendicular to $\mathbf{H}_0$, causing them to spiral out around $\mathbf{H}_0$, pick up energy, and damp the wave. This cyclotron damping is the analog of the Landau damping which was discussed in some detail in Chapter III.

Before leaving our simple dispersion formula [Eq. (42.16)] it is instructive to see how we can modify it to describe two other physically interesting phenomena.

43. Alfven Waves

Many solid state plasmas (see Chapter I) consist of more than a single carrier species. Compensated plasmas, such as those in tungsten, bismuth, and antimony contain equal numbers of holes (positively charged carriers) and electrons [negatively charged carriers (7)]. It suffices to say that these two types of carriers exist because of the complexities of the band structure of these metals. In many cases, the two types of charge carriers may in fact consist of many effective groups of carriers with different masses, etc. The complications are numerous. Fortunately, to discuss Alfven waves qualitatively it is necessary to assume only that we have equal numbers of two different mass carriers with equal and opposite charges.

Assume therefore that we have n_1 electrons with mass m_1 and n_2 holes with mass m_2. The total conductivity σ_+ is linearly proportional to the total current induced in the SSP by the total electric field. For noninteracting carriers, this current is the sume of those due to each carrier,

$$\mathbf{j}_T = \mathbf{j}_1 + \mathbf{j}_2, \qquad (43.1)$$

where

$$\mathbf{j}_1 = \boldsymbol{\sigma} \cdot \mathbf{E} \qquad (43.2)$$

$$\mathbf{j}_2 = \boldsymbol{\sigma} \cdot \mathbf{E}. \qquad (43.3)$$

In this case the total dielectric constant is given by

$$\epsilon_+ = \frac{-1}{\omega} \left[\frac{\omega_{p_1}^2}{(\omega - \omega_{c_1} + i/\tau_1)} + \frac{\omega_{p_2}^2}{(\omega + \omega_{c_2} + i/\tau_2)} \right]. \qquad (43.4)$$

When $\omega_{c_{1,2}} \tau_{1,2} \gg 1$ and $\omega \ll \omega_{c_{1,2}}$ we may expand Eq. (43.4) in powers of $\omega/\omega_{c_{1,2}}$ to obtain,

$$\epsilon_+ = -\frac{\omega_{p_1}^2}{\omega \omega_{c_1}} + \frac{\omega_{p_2}^2}{\omega \omega_{c_2}} + \frac{\tilde{\omega}}{\omega} \left[\frac{\omega_{p_1}^2}{\omega_{c_1}^2} + \frac{\omega_{p_2}^2}{\omega_{c_2}^2} \right] + \cdots$$

$$= \frac{(n_2 - n_1) 4\pi e^2}{\omega H_0} + \frac{\tilde{\omega}}{\omega} \left[\frac{\omega_{p_1}^2}{\omega_{c_1}^2} + \frac{\omega_{p_2}^2}{\omega_{c_2}^2} \right]. \qquad (43.5)$$

In a compensated material when $n_1 = n_2$, the leading term in Eq. (43.5) vanishes. The second set of terms in Eq. (43.5) is now the dominant one. Under these conditions a qualitatively different type of wave propagation occurs. When $\omega \tau \geqslant 1$, Eq. (43.5) yields a propagating mode with velocity,

$$(\omega/k)^2 \equiv V_A^2 = H_0^2 / 4\pi n_0 (m_1 + m_2). \qquad (43.6)$$

The velocity V_A is termed the Alfven speed.

Equation (43.6) has a simple interpretation. We may think of the static line of force, perturbed by the rf field, as vibrating like a stretched string. At frequencies much smaller than the cyclotron frequencies of the carriers, the carriers motion about the lines of force is adiabatic. The particles are constrained to move with the lines of force like beads on a string. Since the net charge of the system is zero, the equation of motion of the string is solely determined by the tension in the string and the mass density weighting it, i.e.,

$$V_A{}^2 = T/\rho. \tag{43.7}$$

To compute the effective tension in the line of force is slightly complicated. The tensile stress in a single line is $H_0{}^2/8\pi$. This line inturn moves in a medium with a compressive stress of $H_0{}^2/8\pi$. The net restoring force on a displaced line can be seen to be $H_0{}^2/4\pi$, i.e., the sum of the tensile and compressive stresses on a single line. Since $\rho = n_0(m_1 + m_2)$, we simply recover Eq. (43.6).

This simple picture also accurately describes the helicon mode. However in this uncompensated situation, we must take into account the Lorentz force ($\mathbf{v} \times \mathbf{H}_0$) exerted by $\mathbf{H}_0$ on the charged string. The same argument which leads to Eq. (43.6) can now be used to derive Eq. (42.18). In this case it is the Lorentz force which is proportional to the tension T. This is the origin of the $H_0^{1/2}$ variation of the velocity.

For compensated materials, the condition $\omega\tau \gg 1$ severely restricts the range of materials in which one can hope to see Alfven wave propagation. By contrast, in the helicon case we merely required $\omega_c\tau$ to be greater than one. Since $\omega \ll \omega_c$, it is much more difficult to propagate Alfven waves.

In metals like tungsten, Alfven propagation has yet to be observed. However, one can achieve the appropriate condition $\omega\tau \gg 1$ in extremely pure bismuth and antimony (7). The reason for this is simply that the parameters characterizing these semimetallic materials are such that we can propagate Alfven waves at relatively high frequencies without running into cyclotron damping effects. Utilizing Eq. (43.5) we find that for Alfven wave propagation the condition $kv_F/\omega_c < 1$, the analog of Eq. (42.25), may be written as

$$(v_F/c)(\omega_p/\omega_c)(\omega/\omega_c) < 1. \tag{43.8}$$

For the semimetals, v_F is comparable to metallic Fermi velocities, ω_p is much smaller and ω_c, for the same field, is much larger since the effective mass of the carriers is small. It is clear that for the same fields, we can work at higher frequencies (i.e., higher $\omega\tau$) than for the compensated metallic materials.

144   VII. THE LOCAL LIMIT

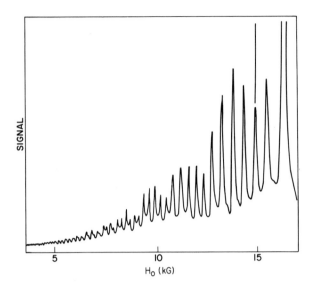

FIG. VII-10. A typical transmission pattern taken in Bi sample in the Alfven wave regime. $d = 2.23$ mm; $T = 1.2°$K; $\omega = 1.02 \times 10^{11}$ Hz.

In Fig. VIII-10 we show some data taken in bismuth by G. A. Williams (8). The experiment is, aside from details, identical with the helicon experiments of Grimes (see Fig. VII-7). The sample, a thin (2.23 mm) slab of single crystal bismuth, which has its trigonal axis oriented perpendicular to the plane of the specimen is driven at a frequency $\omega = 1.02 \times 10^{11}$ Hz. The resulting transmission spectrum may be interpreted, as for the helicons, in terms of the infinite medium dispersion relation, Eq. (43.8). The comparison is displayed in Fig. VII-11. For the case of Bi, there is good agreement between theory (Eq. 43.6) and experiment over the entire range of magnetic fields studied in this particular experiment.

44. Helicon–Phonon Interaction

Since helicons are such slow waves, it occurred to many people that their velocity could be made to match the velocity of transverse sound waves in the metal. This situation is diagrammed in Fig. VII-12. The parameter $v \equiv \omega/k$ is the phase velocity of the coupled modes. The phase velocity of the helicons is proportional to $H_0^{1/2}$ and the velocity of sound (V_s) is roughly independent of a magnetic field. At the matching, or degeneracy point, of the two modes, we would expect some type of

44. HELICON–PHONON INTERACTION

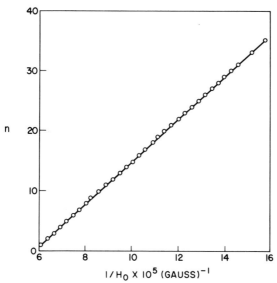

FIG. VII-11. Comparison of the calculated local dispersion relation for a compensated plasma with the data displayed in Fig. VII-10. $d = 2.23$ mm; $T = 1.2°$K; $\omega = 1.02 \times 10^{11}$ Hz.

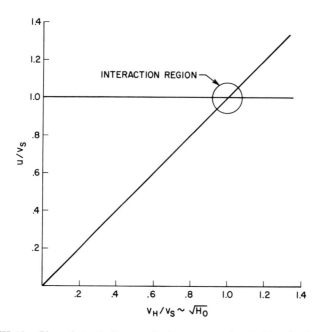

FIG. VII-12. Plot of the helicon and phonon sound velocities in the absence of interactions, as a function of magnetic field.

wave coupling to break the degeneracy. We will go into some detail in describing this coupling. The reasons for this are three fold:

(1) The helicon–phonon coupling is illustrative of all mode-coupling problems in solids where the coupling is via the total self-consistent field.
(2) The results of the calculation are simple and easily understood.
(3) The experiments demonstrating the effects are clean and unambiguous.

In order to describe the coupled helicon phonon system, we need an equation of motion for the ions. This equation will enable us to compute the current which is induced in the ion system by an appropriately polarized electromagnetic field. Given the current, we may add it to the electronic current to obtain a total dielectric constant for the "coupled system." The entire coupling between the two systems exists because both systems respond to the total field.

Let us consider an acoustic wave of angular frequency ω and wave vector $\mathbf{k}$ whose displacement field $\boldsymbol{\xi}(\mathbf{r}, t)$ is of the form

$$\boldsymbol{\xi} = \boldsymbol{\xi}_0 \exp i(\mathbf{k} \cdot \mathbf{r} - \omega t). \tag{44.1}$$

The normal coordinate $\boldsymbol{\xi}$ satisfies an equation of motion which describes acoustic disturbances of the lattice. For a metal having a single transverse sound velocity, this equation of motion is (9)

$$M \, \partial^2 \boldsymbol{\xi} / \partial t^2 = C_l \nabla (\nabla \cdot \boldsymbol{\xi}) - C_t \nabla \times \nabla \times \boldsymbol{\xi} + e\mathbf{E} - (e/c)(\partial \boldsymbol{\xi}/\partial t) \times \mathbf{H}_0 + \mathbf{F}_c . \tag{44.2}$$

Here M is the mass of the ion and C_l and C_t are elastic constants characterizing the velocity of longitudinal $[V_s{}^l = (C_l/M)^{1/2}]$ and transverse $[V_s{}^t = (C_t/M)^{1/2}]$ sound waves. For a material with several distinct transverse sound velocities, such as potassium, Eq. (44.2) must be modified to take into account this anisotropy. For the moment, we neglect this complication and use Eq. (44.2). The force $\mathbf{F}_c$ in Eq. (44.2) is a dissipative term analogous to the $-\mathbf{v}/\tau$ term in Eq. (42.3). Its actual form is unimportant. For our discussion, we can take $\mathbf{F}_c = 0$ as long as $\omega_c \tau \gg 1$.

For a purely transverse sound wave, Eq. (44.2) is equivalent to

$$(\omega^2 - V_s^2 k^2 \pm \Omega_I \omega) \xi_\pm{}^0 = -(e/M) E_\pm \tag{44.3}$$

where $\xi_\pm{}^0 \equiv \xi_x{}^0 \pm i\xi_y$ and $E_\pm = E_x \pm iE_y$. The frequency $\Omega_I = eH_0/Mc$ is the ion cyclotron frequency. The current due to the ion motion is,

$$j_\pm^{\text{ion}} = -n_0 e i \omega \xi_\pm , \tag{44.4}$$

44. HELICON–PHONON INTERACTION

so that (with $\omega_{p_\pm}$ equal to the ion plasma frequency),

$$\frac{4\pi\sigma_I^+}{i\omega} = \omega_{p_I}^2 \frac{1}{(\omega^2 - V_s^2 q^2 + \Omega_1 \omega)}. \tag{44.5}$$

Equation (44.5) implies that the dispersion relation

$$k^2/k_0^2 = \epsilon_+ = (4\pi/i\omega)(\sigma_e^+ + \sigma_I^+) \tag{44.6}$$

takes the form

$$\frac{k^2}{k_0^2} \simeq \frac{\omega_p^2}{\omega \omega_c} + \frac{\omega_{p_I}^2}{(\omega^2 - V_s^2 k^2 + \Omega_1 \omega)} \tag{44.7}$$

if $\omega_c \gg \omega$. Equivalently,

$$(\omega - \omega_H)(\omega^2 - V_s^2 k^2 + \Omega_1 \omega) = \omega^2 \Omega_1, \tag{44.8}$$

with

$$\omega_H = c^2 k^2 \omega_c / \omega_p^2. \tag{44.9}$$

Equation (44.8) may be rewritten in a slightly more convenient form, i.e.,

$$[u/V_s - V_H/V_s][(u/V_s)^2 - 1 + (\Omega_1/\omega)(u/V_s)^2] = (\Omega_1/\omega)(V_H/V_s). \tag{44.10}$$

For potassium, $V_s \simeq 10^5$ cm sec^{-1} and $\omega_p \simeq 10^{16}$. This implies, that for a frequency of 20 MHz, helicon and sound velocities are equal at a field of approximately 10^5 G. In this case the coupling parameter

$$(\Omega_1/\omega)_{V_H = V_s} \simeq 0.2 \tag{44.11}$$

so that there is a splitting in the phase velocity of about 20 % at the cross over point.

In 1964 Grimes and Buchsbaum (9) reported the results of experiments designed to measure the effective coupling between helicons and phonons utilizing single crystal slabs of potassium. In single crystal specimens they launched helicons as illustrated in Fig. VII-5, along one of the crystal axes. They then observed the transmitted rf amplitude at several frequencies. Typical experimental results are shown in Fig. VII-13. These curves are to be contrasted with the pure helicon data shown in Fig. VII-7. The traces shown in Fig. VII-13 display the phenomenon of helicon phonon interaction in potassium. The beat pattern occurs because the signals from the two coupled waves transmitted through the specimen are added to a reference signal which has fixed amplitude and phase.

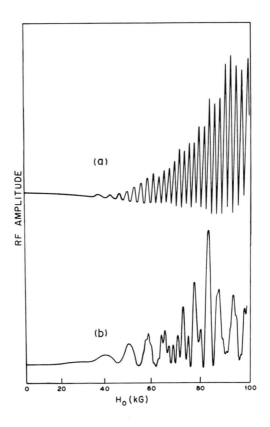

FIG. VII-13. Experimental curves showing the helicon phonon interaction effect. Curves (a) and (b) were obtained in single crystal K slabs 0.048 cm thick at $\omega = 1.98 \times 10^8$ Hz and $\omega = 1.26 \times 10^8$ Hz, respectively. The normal to the surface was oriented along the $\langle 110 \rangle$ axis of the crystal, $\mathbf{k} \parallel \langle 110 \rangle$; $T = 4.2°$K.

Curves (a) and (b) of Fig. VII-13 correspond to wave propagation in the $\langle 110 \rangle$ direction in potassium with $\omega = 1.98 \times 10^8$ Hz and $\omega = 1.26 \times 10^8$ Hz, respectively and at a temperature of 4.2°K. At the higher frequency the coupling between the helicon and sound waves is relatively weak and manifests itself by an increasing separation of the successive maxima as the magnetic field is decreased; i.e., a less-rapid slow-down of the phase velocity of the wave with decreasing magnetic field than that dictated by the "pure" helicon wave formula. Here only the upper branch of the dispersion relations of Fig. VII-12 is discernible. In Fig. VII-13 curve (b) taken at 20 MHz, the coupling is sufficiently strong for both waves to appear, i.e., the beat pattern in the figure can be resolved into two series of peaks corresponding to two coupled branches of the dispersion relation. The series of peaks that are widely spaced at lower fields and become more closely spaced at higher fields belong to a branch of the dispersion relation that resembles a sound wave at low fields and a helicon wave at high fields. The second series of peaks becomes more widely spaced with increasing H_0 and belongs to a branch of the dispersion relation which is changing from a helicon wave to a sound wave.

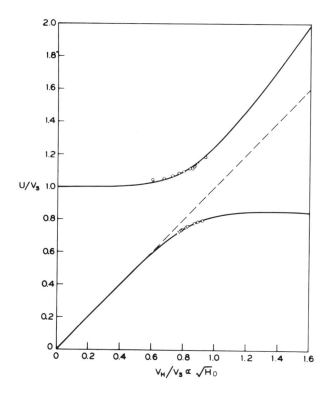

FIG. VII-14. Comparison of the calculated local dispersion curves for the coupled helicon–phonon system with the data displayed in Fig. VII-13. $\mathbf{k} \parallel \langle 110 \rangle$; $\omega = 1.26 \times 10^8$ Hz.

The dispersion relation for coupled helicon and sound waves in a medium with a single transverse sound velocity is given by Eq. (44.8). In order to interpret the results of the experiments in potassium it was necessary to extend Eq. (44.8) to describe the elastically anisotropic case of wave propagation along a twofold axis. Take the magnetic field $\mathbf{H}_0$

in the z direction. Assume it points along the $\langle 110 \rangle$ crystal axis. In this case the equation of motion analogous to Eq. (44.3) becomes

$$(\omega^2 - \tfrac{1}{2}(V_1^2 + V_2^2) k^2 \pm \Omega_1 \omega)\, \xi_\pm^0 - \tfrac{1}{2}(V_1^2 - V_2^2) k^2 \xi_\mp^0 = -e/M E_\pm \,. \tag{44.12}$$

Here V_1 and V_2 are the phase velocities of the two independent shear waves when they propagate along the twofold direction in the absence of a magnetic field. Equation (44.12) expresses the fact that when V_1 and V_2 are unequal the normal modes of propagation are elliptically polarized waves, i.e., ξ_+^0 and ξ_-^0 are coupled together.

Utilizing Eq. (44.12) we can derive the dispersion relation for the coupled system. A comparison of the data, extracted from curves like Fig. VII-13 and processed exactly as described on pp. 138–140 is shown in Fig. VII-14. Velocities are normalized to the sound velocity $V_1 = 7 \times 10^4$ cm sec^{-1} with $V_2 = 1.70 \times 10^5$ cm sec^{-1}. The agreement between theory (solid line) and the experimental points is excellent. No adjustable parameters were used in the comparison.

REFERENCES

1. R. Bowers, C. Legendy, and F. Rose, *Phys. Rev. Lett.* **7**, 339 (1961).
2. A. Libchaber and C. C. Grimes, *Phys. Rev.* **178**, 1145 (1969).
3. For further discussion of the Onsager relations see S. R. DeGroot, "Thermodynamics of Irreversible Processes." North-Holland Publ., Amsterdam, 1961.
4. E. A. Kaner and V. G. Skobov, "Electromagnetic Waves in Metals in a Magnetic Field." Taylor & Francis, London, 1968.
5. G. E. Reuter and E. H. Sondheimer, *Proc. Roy. Soc. (London)* **195**, 136 (1948).
6. S. Schultz and G. Dunifer, *Phys. Rev. Lett.* **18**, 283 (1967).
7. S. J. Buchsbaum and J. K. Galt, *Phys. Fluids* **4**, 1514 (1968).
8. G. A. Williams, *Phys. Rev. A* **139**, 771 (1965).
9. C. C. Grimes and S. J. Buchsbaum, *Phys. Rev. Lett.* **12**, 357 (1964).

VIII. Nonlocal Effects in a Noninteracting Gas

45. The Solution of the Boltzmann–Vlasov Equation

To include nonlocal, i.e., finite-**k** or finite-velocity, effects in our analysis of wave propagation, we need, in addition to Maxwell's equations, an equation which determines the current induced in the medium by a spatially and time varying electric field. For strictly longitudinal fields, we have seen that the classical analogue of the quantum-mechanical Hartree approximation was the collisionless Boltzmann–Vlasov (BV) equation (22.3). More generally, we conjectured and argued qualitatively that if magnetic fields and transverse currents were present, as they will be in all wave propagation problems, that Eq. (22.3) could be generalized to yield Eq. (22.7). We will use Eq. (22.7) to evaluate the nonlocal conductivity tensor $\sigma(\mathbf{k}, \omega)$ and obtain the dispersion properties of the normal mode solutions of Maxwell's equations as discussed in Chapter VII.

The BV equation (22.7) incorporates the Coulomb interactions among electrons into a self-consistent field. The short-range effects of the Coulomb interactions, exchange and correlation, are omitted. These effects are higher order in r_s (see the discussion in Section 2) and would be small in *high-density* metallic plasmas. At real metallic densities there is no a priori reason to ignore these "higher-order" interaction effects, since in the alkalis r_s ranges from 3.4 in Li to 5.57 in Cs. Despite these obvious facts, we will see that the weak coupling picture seems to work exceedingly well in describing many features of wave propagation in metals. When we discuss the Fermi-liquid theory in Chapter X, we will get some idea as to why this picture works so well.

The mass of the charged particles is an important parameter in the BV equation. Frequently an effective mass m^* is used instead of the bare electron mass m_0 appearing in Eq. (22.7). This mass, often called the quasi-particle mass, takes into account band structure effects as well as some of the effects of electron–electron and electron–phonon interactions. However, such a simple modification is not a very accurate way of including the effects of interactions. This point will be discussed later in detail. However, for the time being, we will work with a modified BV equation having replaced m_0 by m^* and inserting a phenomenological collision term which simulates the effects of electron impurity collisions. This equation for the distribution function $n(\mathbf{p}, \mathbf{r}, t)$ is

$$\partial n/\partial t + [\mathbf{v} \cdot \nabla + e(\mathbf{E} + \mathbf{v} \times \mathbf{H}/c) \cdot \nabla_\mathbf{p}]n = -(1/\tau)[n - n^\circ(\mathbf{p})] \quad (45.1)$$

with $\mathbf{v} \equiv \mathbf{p}/m^*$.

The mechanics of solving the BV equation in the infinite medium for weak time-varying fields are elementary. Since we will frequently use the results of such calculations in an analysis of wave propagation phenomenon in metals, it is useful to consider in some detail the solution of Eq. (45.1). To proceed, we linearize Eq. (45.1) in the external driving fields, i.e., we take

$$n(\mathbf{p}, \mathbf{r}, t) = n^\circ(\mathbf{p}) + \delta n(\mathbf{p}, \mathbf{r}, t) \quad (45.2)$$

where δn is first order in the driving fields $\mathbf{E}$ and $\mathbf{H}$. Assuming that these fields are plane waves in space and time, we may write the linearized Fourier transformed version of Eq. (45.1) as

$$i(\mathbf{k} \cdot \mathbf{v} - \omega - i/\tau)\,\delta n + \frac{e}{c}(\mathbf{v} \times \mathbf{H}_0) \cdot \frac{\partial}{\partial \mathbf{p}}(\delta n)$$
$$= -e\mathbf{E} \cdot \frac{\partial n^\circ}{\partial \mathbf{p}} - \frac{e}{c}(\mathbf{v} \times \mathbf{H}) \cdot \frac{\partial n^\circ}{\partial \mathbf{p}}. \quad (45.3)$$

When n° is an equilibrium distribution, the second term on the right-hand side of Eq. (45.3) is identically zero. This follows naturally since n° is only a function of the energy E_p° of the particles so that

$$\partial n^\circ/\partial \mathbf{p} = (\partial n^\circ/\partial E_p^\circ)(\partial E_p^\circ/\partial \mathbf{p}) \equiv \mathbf{v}\, \partial n^\circ/\partial E_p^\circ. \quad (45.4)$$

In other cases this term must be kept, e.g., in a drifting plasma. We will be concerned with perturbation about equilibrium, so we will, in fact, drop this term.

Given $\delta n(\mathbf{p}, \mathbf{k}, \omega)$ one finds $\boldsymbol{\sigma}(\mathbf{k}, \omega)$ by integration,

$$\mathbf{j}(\mathbf{k}, \omega) = e(2\pi)^{-3} \int (\mathbf{p}/m^*)\, \delta n(\mathbf{p}, \mathbf{k}, \omega)\, d^3p \quad (45.5)$$

and the relation

$$\mathbf{j}(\mathbf{k}, \omega) = \boldsymbol{\sigma}(\mathbf{k}, \omega) \cdot \mathbf{E}. \quad (45.6)$$

In order to solve Eq. (45.3), we choose a coordinate system (see Fig. VIII-1) with $\mathbf{H}_0$ along the z axis and $\mathbf{k}$ in the x, z plane making an angle Δ with the z axis. In this coordinate system (for a negatively charged electron) the term

$$-(e/c)(\mathbf{v} \times \mathbf{H}_0) \cdot \partial(\delta n)/\partial \mathbf{p} = \omega_c\, \partial(\delta n)/\partial \varphi \quad (45.7)$$

45. SOLUTION OF BOLTZMANN–VLASOV EQUATION

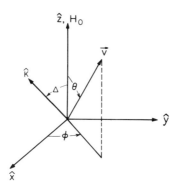

FIG. VIII-1. The coordinate system. The wave vector **k** of the disturbance makes an angle Δ with the z axis and is in the x, z plane. The particle velocity is specified by the spherical coordinates $|\mathbf{v}|$, θ, and ϕ.

and Eq. (45.3) may be written as

$$\omega_c \, \partial(\delta n)/\partial\varphi - i(\omega + i/\tau - kv \cos\Delta \cos\theta - kv \sin\theta \cos\varphi \sin\Delta) \, \delta n$$
$$= e\mathbf{E} \cdot \mathbf{v} \, (\partial n^\circ/\partial E^\circ). \tag{45.8}$$

For a degenerate Fermion system at zero degrees, i.e., a metal

$$\partial n^\circ/\partial E_p^\circ = -\delta(E_p^\circ - E_\mathrm{F}) \tag{45.9}$$

where E_F is the Fermi energy and $E_p^\circ = p^2/2m^*$. The presence of the delta function in the driving term forces **v** to be on the Fermi surface, i.e., $|\mathbf{v}| = p_\mathrm{F}/m^*$. The equation for the distribution function is then simply an equation in the angle the velocity (momentum) vector makes as it moves over the Fermi surface. In fact, Eq. (45.8) is simply a first-order differential equation in the variable φ, the azimuthal angle the velocity (momentum) vector of the electron makes with respect to the magnetic field direction.

For a classical plasma and in some low density plasmas in semiconductors, n° may be Maxwellian. In this case, the magnitude of the velocity **v** is not confined to the Fermi surface as indicated by Eq. (45.9). However, Eq. (45.3) along with the definition of the conductivity, Eq. (45.6) is still valid. The integral on momentum (velocity) in Eq. (45.5) must be carried out over all three variables for the classical plasma (1). The manipulations are hardly more difficult than those we will outline. We will deal exclusively with the quantum (zero temperature) results and merely give the expression for one component of the conductivity tensor of a classical plasma.

Equation (45.8) is a first-order differential equation which can immediately be integrated. Defining the function g by

$$\delta n = -(\partial n°/\partial E_p°)\, g \qquad (49.10)$$

we obtain

$$g = (e/\omega_c)\mathbf{E} \cdot \int_{-\infty}^{\phi} \mathbf{v}' \exp i[a(\phi - \phi') - b(\sin\phi - \sin\phi')]\, d\phi'. \qquad (45.11)$$

Here

$$b = (kv_F \sin\theta \sin\varDelta)/\omega_c \qquad (45.12)$$

and

$$a = (\omega + i/\tau - kv_F \cos\varDelta \cos\theta)/\omega_c. \qquad (45.13)$$

The lower limit on the integral fixes the constant of integration. Minus infinity arises from the condition that $g(\theta, \phi) = g(\theta, \phi + 2\pi)$ and from the requirement that g be bounded at $\phi = -\infty$.

Since the general magnetoconductivity tensor is the basis for our entire discussion of wave propagation phenomenon, we will outline the calculation of σ_{zz}. This calculation is typical of the calculation of any element σ_{ij}.

From Eq. (45.11) we see that for $\mathbf{E}$ in the z direction,

$$g = (e/\omega_c)\, v_F \cos\theta \int_{-\infty}^{\phi} \exp i[a(\phi - \phi') - b(\sin\phi - \sin\phi')]\, d\phi'. \qquad (45.14)$$

In general it is difficult to integrate Eq. (45.14) over the ϕ' variable because of the $\sin\phi'$ in the exponent of the integrating factor. However, one can perform the ϕ' integration and put the conductivity in a physically more transparent form by expanding $\exp[ib \sin\phi]$ as a series of Bessel functions, i.e.,

$$e^{ib \sin\phi} = \sum_{m=-\infty}^{+\infty} J_m(b)\, e^{im\phi}. \qquad (45.15)$$

Inserting this expansion into Eq. (45.14) reduces the ϕ' integration to a sum of exponential integrals which are easily evaluated and yield

$$g = (e/\omega_c) v_F \cos\theta\, e^{-ib \sin\phi} \sum_{m=-\infty}^{+\infty} [i/(a-m)]\, J_m(b)\, e^{im\phi}. \qquad (45.16)$$

The quantity

$$\sigma_{zz} \sim \int g \cos\theta\, d(\cos\theta)\, d\phi. \qquad (45.17)$$

45. SOLUTION OF BOLTZMANN–VLASOV EQUATION

Substituting Eq. (45.16) into (45.17), we see that

$$\sigma'_{zz} \sim \sum_{m=-\infty}^{+\infty} \int \sin\theta \cos^2\theta \, d\theta \, d\phi (a-m)^{-1} e^{im\phi} e^{-ib\sin\phi} J_m(b). \quad (45.18)$$

Expanding $\exp(-ib\sin\phi)$ once again, doing the ϕ integration and putting in the appropriate normalization constants yields the result

$$\sigma_{zz} = iN \sum_{m=0}^{\infty} \int_0^\pi \frac{a J_m^2(b) \cos^2\theta \sin\theta \, d\theta}{(1+\delta_{m0})(a^2-m^2)} \quad (45.19)$$

with

$$N = 3n_0 e^2 / m^* \omega_c. \quad (45.20)$$

The calculation of the remaining components of the conductivity tensor parallels the simple calculation outlined above. In all cases there is an integral over the Fermi surface and an infinite, rather rapidly convergent sum to perform. Various components of the conductivity differ only in the function of θ multiplying the Bessel functions and by the fact that derivatives of Bessel functions may also enter. The derivatives arise since

$$\int e^{i(b\sin\phi - a\phi)} \sin\phi \, d\phi = \frac{1}{i} \frac{\partial}{\partial b} \int e^{i(b\sin\phi - a\phi)} \, d\phi$$

$$= \sum_{m=-\infty}^{+\infty} \frac{J_m'(b)}{a-m} e^{i(m-a)\phi}. \quad (45.21)$$

The remaining components of the conductivity tensor are (2),

$$\sigma_{xx} = iN \sum_{n=0}^{\infty} n^2 \int_0^\pi \frac{a}{b^2} \frac{J_n^2(b) \sin^3\theta \, d\theta}{(a^2-n^2)}, \quad (45.22)$$

$$\sigma_{yy} = iN \sum_{n=0}^{\infty} \int_0^\pi \frac{a J_n'(b)^2 \sin^3\theta \, d\theta}{(1+\delta_{n0})(a^2-n^2)}, \quad (45.23)$$

$$\sigma_{zz} = iN \sum_{n=0}^{\infty} \int_0^\pi \frac{a J_n^2(b) \cos^2\theta \sin\theta \, d\theta}{(1+\delta_{n0})(a^2-n^2)}, \quad (45.24)$$

$$\sigma_{xy} \equiv -\sigma_{yx} = N \sum_{n=0}^{\infty} \int_0^\pi \frac{a^2}{b} \frac{J_n(b) J_n'(b) \sin^3\theta \, d\theta}{(1+\delta_{n0})(a^2-n^2)}, \quad (45.25)$$

$$\sigma_{yz} = -\sigma_{zy} = -N \sum_{n=0}^{\infty} \int_0^\pi \frac{a J_n(b) J_n'(b) \sin^2\theta \cos\theta \, d\theta}{(1+\delta_{n0})(a^2-n^2)}, \quad (45.26)$$

$$\sigma_{xz} = \sigma_{zx} = iN \sum_{n=0}^{\infty} n^2 \int_0^\pi \frac{J_n^2(b) \sin^2\theta \cos\theta \, d\theta}{b(a^2-n^2)}. \quad (45.27)$$

Although Eq. (45.22)–(45.27) look complicated they have several simple features.

(1) For finite k with $\Delta = \pi/2$, every component of the conductivity tensor exhibits an infinite set of resonances at multiples of the cyclotron frequency ($\omega = n\omega_c$, $n = 1, 2, 3,...$).

(2) At long wavelengths, or more explicitly, as $b \to 0$, it is clear that the order (power of b) which dominates a given singularity ($\omega = n'\omega_c$) increases with increasing n'.

(3) There is an infinite set of branch cut singularities present in each term of the conductivity tensor. The boundaries of these regions are determined by the conditions

$$\omega - n\omega_c = kv_F \cos \Delta, \qquad n = 1, 2,... . \qquad (45.28)$$

These cuts mark the onset of Doppler-shifted cyclotron absorption. The fans described by Eq. (45.28) are shown in Fig. VIII-2. When the angle

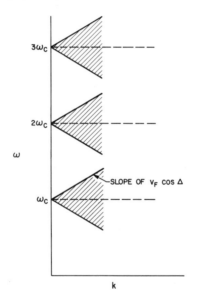

FIG. VIII-2. A plot of the fans marking the boundary region for cyclotron damping. Inside the shaded regions there is cyclotron damping.

of propagation is strictly perpendicular to the field $\mathbf{H}_0$, i.e., $\cos \Delta = 0$ the fans collapse to a set of lines. In this case there is no resonant absorption of energy except for a series of delta functions at $\omega = n\omega_c$.

(4) The components σ_{xz} and σ_{zy} are clearly zero for $\mathbf{k} \parallel \mathbf{H}_0$, i.e., $b = 0$. For $\mathbf{k} \perp \mathbf{H}_0$, these components are still zero because the angular weighting factor $\sin^2 \theta \cos \theta$ is odd about $\pi/2$. When there is no θ dependence in the denominator, as is the case for $\Delta = \pi/2$, the remainder of the integrand is even. The integral over θ gives zero.

In the classical plasma, the results for the conductivity tensor are very similar to those given here for the degenerate plasma (1). In fact for $\mathbf{k} \perp \mathbf{H}_0$, unlike the degenerate case, *all* the integrals may be done explicitly and one finds for example that

$$\sigma_{xx} = -i \frac{n_0 e^2 \tilde{\omega}}{m^* \omega_c^2} \frac{1}{\lambda} \left\{ [e^{-\lambda} I_0(\lambda) - 1] + 2a^2 \sum_{n=1}^{\infty} \frac{e^{-\lambda} I_n(\lambda)}{a^2 - n^2} \right\}. \quad (45.29)$$

Here

$$\lambda = k^2 v_{\text{th}}^2 / \omega_c^2 \quad (45.30)$$

where

$$v_{\text{th}}^2 = kT/m^* \quad (45.31)$$

and $I_n(\lambda)$ is the Bessel function of the second kind.

46. Helicons in the Nonlocal Regime

In Chapter VII we considered the properties of helicons in the local regime. In the next two sections we will utilize the results of Section 45 to analyze the behavior of helicons in the regime where nonlocal effects are important.

When $\mathbf{k} \parallel \mathbf{H}_0$, $b = 0$ and the infinite sum of Bessel functions collapses to either a single constant term or to zero. The θ integration can then be performed and we find that

$$\epsilon_+ \equiv \epsilon_{xx} + i\epsilon_{xy} \cong \frac{4\pi}{i\omega} [\sigma_{xx} + i\sigma_{xy}] = \frac{3}{4} \frac{\omega_p^2}{\omega k v_F} \left[(1 - z^2) \ln\left(\frac{z+1}{z-1}\right) + 2z \right] \quad (46.1)$$

with

$$z = -(\omega - \omega_c + i/\tau)/k v_F. \quad (46.2)$$

As $k \to 0$, i.e., $|z| \to \infty$ we may expand the logarithm in Eq. (46.1) as a power series in $1/z$. The leading term in this power series expansion is just the local dielectric constant, Eq. (42.15), found utilizing the simple uniform field equation of motion technique. As z gets smaller, there are corrections to the local dispersion relation Eq. (42.15) of the form

$$\frac{k^2}{k_0^2} = -\frac{\omega_p^2}{\omega(\omega - \omega_c + i/\tau)} \left[1 + \frac{1}{5z^2} + \cdots \right]. \quad (46.3)$$

When $\omega_c\tau \to \infty$, it is clear from Eq. (46.1) that as long as $x > 1$ the dielectric constant is real. As x passes through unity we pick up an extra $i\pi$ from the logarithm. This $i\pi$ is the mathematical statement that cyclotron damping is present. In other words when $|kv_F/(\omega_c - \omega)| > 1$, there will be some electrons on the Fermi surface having velocities in the z direction such that they experience a static electric field due to the wave.

For a spherical Fermi surface, the onset of cyclotron damping measures the maximum velocity of the electrons in the z direction. Analysis shows that, in materials with more complicated Fermi surfaces, Doppler-shifted cyclotron resonance occurs when

$$k[v_z/\omega_c(p_z)]_{\max} \equiv [ck(\omega, H_0)/2\pi eH_0] \mid \partial S(E_F, p_z)/\partial p_z \mid_{\max} = 1. \quad (46.4)$$

In Eq. (46.4) we have expressed $|v_z/\omega_c(p_z)|$ in terms of the derivative of the cross-sectional area $S(E_F, p_z)$ of the Fermi surface.

The details of how one derives Eq. (46.4) are not important here. They may be found in a number of elementary discussions of electronic motion in a magnetic field, e.g. Refs. (1) and (2) of Chapter VI. Several people (3, 4), have suggested utilizing the onset of this absorption to make a point by point mapping of the Fermi surface. To date this method has not proved practically useful in mapping the shape of rather complicated Fermi surfaces (4). However, it is quite helpful in self-consistently checking the predicted behavior for simple Fermi surfaces. The main problem in such measurements seems to be one of accuracy. In most materials the edges are not very sharp and the fact that k as well as $(\partial S/\partial p_z)_{\max}$ comes into Eq. (46.4) makes interpretation of the data difficult.

In Fig. VIII-3 we compare the helicon data in K shown in Fig. VII-7

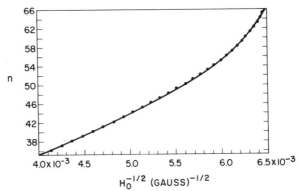

FIG. VIII-3. A plot of the non-local dispersion relation. The experimental points are obtained from the transmission spectra in K displayed in Fig. VII-7. $d = 0.464$ mm; $T = 4.2°$K; $\omega = 1.26 \times 10^8$ Hz.

with the theoretical dispersion relation (solid line) computed using the more exact dielectric constant given in Eq. (46.1). The dispersion is plotted for $|x| > 1$, i.e., above the helicon absorption edge where absorptive effects are small since

$$|\text{Im}(\epsilon_+)/\text{Re}(\epsilon_+)| \sim 1/\omega_c\tau \ll 1. \tag{46.5}$$

The nonlocal theory is in excellent agreement with the experiment. We wish to stress the fact that no adjustable parameters were used in this theoretical fit to the data.

Equation (46.1) implies that the position of the helicon absorption edge, at low frequencies, is given by

$$kv_\text{F}/\omega_c = 1 \tag{46.6}$$

with

$$k^2/k_0{}^2 = \epsilon_+(\mathbf{k}, \omega). \tag{46.7}$$

Utilizing Eq. (46.7), Eq. (46.6) is plotted in Fig. VIII-4 along with the

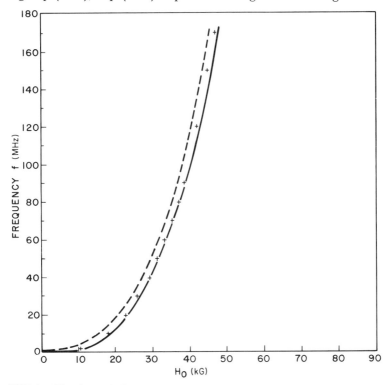

Fig. VIII-4. The theoretical position of the absorption edge [Eq. (46.6)] is plotted as solid line. The points are the experimental data taken in various samples of pure K. The dashed line refers to Eq. (46.12).

data points taken in K by Libchaber and Grimes (5) at a number of different frequencies. The agreement is of the order of one percent, which is within the uncertainty in the dc field measurement.

In recent years Overhauser (6, 7) has suggested that the ground state of the interacting electron gas in the alkalis may in fact be quite different from the normal fermion ground state we have used here. He originally proposed this anomalous ground state in order to explain some very unusual infrared optical reflectivity experiments in these materials (8). In his paper Overhauser suggested that the ground state of the electron gas was a spin or charge density wave. He postulated that the spin or charge density wave had a wave vector **Q** associated with it which was not commensurate with a reciprocal lattice vector. We do not wish to go into a detailed discussion of the validity of this picture or how it arises. We would like to point out however that the position of the helicon transmission edge in potassium is significantly different in the Overhauser model than in the normal ground state. This discussion is useful, even though it is now believed that the ground state of materials like potassium are not in an Overhauser-like ground state (9, 10) because it illustrates how one can at least in principle look at Fermi surfaces which are slightly more complicated than spherical.

The Fermi surface of the Overhauser ground state is a lemon-shaped object. The tip of the lemon extends out to the point **Q**/2 in momentum space. At this point a gap G is opened in the band structure. The Fermi surface is still rotationally symmetric about the z axis (the direction of **Q**) and the energy momentum relation of the electrons is, for $k_z > 0$, given by

$$\epsilon = p^2/2m^* + \mu(\tfrac{1}{2}Q - p_z) - [\mu^2(\tfrac{1}{2}Q - p_z)^2 + \tfrac{1}{4}G^2]^{1/2} \qquad (46.8)$$

with $\mu = Q/2m^*$. The maximum velocity of electrons along the z axis is in turn given by

$$v_{\max} = (\hbar Q/2m^*)[1 - (G/\mu Q)^{2/3}]^{3/2}. \qquad (46.9)$$

The gap value G and the wave vector Q of the spin density wave are parameters in the theory whose values are set by fitting the infrared optical data of Mayer and co-workers (8). For potassium, Overhauser chose the values $G = 0.62$ eV and $Q = 1.42$ Å^{-1}. Substituting the above values into Eq. (46.9), we find that

$$v_{\max} = 0.61 \times 10^8 \text{ cm/sec.} \qquad (46.10)$$

For the free fermion ground state,

$$v_F = 0.71 \times 10^8 \text{ cm/sec.} \qquad (46.11)$$

The two maximum velocities in the z direction differ by something approximating 15 %.

The *transmission* edge is still given by

$$kv_{max}/\omega_c = 1. \qquad (46.12)$$

The value of k in Eq. (46.12) is determined from a solution of the dispersion relation, Eq. (46.7). In this case, however, the distribution function and velocity are obtained from the modified energy–momentum relation given in Eq. (46.8). A numerical plot of Eq. (46.12) is shown in Fig. VIII-4 as the dashed line. It is distinctly not as good a fit to the transmission data as the solid free electron curve.

47. Helicon Propagation at an Angle to the Magnetic Field

We have seen that when a helicon wave propagates in a metal (with a spherical Fermi surface) along a static magnetic field, it can be damped by two processes, collisional damping (finite $\omega_c \tau$) and absorption caused by resonant cyclotron damping. The cyclotron damping occurs only under suitable conditions, namely, when there exist carriers within the metal whose velocity along the magnetic field is such that they experience Doppler-shifted cyclotron resonance.

When the helicon wave propagates at an angle to the magnetic field $(0 < \Delta < \pi/2)$, it is no longer a purely transverse wave and two other sources of damping can exist. The existence of a (small) longitudinal electric field leads to so-called Landau damping (*11*). (See the discussion in Section 21). Like cyclotron damping, Landau damping is a collisionless resonant damping mechanism. Electrons traveling at approximately the phase velocity of the wave, i.e., $\omega/(\mathbf{k} \cdot \mathbf{v}) \simeq 1$ experience a longitudinal static electric field. Landau damping is the dominant collisionles damping mechanism for longitudinal *plasma waves*. However, this particular form of damping can be shown to be negligbly small for helicons as long as the phase velocity of the helicon is small compared to the Fermi velocity of the carriers.

In addition to the Landau damping which appears for nonparallel propagation there is still another collisionless mechanism available for damping the wave. For oblique propagation, the magnetic field of the wave is now such that when added to the static magnetic field it causes the carriers to experience a moving periodic mirror field. "Mirror field" is an expression used by gas plasma physicists to describe a magnetic field configuration that will reflect electrons (*12*). In the simplest case it is a nonuniform field that gets more intense as the particle moves along the z axis, the direction of the field. The variation of magnetic fields in the z direction produces a z component of force on an electron moving in this field. If the field varies slowly enough, this force is given by

$$F_z^m \simeq \left(\frac{mv_\perp^2}{H_0}\right)\left(\frac{\partial H_z}{\partial z}\right). \tag{47.1}$$

For static fields, this z component of force eventually reflects a freely moving particle. In the dynamic or moving case, it leads to a collisionless damping of the wave similar to Landau damping. It has been called magnetic Landau damping (13, 14). Since the helicon wave is practically all magnetic field, this turns out, under most experimental situations, to be the dominant collisionless damping mechanism.

Experimental measurements of helicon propagation, quite dramatically, exhibit the effects of the additional collisionless damping which is present. In Fig. VIII-5 we show a series of experimental traces on helicon propagation in K taken by Libchaber and Grimes (5). As the field is tilted, the amplitude of the signal decreases. This decrease in amplitude is due to the existence of longitudinal fields and the resonant collisionless damping of the wave.

The analysis of off-angle helicon propagation is a straightforward but tedious exercise in the use of the conductivity tensor [Eqs. (45.22)–(45.27)]. We refer the interested reader to references (13) and (14) for the details of the calculation. The results are partially contained in the complex dispersion relation,

$$k^2 = k_H^2[1 + i\Gamma], \tag{47.2}$$

where

$$k_H^2 = \omega_p^2 \omega / (c^2 \omega_c \cos \Delta)$$

and

$$\Gamma = (\omega_c \tau \cos \Delta)^{-1}[1 - \tfrac{3}{2}(1 - \pi^2/16) \sin^2 \Delta] + (3\pi/16) kR_c \sin^2 \Delta. \tag{47.3}$$

Equation (47.3) is derived assuming that,

(1) $\omega_c \tau \gg 1$, (2) $kR_c \ll 1$, (3) $(kR_c)(\omega_c \tau)$.

The $(\omega_c \tau)^{-1}$ in Eq. (47.3) is the collisional term while the $(3\pi/16) kR_c \sin^2 \Delta$ characterizes collisionless magnetic Landau damping. Equation (47.2) accurately describes the off-angle data shown in Fig. VIII-5. In Fig VIII-6 we show a plot of $\Gamma(\Delta) - \Gamma(0)$ extracted from transmission data for a fixed Δ as a function of kR_c (5). The variation of the collisional $(\omega_c \tau)^{-1}$ term is negligible, and the data fall accurately on the straight line $(3\pi/16) kR_c \sin^2 \Delta$.

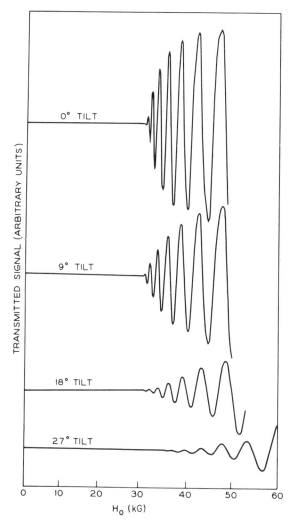

Fig. VIII-5. The effect of magnetic field tipping on the helicon transmission spectrum in K. $d = 0.23$ mm; $T = 1.2°$K; $\omega = 3.14 \times 10^8$ Hz.

There have been several relatively extensive experimental investigations of off-angle helicon propagation (5, 15). In all cases the phenomenon can be understood within the framework of the simple noninteracting spherical Fermi-surface picture of the electron gas. In fact, all of the results we have discussed in this section relating to helicon propagation in metals are, to within the accuracy of the experiments, in

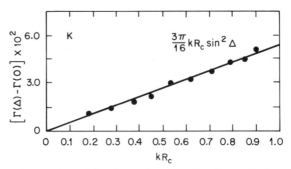

FIG. VIII-6. Comparison of the observed and predicted contribution to the magnetic Landau damping of helicon waves for an 18° tilt angle.

excellent agreement with the noninteracting effective mass approximation for the electron gas.

We have not really learned anything new about the properties of the electrons in simple metals. However, many of the techniques, arguments, and physical insights we have gained here will enable us to understand more easily the more complicated wave phenomenon in the alkalis. In this sense helicons are the launching pad for the rest of our discussion.

48. Cyclotron Wave Propagation in a Noninteracting Gas

We now go on to discuss another type of wave propagation, the so-called cyclotron waves (CW). These waves, briefly mentioned in the introductory remarks (Section 39), propagate most easily at right angles to the field ($\mathbf{k} \perp \mathbf{H}_0$) in the vicinity of cyclotron resonance ($\omega \simeq n\omega_c$). Their propagation characteristics depend on the nonlocal (finite k) properties of the conductivity tensor. They do not propagate, as helicons do, in a medium with a local dielectric constant. These waves are important because, as we will see, they are more sensitive probes of the dynamics of the electron gas than are the helicons.

The first evidence for such wavelike excitations capable of relatively undamped propagation in the vicinity of cyclotron resonance was observed in potassium (16). The experimental configuration is shown in Fig. VIII-7. The sample in the form of a thin slab of metal encased, for protection, in sheets of an inert hydrocarbon is placed inside a microwave cavity. Due to lack of electrical contact between the specimen and the cavity wall, microwave currents flow on both faces of the sample. While not of rigorously equal amplitude, the currents on the surface are in opposite direction (antisymmetric excitation). This type of configura-

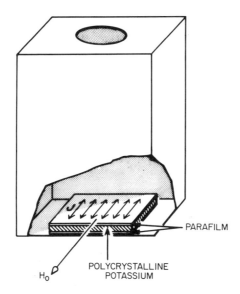

FIG. VIII-7. The experimental configuration utilized in the first experiments on CW.

tion is extremely sensitive to any magnetic-field-dependent transmission of microwave energy from one sample face to the other and thus is an ideal geometry for observing wave propagation.

A typical spectrum displaying the field differential power absorption for a thin, relatively flat potassium specimen is shown in Fig. VIII-8. The sample was 0.13-mm thick and was made from pure bulk material. The dc magnetic field was parallel to the surface of the specimen and was oriented so that the ac currents flowed parallel to it ($\mathbf{J} \parallel \mathbf{H}_0$). The driving frequency $\omega \cong 7.54 \times 10^{10}$ Hz. To avoid undue complications in the spectrum, that portion in the interval $1 \leqslant \omega_c/\omega < 1.9$ is shown.

There are several distinctive features of the spectrum.

(1) The large negative peak in dA/dH_0 near the resonance field ($\omega = \omega_c$) expected for a carrier with an effective mass ratio $m^*/m_0 = 1.21$ is due to single-particle Azbel–Kaner cyclotron resonance (AKCR). The shape and position of the AKCR signal is independent of the geometry of the specimen. While we will not discuss the mathematical details of this resonance, its physical origin is well known (17) and easily described, as follows. Since the magnetic field is parallel to the surface of the specimen, the periodic part of the electrons' motion is a circle perpendicular to $\mathbf{H}_0$. This is diagrammed in Fig. VIII-9. Because the skin depth (shaded region in Fig. VIII-9) is much smaller than the orbit diameter, the electron will be subjected to a short pulsed rf field each time it enters

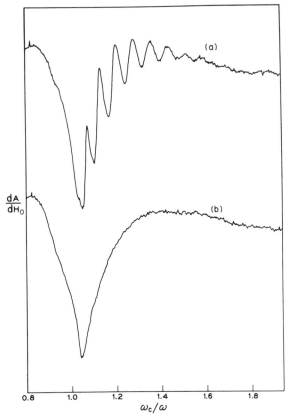

FIG. VIII-8. Field derivative of the microwave power absorbed by a slab of pure potassium metal versus magnetic field. The experiment was performed in the $\mathbf{J} \parallel \mathbf{H_0}$ polarization. $d = 0.13$ mm; $T = 1.4°$K; $\omega = 7.54 \times 10^{10}$ Hz.

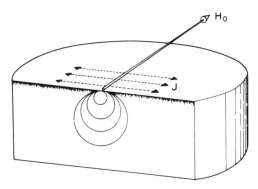

FIG. VIII-9. Schematic of the physics and experimental configuration in Azbel–Kaner cyclotron resonance. Resonances at $\omega = n\omega_c$, $\omega_c = eH_0/m^*c$; observable at $\omega_c \tau > 1$.

the skin region. If $\omega \neq n\omega_c$, the average, over many passes, of the power absorbed (gained) by a typical electron is zero. On the other hand, if the field has the same phase ($\omega = n\omega_c$) each time the electron enters the skin depth, it will on the average either gain or lose energy to the field depending on its initial phase. The experimental arrangement is analogous to a little cyclotron wherein the skin region acts as the accelerating gap. This was the effect which Grimes and Kip (*18*) utilized to determine the effective mass of the carriers in sodium and potassium.

(2) The "extra" oscillations in dA/dH (Fig. VIII-8) on the high-field side of AKCR occur only for thin *plane parallel* specimens. A similar purposely wedge-shaped slab shows no such additional structure (see Fig. VIII-8b). By examining samples of differing thickness one may qualitatively establish that the oscillations result from interference between the primary surface currents and much weaker currents transmitted across the sample by a propagating mode whose wavelength is controlled by the magnetic field.

In Fig. VIII-10 we show a *complete* spectrum taken in reflection under

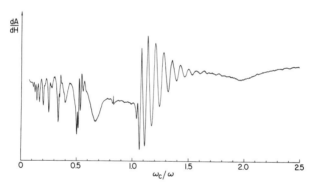

FIG. VIII-10. Field derivative of the microwave power absorbed by a slab of pure potassium metal versus magnetic field. The experiment was performed in the $\mathbf{J} \parallel \mathbf{H}_0$ polarization. $d = 0.141$ mm; $T = 1.4°$K; $\omega = 7.54 \times 10^{10}$ Hz.

conditions almost identical with those shown in Fig. VIII-8. The specimen is a parallel thin specimen of K placed near the center of the cavity in order to supress AKCR. The oscillations on the high-field side of each subharmonic are clearly visible. The small bit of structure between $\omega = \omega_c$ and $\omega = 2\omega_c$ is the spin resonance which will be discussed in great detail in Chapter IX.

The "extra" oscillations in Figs. VIII-8 and VIII-10 are the signature of the CW and are the subject of the discussion in this section. Experiments show that similar structure exists in limited bands on the high-field side of all harmonics $\omega = n\omega_c$ and that similar but distinct propagation occurs for $\mathbf{J} \perp \mathbf{H}_0$ as well.

At microwave frequencies the variations of the magnetic-field induced structure with sample thickness and experimental frequency are consistent with the concept of "standing" wave excitation of CW. The oscillations of the power absorption as in the helicon transmission experiments results from the varying phase of the transmitted currents relative to the primary driving currents. One-sided transmission experiments (as in the helicon case) would be expected to show the variation in phase of the transmitted field directly.

Figure VIII-11 is an experimental spectrum taken in transmission

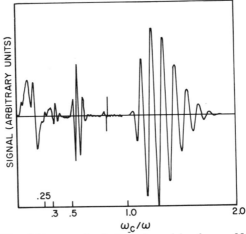

FIG. VIII-11. The field transmitted through a slab of pure Na versus magnetic field. The experiment was performed in the $\mathbf{J} \parallel \mathbf{H}_0$ geometry. $d = 0.3$ mm; $T = 1.4°$K; $\omega = 5.71 \times 10^{10}$ Hz.

by Schultz and Dunifer (20). The specimen in this case is a slab of Na approximately 0.165-mm thick. As in Figs. VIII-8 and VIII-10, the dc magnetic field was oriented so that ac currents flowed parallel to it ($\mathbf{J} \parallel \mathbf{H}_0$). The driving frequency $\omega \simeq 5.7 \times 10^{10}$ Hz. The bands of transmitted power on the high-field side of the cyclotron harmonics and the oscillations in phase with the transmitted field are clearly visible. The sharp structure between $\omega = \omega_c$ and $\omega = 2\omega_c$ is, as in Fig. VIII-10, the spin resonance. The two spectra, reflection and transmission, are remarkably similar.

The transmission experiments have two distinct advantages over the reflection or standing wave experiments. Figure VIII-11 shows clearly that the AKCR signal is completely surpressed. In addition, the field configuration, one-sided excitation, is completely specified. The field-dependent phase variation of the waves yields the dispersion or variation of wavelength with field and frequency. However, there is a real, very important, difficulty connected with the transmission experiments, which does not seem to plague the two-sided reflection experiments. The reader will note that in Fig. VIII-11 the field intensity turns on gradually at cyclotron resonance. This behavior makes it difficult to accurately measure the properties of the first few phase wiggles, i.e., the properties of the long-wavelength modes. We will see, that in a very real sense, these long-wavelength modes are the most interesting aspects of the spectrum, so that there is an advantage to the two-sided reflection experiments, because in them the structure turns on suddenly near cyclotron resonance.

For the CW case the two-sided reflection experiments were the original experiments. However, unlike the helicon case these standing CW experiments are easy to interpret because of the simplified geometry in which they were done. To date the two-sided reflection experiments in K are the most complete available. We will discuss the results of these "quasi" transmission experiments in some detail. We will not analyze in any detail the CW data available from the more conventional transmission experiments.

Once we recognize that the structure in the microwave power absorption spectrum is due to some type of standing wave resonance in the specimen, we must conclude that the observed wavelengths, as in Section 42, are typically less than or approximately equal to the sample thickness, i.e., $0.1 < \lambda/L < 1$. Since $L \sim 10^{-2}$ cm and $R_c \sim v_F/\omega_c \sim 10^{-3}$ cm for Fermi velocities, $v_F \sim 10^8$ cm sec^{-1} and magnetic fields of a few thousand Gauss we conclude that for the waves observed in these experiments $kR_c \cong 1$.

Figure VIII-12 shows the magnitude of the important physical parameters involved in the CW experiments (21). The skin depth is included to show the region in which most of the microwave energy is concentrated. The skin depth is fundamentally an unknown region in which the rf current flows and excites the waves. The ratio $\delta_s/L \sim 10^{-3}$ is clearly a measure of the uncertainty involved in the application of an infinite-medium calculation to the actual boundary value problem. In some sense the "electromagnetic" thickness of the specimen is uncertain to a length which is of order δ_s. When we identify, as we shall, standing-wave resonances with wave vectors $k = 2n\pi/L$ we really are uncertain about what value to use for L, i.e., $L \pm \mathcal{O}(\delta_s)$.

The idea that wave propagation might occur near AKCR was first discussed by Kaner and Skobov (22). However, they only envisaged

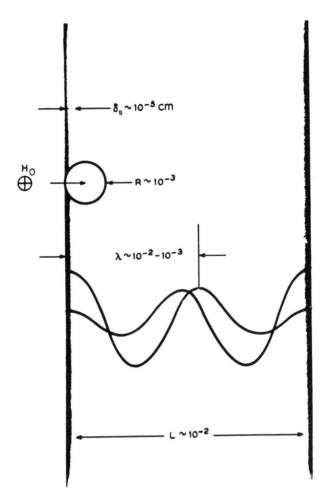

FIG. VIII-12. A schematic of the important physical parameters characterizing the CW experiments.

coupling to waves with wavelengths comparable to the anomalous skin depth δ_s, i.e., much smaller than the cyclotron orbit radii ($kR_c \gg 1$). While little theoretical distinction exists between the two limits $kR_c \leqslant 1$ and $kR_c > 1$, it is important to recognize that only the long-wavelength limit has proved to be experimentally accessible. We will discuss in some detail the behavior of CW in the physically interesting long-wavelength limit $kR_c \cong 1$ and refer only briefly to the large kR_c limit of these waves.

Although there has been some fragmentary work on CW propagation at arbitrary angles to the magnetic field the contact between theory and experiment is tenuous (23, 24). Theoretical analysis shows that for arbitrary angles of propagation, the presence of the ($kv_F \cos \Delta$) term gives rise to two distinctly different physical effects both of which could have a profound effect on the propagation characteristic of wavelike disturbances. The Doppler-like ($kv_F \cos \Delta$) term introduces finite regions of Landau damping, magnetic Landau damping, and Doppler-shifted cyclotron damping. Such collisionless damping will act to destroy propagation in otherwise "open" regions of the ω vs. k plane. On the other hand, this type of term could permit a phase reversal, i.e., change in sign of selected components of the conductivity tensor in certain regions of the ω vs. k plane. There is some evidence, both theoretical (24) and experimental (25), that this type of effect can lead to propagation in regions which, for $\mathbf{k} \perp \mathbf{H}_0$, would be cut off.

The algebraic complications associated with an analysis of oblique propagation are formidable, and to date no new physical phenomena have been revealed in an examination of the off-angle propagation data. Hence, we will only consider the case $\Delta = \pi/2$. Since several components of σ_{ij} are zero when $\Delta = \pi/2$ (see Section 45, p. 157), this restriction enormously simplifies the algebra. In addition, the form of the remaining components are simplified by the absence of the $kv_F \cos \Delta$ term in the denominator of the various terms in the conductivity sums.

The dispersion relation for waves propagating perpendicular to the magnetic field in a infinite metal system is given by

$$\begin{vmatrix} \epsilon_{xx} & \epsilon_{xy} & \epsilon_{xz} \\ \epsilon_{yx} & \epsilon_{yy} - k^2/k_0^2 & \epsilon_{yz} \\ \epsilon_{zx} & \epsilon_{zy} & \epsilon_{zz} - k^2/k_0^2 \end{vmatrix} = 0. \quad (48.1)$$

Since $\epsilon_{xz} = \epsilon_{zx} = \epsilon_{yz} = \epsilon_{zy} = 0$, Eq. (48.1) is equivalent to

$$(k^2/k_0^2 - \epsilon_{zz})(k^2/k_0^2 - (\epsilon_{yy} + \epsilon_{xy}^2/\epsilon_{xx})) = 0. \quad (48.2)$$

The root at

$$k^2/k_0^2 = \epsilon_{zz} \quad (48.3)$$

is called the ordinary wave. It is purely transverse in character, i.e., $\mathbf{E} \parallel \mathbf{z}$. The other root

$$k^2/k_0^2 = \epsilon_{yy} + \epsilon_{xy}^2/\epsilon_{xx} \quad (48.4)$$

is known as the extraordinary wave. It differs from the ordinary wave in that it is not purely transverse but can have a longitudinal component of electric field associated with it. In metals, however, the longitudinal component is small because the plasma frequency is extremely high relative to the applied frequency.

The ordinary and extraordinary modes have been studied in great detail in high-temperature gas plasmas where the physical parameters of the plasma are completely different from those of a metal (26, 27). We will discuss this well-known regime briefly and then go on to a more thorough discussion of CW in metallic plasmas. The experimental frequencies of excitations in gas plasmas are usually in the neighborhood of the plasma frequency. At densities of 10^{10} the plasma frequency, in turn, is of the order of the cyclotron frequencies at fields of the order of a few thousand gauss. On the other hand, in a metal, the plasma frequency is essentially infinite compared to ω_c. The approximate equality of all the characteristic frequencies in the gas plasma implies that the dielectric tensor is of the order of unity and hence that the displacement current (unlike the situation in metals) [see Eq. (40.13)] is nonnegligible.

In high-temperature gas plasmas the waves generally divide up into two distinct classes (27). There are those which are electromagnetic in character $\omega \sim ck$, i.e., propagate at a velocity near the velocity of light and those whose characteristic frequency is of the order of the plasma frequency propagating at typical particle velocities. This rather neat division is almost obvious from the dispersion relations [Eqs. (48.3), (48.4)]. The ordinary wave is clearly electromagnetic in character. When $\omega > \omega_p$, it has a velocity "near" the velocity of light. For this mode we can neglect the dispersion in $\epsilon_{zz}(\mathbf{k}, \omega)$. (Letting $k \to 0$ and $c \to \infty$ are equivalent.) In this long-wavelength limit the mode is characterized by a dispersion relation of the form

$$k^2/k_0^2 = 1 - \omega_p^2/\omega^2 \equiv \epsilon_{zz}(0, \omega). \tag{48.5}$$

There is another, almost transverse, mode satisfying Eq. (48.4), i.e.,

$$k^2/k_0^2 \simeq \epsilon_{yy}(0, \omega) + \epsilon_{xy}^2(0, \omega)/\epsilon_{xx}(0, \omega) \tag{48.6}$$

where the ϵ_{ij} are given by Eqs. (45.22)–(45.27).

The third mode which arises from an approximate solution of Eq. (48.4) is quite different. It is almost longitudinal; it has a phase velocity not very different from the thermal velocity of the particles and it approximately satisfies the dispersion relation

$$\epsilon_{xx} = 0. \tag{48.7}$$

There are several ways to arrive at Eq. (48.7). The simplest is to note

48. CYCLOTRON WAVE PROPAGATION

that the left-hand side of Eq. (48.4) is for slow waves essentially infinite

$$k^2/k_0^2 = c^2/(\omega/k)^2 \gg 1. \tag{48.8}$$

An infinity in Eq. (48.4) can only come from a zero in ϵ_{xx}; thus Eq. (48.7).

The solution of Eq. (48.7) for a Maxwellian or a degenerate plasma is given by (26)

$$\omega = [\omega_p^2 + \omega_c^2 + \mathcal{O}(k^2 v_{th}^2/\omega_c^2)]^{1/2} \tag{48.9}$$

$$\omega = 2\omega_c + \mathcal{O}(k^2 v_{th}^2/\omega_c^2) \tag{48.10}$$

$$\omega = 3\omega_c + \mathcal{O}(k^4 v_{th}^4/\omega_c^4), \quad \text{etc.} \tag{48.11}$$

The first mode in this group is called the upper hybrid, and the infinite set near the cyclotron harmonics are conventionally called the Bernstein modes (27). The longitudinal modes show up, much in the same manner as the CW do, as structure in the microwave absorption spectrum near the cyclotron harmonics in gas plasmas (28).

Because of the large difference in plasma frequency between a metal and a gas plasma, the upper hybrid and Bernstein modes are quite different in character from the CW. The Bernstein modes [Eqs. (48.10) and (48.11)] are almost pure longitudinal waves. The CW for both polarizations (ordinary and extraordinary) are almost transverse. There are *no* low-frequency longitudinal electromagnetic waves in metals.

Neglecting the displacement current the dispersion relation for the ordinary wave in a metal is simply

$$k^2/k_0^2 = 4\pi\sigma_{zz}(k, \omega)/i\omega \equiv -(\omega_p^2/\omega^2) F(kR_c, \omega_c/\omega, \omega_c\tau). \tag{48.12}$$

The explicit form for F may be obtained from Eq. (45.24). One of the important characteristics of F is that for $kR_c \sim 1$ it too is typically of order one. Although there are resonances present in the various terms, finite collision times (particularly in solids) prevent the resonance terms from getting much bigger than ten to a hundred. There is certainly no large number of the order of $(\omega_p/\omega)^2 \simeq 10^{10}$ present in F.

In the physically interesting regime where the CW are observed $\omega_p^2/\omega^2 \sim 10^{10}$ whereas $k^2/k_0^2 \sim 10^5$. Due to this state of affairs, i.e., high plasma frequency, the conduction electron response dominates all other considerations. Not only may we neglect the displacement current, but we may also, in calculating the dispersion relation, neglect k^2/k_0^2. We remind the reader that in the Bernstein mode regime k^2/k_0^2 was taken to be infinite. In this case, we assume it to be *zero*. For the metallic case,

Eqs. (48.3) and (48.4) may be accurately solved by seeking the zeros in the appropriate components of the conductivity tensor, i.e.,

$$\sigma_{zz} = 0 \quad (\mathbf{J} \parallel \mathbf{H}_0) \tag{48.13}$$

and

$$\sigma_{yy} + \sigma_{xy}^2/\sigma_{xx} = 0 \quad (\mathbf{J} \perp \mathbf{H}_0). \tag{48.14}$$

The physical content of these equations is that wave propagation, can occur only if for some combination of experimental frequency, wavelength in the medium, and magnetic-field value, the conduction current is essentially zero. In fact, of course, the current associated with a wave in a metal is not rigorously zero. However, the dispersion properties of the CW are determined (to order k^2c^2/ω_p^2) by assuming this to be true.

The apparently paradoxical requirement that the total conductivity vanish is only achievable because of the nonlocal nature of σ, i.e., its $\mathbf{k}$ dependence. This nonlocality implies that the total current at a point in the medium is made up of a local contribution, due to the electric field at that point, plus a nonlocal contribution arising from electrons arriving at that point with velocity increments acquired at other points and earlier times. It is only when cancellation between the local and nonlocal currents occurs that wave propagation becomes possible. This state of affairs is diagrammed in Fig. VIII-13.

It is not immediately obvious from such a diagram when a null in the current occurs. The transport equation solution for σ_{zz} [Eq. (45.24)]

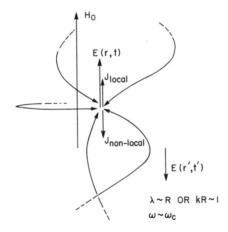

FIG. VIII-13. A schematic of the nonlocal contributions to the current induced by fields at nearby points in space.

48. CYCLOTRON WAVE PROPAGATION

will give us this information. It is, however, immediately clear that the null condition will depend critically on the relationship between (kR_c) and $n\omega_c - \omega$. As the electron traverses its circular orbit, perpendicular to the magnetic field its phase relationship relative to the electric field is determined by the space and time variation of the field across the orbit, which in turn depends on kR_c and ω_c/ω.

In order to examine this relationship more closely, consider an electron which starts (at $t = 0$) at the origin. At a later time t it will be found at a distance x (measured along $\mathbf{k}$) given by the expression

$$x(t) = R_c[\sin(\omega_c t + \phi) - \sin \phi]. \tag{48.15}$$

The perturbed momentum of the electron obeys the equation of motion

$$\dot{\mathbf{p}} - (e/c)(\mathbf{v} \times \mathbf{H}_0) = e\mathbf{E}[x(t)]. \tag{48.16}$$

In the limit $kR_c \ll 1$ we may expand $\mathbf{E}[x(t)]$ in a Taylor series about the point $x = 0$ to find

$$\mathbf{E}[x(t)] = \mathbf{E}(0)[1 + ikx(t) + \tfrac{1}{2}(ikx(t))^2 + \cdots]. \tag{48.17}$$

Thus, the field "seen" by the electron is modulated by the cyclotron motion of the carriers. The effective field due to this motion contains components at all frequencies $\omega' = \omega \pm n\omega_c$. The amplitude at ω' is determined by the power of k retained in the series expansion. For $\mathbf{E} \parallel \mathbf{H}_0$, Eq. (48.16) tells us that the current is resonant (infinite in the absence of τ^{-1}) when $\omega' = 0$.

The Taylor series expansion Eq. (48.17) implies that the field e_n with frequency $\omega \pm n\omega_c$ has an amplitude of order k^n. A priori then, we might expect an induced current of the same order (in the wave vector $\mathbf{k}$ of the field). However, the current is related to an average over all electrons in the gas. This averaging process is equivalent to averaging over the initial phase ϕ appearing in Eq. (48.14). For $\mathbf{E} \parallel \mathbf{H}_0$, the first component of the field at $\omega \pm n\omega_c$ which is nonvanishing, i.e., dc in the phase ϕ has the form

$$e_n \sim k^{2n} e^{in(\omega_c t + \phi)} e^{-in\phi} e^{-i\omega t}. \tag{48.18}$$

Correspondingly the current induced by this field has an amplitude given by

$$I_n \sim \frac{(kR_c)^{2n}}{\omega - n\omega_c} E(0). \tag{48.19}$$

To carry out this argument in detail, i.e., to obtain the proportionality constant in Eq. (48.19), it is necessary to properly average over all possible electron trajectories. The most convenient and unambiguous way to do this is to solve the linearized BV equation [Eq. (45.3)]. The arguments [first used by Nozières (29)] presented on the preceding few pages give us a nice qualitative way of understanding the origin and magnitude of the various resonant terms in the conductivity tensors.

Quantitatively the Bessel functions in Eqs. (45.22)–(45.25) may be expanded and the integrals over angle carried out in a straightforward manner. The results expressed as power series in $(kR_c)^2$ are (19)

$$\sigma^\circ_{zz} = iNa \sum_{n=1}^{\infty} \frac{(kR_c)^{2(n-1)}}{(2n-1)(2n+1)(a^2-1)(a^2-4) \cdots [a^2-(n-1)^2]} \quad (48.20)$$

$$\sigma^\circ_{yy} = \frac{iN}{a} \sum_{n=1}^{\infty} \frac{[a^2+2(n-1)n^2](kR_c)^{2(n-1)}}{(2n-1)(2n+1)(a^2-1)(a^2-4) \cdots (a^2-n^2)} \quad (48.21)$$

$$\sigma^\circ_{xx} = \frac{iN}{a} \sum_{n=1}^{\infty} \frac{(kR_c)^{2(n-1)}}{(2n+1)(a^2-1)(a^2-4) \cdots (a^2-n^2)} \quad (48.22)$$

$$\sigma^\circ_{xy} = -\sigma^\circ_{yx} = N \sum_{n=1}^{\infty} \frac{n(kR_c)^{2(n-1)}}{(2n+1)(a^2-1)(a^2-4) \cdots (a^2-n^2)}. \quad (48.23)$$

The resonances (infinities) in the components of the conductivity tensor occur at the cyclotron harmonics ($\omega = n\omega_c$) and have coefficients in the long-wavelength limit $kR_c \to 0$ proportional to $(kR_c)^{2n}$ in the case of σ°_{zz} and $(kR_c)^{(2n-1)}$ for the others.

Near the fundamental Azbel–Kaner resonance the long-wavelength limits of the dispersion relations Eqs. (48.13) and (48.14) are

$$\sigma^\circ_{zz} = \frac{in_0 e^2}{m^*\omega} \left[1 + \frac{(kR_c)^2}{5(a^2-1)} + \cdots\right] = 0 \quad (48.24)$$

$$\sigma^\circ_{yy} + \frac{(\sigma^\circ_{xy})^2}{\sigma^\circ_{xx}} = \frac{in_0 e^2}{m^*\omega(a^2-1)} \left[1 + \frac{1}{350} \frac{(kR_c)^4}{(a^2-1)} + \cdots\right]. \quad (48.25)$$

For $a^2 < 1$ ($\omega_c > \omega$), Eq. (48.24) has a solution *quadratic* in kR_c, where as Eq. (48.25) predicts a transverse "extraordinary wave" with a *quartic* dependence on kR_c for $\omega_c > \omega$. It is of interest to note that the coefficient 1/350 of the quartic term is remarkably small. One is tempted to conclude that the electron gas resists exhibiting an appreciable singularity at the fundamental cyclotron resonance, a state of affairs noted earlier by Smith, Hebel, and Buchsbaum (30) in a treatment of nearly local conductivity. This fact may be qualitatively understood as follows. The equation $j_x = 0$ implies that

$$E_x/E_y = -\sigma_{xy}^\circ/\sigma_{xx}^\circ. \tag{48.26}$$

To zeroth order in $(kR_c)^2$ Eq. (48.26), in turn, implies that

$$E_x/E_y = i\omega_c/\omega. \tag{48.27}$$

At cyclotron resonance ($\omega = \omega_c$) the field, to order $(kR_c)^2$, is circularly polarized with the sense of polarization *opposite* to that required for resonant interaction with electrons precessing in the static magnetic field. This is the physical reason for the vanishing $(kR_c)^2$ term in the dispersion relation for the extraordinary wave near the first harmonic.

The situation is quite different for the second subharmonic when $\mathbf{J} \perp \mathbf{H}_0$. Here

$$\sigma_{yy}^\circ + \frac{(\sigma_{xy}^\circ)^2}{\sigma_{xx}^\circ} \sim 1 + \frac{6}{5} \frac{(kR_c)^2}{(a^2 - 4)}, \tag{48.28}$$

has a singularity quadratic in kR_c analogous to the $\mathbf{J} \parallel \mathbf{H}_0$ case for $\omega_c \gg \omega$.

The dispersion relations are also relatively simple in the large k limit (23) as well. For large values of kR_c, we may use the asymptotic form of $J_n(kR_c \sin \theta)$ to evaluate the conductivity integrals. In this limit

$$J_m(u) \cong (2/\pi u)^{1/2} \cos(u - \pi/4 - m\pi/2) \tag{48.29}$$

so that, for example,

$$\sigma_{zz} \cong \frac{2}{\pi kR_c} iN \sum_{m=0}^{\infty} \int_0^\pi \frac{a}{(a^2 - m^2)(1 + \delta_{m0})} \cos^2(kR_c \sin \theta - \pi/4 - m\pi/2)$$
$$\times \cos^2 \theta \, d\theta. \tag{48.30}$$

The integrand in Eq. (48.30) can be rewritten as

$$\cos^2(kR_c \sin \theta - \pi/4 - m\pi/2) = \tfrac{1}{2}[1 + \cos(2kR_c \sin \theta - \pi/2 - m\pi/2)]. \tag{48.31}$$

To leading order in kR_c we may neglect the second term on the right-hand side of Eq. (48.31). It is a rapidly oscillating term which, the conventional method of stationary phase tells us, is small, i.e.,

$$\int_0^\pi d\theta \cos^2 \theta \cos(2kR_c \sin \theta - \pi/2 - m\pi/2) \sim \mathcal{O}(1/kR_c). \tag{48.32}$$

Thus in the limit $kR_c \gg 1$

$$\sigma_{zz} \cong \frac{iN}{2(kR_c)} \sum_{m=0}^{\infty} \frac{a}{(a^2 - m^2)} \frac{1}{(1 + \delta_{m0})}, \qquad (48.33)$$

and the dispersion relation, Eq. (48.13), becomes

$$\frac{k^2}{k_0^2} = \sum_{m=0}^{\infty} \frac{a}{a^2 - m^2} \left(\frac{2\pi N}{kR_c \omega} \right) \frac{1}{(1 + \delta_{m0})}. \qquad (48.34)$$

(Note that when k is large we must retain the k^2/k_0^2 term.) Near a specific resonance ($m = n$), Eq. (48.34) takes the form

$$\omega_n(k) = n\omega_c \left[1 - \frac{3}{4} \left(\frac{\omega_p}{\omega_c} \frac{v_F}{c} \right)^2 (kR_c)^{-3} \right] - \frac{i}{\tau}. \qquad (48.35)$$

To summarize our asymptotic results in the neighborhood of $\omega = \omega_c$ for $\mathbf{J} \parallel \mathbf{H}_0$, we combine Eq. (48.35) with Eq. (48.24). The long-wavelength behavior of the CW is given by

$$\omega = \omega_c [1 + \tfrac{1}{10}(kR_c)^2]. \qquad (48.36)$$

In the short-wavelength limit ($kR_c \to \infty$) Eq. (48.35) yields

$$\omega = \omega_c \left[1 - \frac{3}{4} \left(\frac{\omega_p}{\omega_c} \frac{v_F}{c} \right)^2 (kR_c)^{-3} \right]. \qquad (48.37)$$

In the intermediate regime ($kR_c \cong 1$) we must solve the dispersion relation, Eq. (48.13), numerically. The numerically computed solution of Eq. (48.3) in the region $\omega_c/\omega \geq 1$ with $\omega_p/\omega = 0.85 \times 10^5$ and $v_F/c = 3 \times 10^{-3}$ is shown in Fig. VIII-14 (19). The parameters are typical of microwave resonance in Na and K. A similar computation using the approximation of Eq. (48.36) does not differ appreciably from this curve until kR ($R \equiv v_F/\omega$) values of approximately two are reached. For $kR \cong 6$, the dispersion curve oscillates, a behavior which is characteristic of a sharp Fermi distribution. The oscillations in the attenuation of sound waves in pure metals as a function of a swept magnetic field (magnetoacoustic effect) have the same origins (2). At still higher values of kR ($kR \cong 10^2$, see Fig. VIII-15), the dispersion curve monotonically approaches the $\omega_c/\omega = 1$ axis. In this regime the dispersion relation is well approximated by the asymptotic result given by Eq. (48.37). Note that over the experimentally accessible region ($0 < kR \cong 6$) the dispersion curve is single valued for $\omega_c/\omega < 1.77$, whereas in the interval $1.77 < \omega_c/\omega < 1.95$ it becomes multiple valued, i.e., for a given magnetic field several waves of distinct k may be excited.

48. CYCLOTRON WAVE PROPAGATION

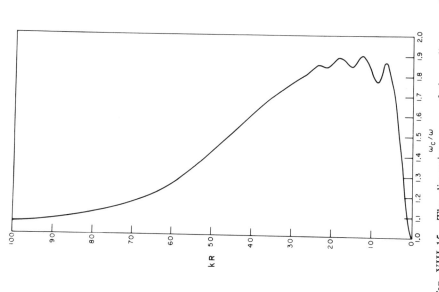

Fig. VIII-15. The dispersion curve of the ordinary mode ($J \parallel H_0$) CW associated with the fundamental cyclotron resonance of a noninteracting electron gas. $\omega_p/\omega = 0.85 \times 10^5$; $v_F/c = 3 \times 10^{-3}$.

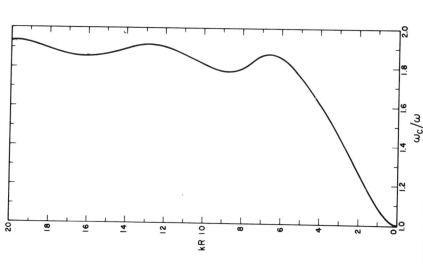

Fig. VIII-14. The dispersion curve of the ordinary mode ($J \parallel H_0$) CW associated with the fundamental cyclotron resonances of a noninteracting electron gas. $\omega_p/\omega = 0.85 \times 10^5$; $v_F/c = 3 \times 10^{-3}$.

180   VIII. NONLOCAL EFFECTS IN A NONINTERACTING GAS

In order to compare the theoretical results with the experimental data obtained in standing-wave type of experiments, it is assumed that the successive oscillations in the experimentally observed dA/dH_0 correspond to inclusion within the sample of successive full wavelengths of the propagating mode. In the case of antisymmetric excitation standing-wave resonances would be expected (16) to occur when

$$(n + \tfrac{1}{2})\lambda = L \qquad n = 0, 1, 2,\dots . \qquad (48.38)$$

Equation (48.38) implies that the current due to the propagating waves has the correct phase to interfere constructively with the field on both

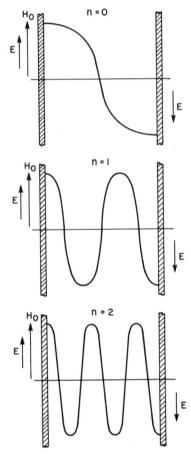

Fig. VIII-16. A schematic of the standing wave modes which are resonant in the antisymmetric two sided excitation experiment. The skin regions are shown as the narrow cross-hatched strips.

48. CYCLOTRON WAVE PROPAGATION

sides of the sample. The state of affairs, i.e., the phase relations for the first few modes, is shown in Fig. VIII-16.

Initially it was believed that the spatial resonances (diagrammed in Fig. VIII-16) would produce a net increase in the power absorption of the sample due to the existence of maximum energy dissipation in the bulk. This is indeed the case for a wave such as the helicon for which there is only a small skin-effect. However, this is not true of the nonlocal mode under discussion. A strong skin-effect exists, i.e., only a small part of the total electromagnetic energy entering the metal is coupled into the observed propagating mode. Under these circumstances the increase in bulk dissipation occurring when Eq. (48.38) is fulfilled is dominated by a decrease in surface dissipation due to transmitted currents aiding the driving currents in shielding the metal. The argument may be seen as analogous to the reduction in dissipation that occurs in the Azbel–Kaner resonance itself where the surface currents act to reduce the overall loss at resonance despite increased dissipation beneath the zero-field skin depth. This argument qualitatively explains why the somewhat sharpened oscillations seen in Fig. VIII-8 near the resonance empirically correspond to local *minima* in the total power absorption. For the most part, however, the oscillations are quite sinusoidal and one may better assign the relative extrema of dA/dH to

$$kL = 2\pi(n + \tfrac{1}{2} \pm \tfrac{1}{4}), \qquad n = 0, 1, 2,\ldots . \tag{48.39}$$

Here the plus and minus signs correspond, respectively, to derivative maxima and minima.

The arguments leading to the $\tfrac{1}{2}$ and $\tfrac{1}{4}$ in Eqs. (48.38) and (48.39) are somewhat qualitative. On the other hand, the fact that successive maxima or minima in the surface impedance correspond to inclusion of an additional full wavelength is quite clear. Ambiguity as to the acutal wavelength of the mode, i.e., the assignment of n and the inclusion of the $\tfrac{1}{2}$ or $\tfrac{1}{4}$ may arise in any given experimental curve. If a large number of curves for a given sample are obtained, e.g., by varying the frequency continuously, this ambiguity is removed by simply requiring that all the curves lie on the same *universal* ω_c/ω vs. kR curve.

Measuring the sample thickness L, assigning the derivative extrema of the upper trace in Fig. VIII-8 according to Eq. (48.39) and finally using the Grimes–Kip value of $m^*/m_0 = 1.21$ for K leads to the experimental points shown in Fig. VIII-17. The solid curve is simply the solution of Eq. (48.13) ($\omega_c\tau \to \infty$) already plotted in Fig. VIII-14. It is important to keep in mind the fact that $\omega_c\tau \cong 30$ in most of these specimens.

The comparison of theory and experiment in Fig. VIII-17 is quite

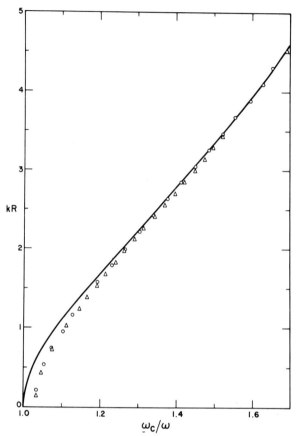

Fig. VIII-17. Comparison of the observed (○ $\omega = 7.54 \times 10^{10}$ and △ $\omega = 1.09 \times 10^{11}$) and predicted (solid line) dispersion of the CW (**J** ∥ **H**$_0$) near the first harmonic of the cyclotron frequency ($\omega = \omega_c$) in K.

good, though a small discrepancy exists at the $kR \to 0$ intercept. Here the experimental points tend to intercept at somewhat higher (approximately 2 %) magnetic field values than expected. This deviation is a real effect and will be discussed in detail later. At higher kR values the general features of the theoretical dispersion appear to be reproduced.

A large number of reflection experiments have failed to reveal clear evidence of CW waves for ω_c/ω appreciably greater than 1.7. This "washing out" is probably due to excitation of several distinct k vectors, since the dispersion curve becomes multiple valued beyond $\omega_c/\omega \simeq 1.7$. Multiple excitation would lead to interference between the waves and a severe weakening of structure in the total power absorption. However, a weak, but well-defined "break" in the experimental trace in Fig. VIII-10 occurs for $\omega_c/\omega \simeq 1.9$. Presumably this reflects the "cutting off" of the metal as the limit of the propagation region is reached.

For finite $\omega_c\tau$, propagation is permissible well into the infinite $\omega_c\tau$ cutoff region. On the one hand the two-sided reflection experiments of Walsh (21) seem to be very insensitive to collisions. The observed dispersion fits the $\omega_c\tau \to \infty$ theory quite accurately and the infinite $\omega_c\tau$ cutoff in propagation is clearly in evidence. On the other hand, the transmission experiments are apparently quite sensitive to collisional effects. There is definite propagation beyond $\omega_c/\omega = 1.9$, and Schultz (25) and co-workers claim that they must include finite $\omega_c\tau$ effects in the

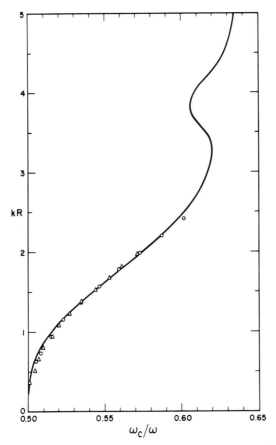

Fig. VIII-18. Comparison of the observed ($\bigcirc$ $\omega = 7.54 \times 10^{10}$ and $\triangle$ $\omega = 1.09 \times 10^{11}$ Hz) and predicted (solid line) dispersion of the CW (J $\parallel$ $\mathbf{H}_0$) near the second harmonic of the cyclotron frequency ($\omega = 2\omega_c$) in K.

analysis of their transmission experiments to obtain good agreement with the theory.

The only firm conclusion that one can come to is to observe that the data must remain consistent with themselves as one goes to higher and higher frequencies, i.e., increasing $\omega_c\tau$. Walsh's data are taken at two frequencies for different specimens (i.e., different $\omega_c\tau$). "The center of the standing-wave resonance" gives an accurate measure of the infinite $\omega_c\tau$ dispersion relation, and there seems to be no need to invoke finite $\omega_c\tau$ effects.

Qualitatively similar evidence of CW propagation is also found on the high-field side of the $\omega = 2\omega_c$ resonance in the $\mathbf{J} \parallel \mathbf{H}_0$ geometry. A comparison of the observed dispersion for K and that computed from Eq. (48.13) near the second cyclotron harmonic ($\omega = 2\omega_c$) is shown in Fig. VIII-18. The overall agreement is very satisfactory. In particular there is *no* discrepancy in the $kR \to 0$ intercept in contrast to that found near the fundamental resonance. It is worth pointing out that if one fixes up the intercept on the first harmonic by shifting the mass m^*/m_0 slightly, then one would seriously affect the agreement between theory and experiment on the second harmonic. The discrepancy in the intercept near $\omega/\omega_c = 1$ and the lack of one near $\omega/\omega_c = 2$ has been attributed to electron–electron interaction effects. This point will be discussed in some detail when we consider the interacting Fermi liquid (Chapter X).

High-frequency wave propagation has been studied in the "extraordinary" or $\mathbf{J} \perp \mathbf{H}_0$ geometry as well (21, 31). A typical experimental trace showing the derivative of the surface impedance of a thin slab of K as a function of magnetic field is shown in Fig. VIII-19. (See Fig. VIII-7 for the experimental configuration.) The structure due to the excitation of waves near the Azbel–Kaner resonance is clearly evident, although the pattern is somewhat more complex than a similar trace (see Figs. VIII-8 and VIII-10) taken in the $\mathbf{J} \parallel \mathbf{H}_0$ geometry.

The simple theory, i.e., the numerically computed solution of Eq. (48.14) is shown in Fig. VIII-20 (32). There are several interesting features of the computed dispersion curves:

(1) The *single* mode which propagates on the high-field side of the first Azbel–Kaner subharmonic exhibits its weak $(kR_c)^4$ dependence [Eq. (48.25)]. At an $\omega_c/\omega \simeq 1.4$ the dispersion curve becomes multivalued and highly oscillatory. The oscillations in the dispersion relation are significantly more pronounced than in the $\mathbf{J} \parallel \mathbf{H}_0$ geometry (see Fig. VIII-15). Propagation above the first harmonic ceases at $\omega_c/\omega \simeq 3.0$.

(2) There are two distinct modes propagating near each subharmonic.

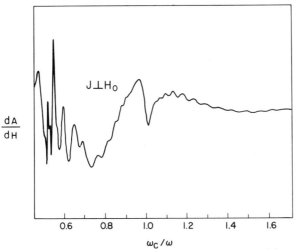

FIG. VIII-19. Field derivative of the microwave power absorbed by a slab of pure potassium metal versus magnetic field. The experiment was performed in the $\mathbf{J} \perp \mathbf{H_0}$ polarization. $d = 0.150$ mm; $T = 1.4°$K, $\omega = 1.08 \times 10^{11}$ Hz.

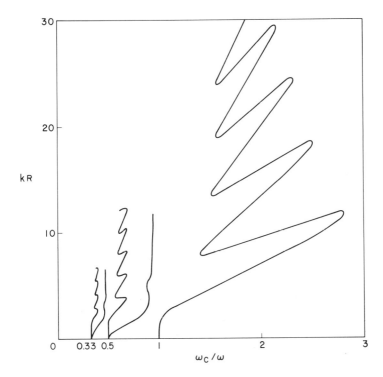

FIG. VIII-20. Theoretical dispersion curves for extraordinary wave propagation in metals, $\omega_p/\omega = 0.85 \times 10^5$; $v_F/c = 3 \times 10^{-3}$.

An expanded trace showing the experimental data for $\omega_c/\omega \geq 1$ is shown in Fig. VIII-21. The experiments accurately confirm the

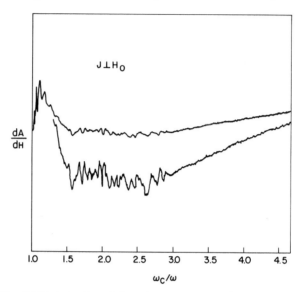

FIG. VIII-21. Field derivative of the microwave power absorbed by a slab of pure potassium metal versus magnetic field. The experiment was performed in the $\mathbf{J} \perp \mathbf{H}_0$ polarization. $d = 0.16$ mm; $T = 1.4°\text{K}$; $\omega = 7.35 \times 10^{10}$ Hz.

predictions of this simple analysis. Above the fundamental resonance a single wave is observed from $\omega_c/\omega = 1.0$ to 1.4, followed by a region of complex beat patterns extending to a well-defined cutoff at $\omega_c/\omega = 2.9$.

Near the second harmonic a careful analysis of the pattern shows that there is in fact a simple beat between two distinct waves. The most dispersive and strongly coupled of these waves corresponds to the long-wavelength (i.e., lowest) branch of the dispersion curves shown in Fig. VIII-20.

The accumulated data for the $\mathbf{J} \perp \mathbf{H}_0$ geometry are shown in Fig. VIII-22. The agreement with the simple theory is very good; however, there is a small disagreement at the intercept of the branch near $\omega = 2\omega_c$. We will see that this discrepancy is real, predicted by theory, and consistent with the intercept error for the $\mathbf{J} \parallel \mathbf{H}_0$ case.

That the phenomenon of CW propagation near AKCR is not uniquely confined to K is demonstrated in Fig. VIII-23 where full AKCR traces of a thin Na slab are shown for both $\mathbf{J} \parallel \mathbf{H}_0$ and $\mathbf{J} \perp \mathbf{H}_0$ polarizations (21). Essentially all the signals on these traces result from wave propagation. The single-particle AKCR signals are extremely weak in Na due to the

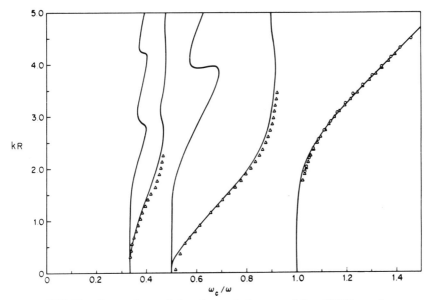

FIG. VIII-22. Comparison of the observed (○ $\omega = 7.5 \times 10^{10}$ Hz and △ $\omega = 1.09 \times 10^{11}$ Hz) and predicted (solid line) dispersion of the extraordinary ($\mathbf{J} \perp \mathbf{H}_0$) CW in K.

fact that the low-temperature martensitic phase transformation ($bcc \to hcp$) creates a large number of small-angle scattering centers. Such scattering hardly affects the dc resistivity ratio, but is disastrous for phenomena such as AKCR where individual electron paths must be very precisely (to order $\delta/R_c \cong 10^{-2}$) defined. The true collective nature of the plasma wave, however, apparently makes it rather insensitive to such individual particle straggling. As we have seen in Section 41, the long-wavelength helicon mode is also unaffected by small-angle scattering in Na.

In Fig. VIII-24 we show the rather meager Na data plotted for the $\mathbf{J} \parallel \mathbf{H}_0$ geometry near the first and second cyclotron subharmonics. The data are not as extensive nor is the agreement as good as in the K case. The fact that the intercept near the first harmonic is not precisely at $\omega_c/\omega = 1$ is real, i.e., similar to the situation in K. It is due to the breakdown of the one-electron model.

Using the transmission technique, Schultz and co-workers have seen clear evidence for CW propagation in an array of metals including Na, K, Rb, Cs, Cu, Al, and Ag (25). A typical transmission trace for Na was shown in Fig. VIII-11. The phase oscillations of the signal gives us, as in the helicon case, the dispersion properties of the waves. The envelope

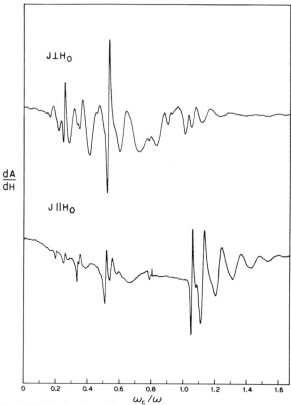

FIG. VIII-23. Field derivative of the microwave power absorbed by a slab of pure Na metal versus magnetic field. $d = 0.08$ mm; $T = 1.4°$K; $\omega = 7.5 \times 10^{10}$ Hz.

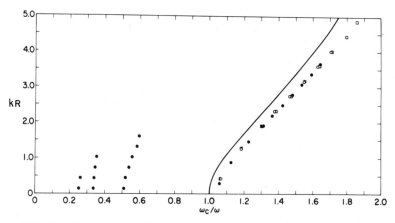

FIG. VIII-24. Comparison of the observed (○ $\omega = 7.50 \times 10^{10}$ Hz, ● $\omega = 1.08 \times 10^{11}$ Hz, and □ $\omega = 7.51 \times 10^{10}$ Hz) and predicted (solid line) dispersion curve in Na ($\mathbf{J} \parallel \mathbf{H_0}$) for extraordinary wave dispersion.

is related to finite $\omega_c\tau$ or damping effects and ultimately determined by the coupling of the waves to the skin-effect field.

A typical trace for a single crystal of Cu is shown in Fig. VIII-25. The ⟨110⟩ plane of Cu is coincidental with the face of the specimen. The

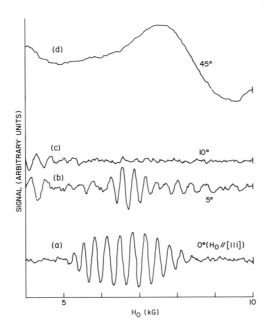

FIG. VIII-25. The field transmitted through a ⟨110⟩ slab of single crystal Cu. $d = 0.4$ mm; $\omega = 5.71 \times 10^{10}$ Hz.

patterns in Fig. VIII-25 are much more complicated than those exhibited by the alkalis for which the Fermi surface is a sphere. However, when the field $\mathbf{H}_0$ points along a ⟨111⟩ axis, the transmission spectrum is greatly simplified and similar to that obtained in Na (see Fig. VIII-20).

There has been no quantitative computation of the properties of CW in materials with anisotropic Fermi surfaces, although some calculations on ellipsoidal Fermi surfaces have been made (32). For extremely complicated Fermi surfaces, we would a priori expect a washing out of the "resonance" in the dielectric constant. Crudely speaking, ω_c takes on a continuum of values. Critical points or regions on the Fermi surface where many electrons have the same ω_c can, however, lead to anomalies in the dielectric constant and could result in propagation as in the Cu case. This seems to be the situation for the trace taken in Fig. VIII-25. For the particular orientation of the magnetic field

($\mathbf{H}_0 \parallel \langle 111 \rangle$), many of the electronic orbits are on closed spherical portions of the Fermi surface. This is the reason for the similarity in the Na and Cu traces. The effective mass of the carriers on this spherical portion of the Fermi surface is known from AKCR experiments (33) to be about $1.4m_0$. This value corresponds to a cyclotron resonance field $H_c \cong 5 \times 10^3$ G at an $\omega = 5.71 \times 10^{10}$ Hz. It is clear from the experimental trace that 5×10^3 G is just about where propagation begins. Calculations on more complicated Fermi surfaces would be extremely useful since they could enormously extend the range of materials in which *useful* CW propagation has been observed.

This essentially concludes our discussion of helicon and CW propagation in an infinite medium containing a high density of noninteracting electrons. Figure VIII-26 summarizes in a qualitative fashion the kind

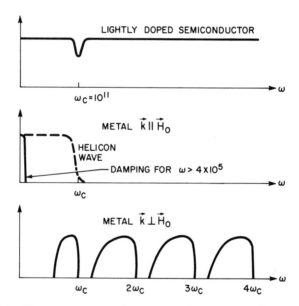

FIG. VIII-26. The power transmission spectrum for an opaque (!!) metal ($H_0 = 5$ kG).

of power transmission spectrum (neglecting magnetism) expected for a macroscopic slab of alkali metal placed in a static field $\mathbf{H}_0$. A schematic of the analogous spectrum for a semiconductor containing a few free carriers is shown for comparison purposes.

"Classical" theories describing the response of metals to microwave radiation demand that a slab of metal exclude all electromagnetic fields whose frequencies lie below the plasma frequency. Our discussion in the last two chapters now makes it clear that this opaque slab of pure metal is, in the presence of a sizable dc magnetic field, quite transparent to a rather broad spectrum of low-frequency ($\omega \ll \omega_p$) radiation. The dispersion properties of these bulk waves, particularly the CW, depend on the detailed dynamics of the conduction electron medium in which they propagate. In the noninteracting picture the kinetics of single particles are the essential ingredients. In the interacting picture we will show (Chapter X) how the potential energy of interaction between electrons begins to play a role and how the properties of CW and the yet to be discussed spin waves are modified by these couplings.

49. The "Problem" of Boundaries

Before going on to a discussion of the weak magnetism of the conduction electrons, we will briefly discuss the question of boundaries. Such a discussion is somewhat outside of the main content of this book. However, some qualitative understanding of the role boundaries play in determining the actual transmission and reflection properties of a slab of metal is useful for a complete understanding of the electromagnetic properties of metals.

In principle one can solve the boundary value problem and compute the reflection or transmission spectrum for the simple metals. To carry this program out we would need, in addition to the BV equation and Maxwell's equation, a model for the surface. More precisely we need to know the character of the scattering suffered by the electrons as they strike the surface of the specimen. In practice, even with ideal surfaces the exact solution of the boundary value problem is either exceedingly difficult or impossible. Although solutions have been given in a few special (*34–36*) cases, one is still left with the nagging thought that the surface is not ideal and that a more realistic picture of the scattering could lead to quantitatively different results (*37*).

We do not intend to go into any of the details associated with a formal solution of the boundary value problem. The interested reader is referred to references (*34*) to (*37*). However, it is fairly easy to write down a solution to the boundary value problem in one very special case (*38*). This solution, which we will generate, is rigorous when the geometry is planar, i.e., the sample is a slab and when the field $\mathbf{H}_0$ is perpendicular to the slab (see Fig. VIII-5). It contains most of the physically interesting phenomena associated with the boundary.

The analysis of this particular configuration completely ignores one rather interesting and physically relevant effect which is worth mentioning. The existence of surface "particlelike" excitations are completely excluded in any analysis of the field "normal" geometry. If the field were parallel to the surface of the metal, then there would be a small group of electrons which could skip (if the surface is specular) along the surface. The characteristic frequency of this motion is not the cyclotron frequency. It has been shown that this particular type of boundary-related skipping leads to a series of low-field resonances in the surface impedance of the slab which are useful for studying the particular properties of Fermi surfaces and the nature of the surface itself (39).

Keeping in mind these shortcomings we consider a one-component electron gas uniformly filling the space bounded by the planes $z = 0$ and $z = L$. The electron gas is immersed in a uniform static magnetic field $\mathbf{H_0}$ oriented in the positive z direction. We wish to compute the reflection and penetration properties of a transverse, circularly polarized wave normally incident from the left.

As a result of the random velocity of the electrons, the plasma conductivity, as we know, is a nonlocal integral operator; that is to say, a field with a β component at some point $\mathbf{r}'$ in the plasma produces a current with an α component at another point $\mathbf{r}$ in the plasma with magnitude $\sigma_{\alpha\beta}(\mathbf{r}, \mathbf{r}', \omega) \, E_\beta(\mathbf{r}')$, i.e.,

$$j_\alpha(\mathbf{r}, \omega) = \int \sigma_{\alpha\beta}(\mathbf{r}, \mathbf{r}', \omega) \, E_\beta(\mathbf{r}', \omega) \, d^3r'. \tag{49.1}$$

The fact that the relation between current and field is an integral one greatly complicates the solution of the boundary value problem. In the presence of boundaries $\sigma_{\alpha\beta}(\mathbf{r}, \mathbf{r}', \omega)$, in general, is a function of the vectors $\mathbf{r}, \mathbf{r}'$ separately, and the integral in Eq. (49.1) extends only over a bounded region. The exact functional dependence of $\sigma_{\alpha\beta}$ will be determined by the geometry of the medium, the nature of the electron trajectories on impact with the boundary, and on the configuration of the external fields. Under such conditions Fourier transform methods do *not* simplify the analysis. However, if the electrons are specularly reflected from the bounding surfaces and if the magnetic field is normal to the boundary, then, and only then, it is possible to show that the finite-medium problem is equivalent to an infinite-medium excited by a periodic array (with period $2L$) of current sheets (see Fig. VIII-27). In this "equivalent infinite-medium problem" the electric field is even and the rf magnetic field odd about the planes $z = 0, L, 2L,...$.

Once the equivalence between the two problems is established, the boundary value problem is easily solved. The equivalent infinite medium is translationally invariant. In it the conductivity $\sigma_{\alpha\beta}(\mathbf{r}, \mathbf{r}', \omega)$

49. THE "PROBLEM" OF BOUNDARIES

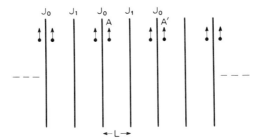

FIG. VIII-27. The equivalent infinite medium with current sheets J_0 at the planes $(2n)L$ and current sheets J_1 at the planes $(2n+1)L$.

depends only on the difference of vectors **r** and **r**′ and the integral in Eq. (49.1) becomes an algebraic relation in Fourier space, i.e.,

$$j_\alpha(\mathbf{k}, \omega) = \sigma_{\alpha\beta}(\mathbf{k}, \omega) E_\beta(\mathbf{k}). \tag{49.2}$$

The solution of the boundary value problem is uniquely specified by the values of the rf magnetic field at the bounding planes $z = 0$ and $z = L$ and by the wave equation satisfied by the electric field in the region $0 < z < L$. The strength of the current sheets at the planes $z = 0$ and $z = L$ in the equivalent infinite medium are fixed to give the correct values of the rf magnetic field for the finite-medium problem. Since it is sufficient to express all results in terms of the unknown strength of the rf magnetic field at the plane $z = 0$ (a simple scaling factor), we need only a single condition on the fields to determine the strength of the current sheet at the plane $z = L$. This condition is supplied by the requirement that there exist only an outgoing wave at this surface. This fixes the ratio $E(L_-)/H(L_-)$ in the boundary-value problem to be equal to one. The subscript on L indicates that the ratio $E(z)/H(z)$ is to be evaluated as $z \to L$ from below.

In order to establish the equivalence of the wave equations in the regions $0 < z < L$, we must show that the currents induced in the medium by the equivalent fields are identical. This is easily achieved by considering an applied delta-function at a point A (Fig. VIII-28) in the original boundary value problem. The equivalent field (the field which is equal to this localized field on the interval $0 < z < L$) in the equivalent infinite-medium problem is a symmetric periodic array of delta-functions (see Fig. VIII-27).

The electric field perturbs the distribution function and creates a local current that is transported to another point in the medium by electrons moving in the static magnetic field $\mathbf{H}_0$. To see if the current

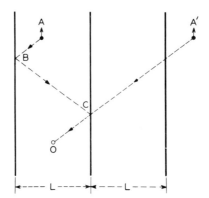

FIG. VIII-28. The equivalence of the path $ABCO$ in the finite medium case with the path $A'CO$ in the fictitious but equivalent infinite medium case.

at a point O (Fig. VIII-28) is the same in the two problems we must examine the electron paths which pass through the point O.

When the static magnetic field is perpendicular to the slab and specular reflection exists, there is a one-to-one correspondence between reflected electron trajectories in the finite medium and direct trajectories from the repeated fields in the periodic infinite medium problem. In the finite-medium problem electrons with velocity **v** which start out at A and follow the path $ABCO$ contribute to the current an amount equal to those electrons which start out with velocity **v** at the point A' in the infinite medium (see Fig. VIII-28). Thus, the equivalence between the currents is established for this special geometry.

The rf magnetic fields in the equivalent infinite medium are equal, by construction, to those in the finite medium. The electric field satisfies the same wave equation for the region $0 < z < L$, since the current induced in both media for equivalent fields are identical. We must then conclude from the uniqueness of the problem that the solutions to the two problems are identical for $0 < z < L$.

For the equivalent infinite medium, and for a right-handed, circularly polarized wave $[E \equiv E_x + iE_y]$, we must solve the scalar "Helmholtz-like" equation

$$\frac{\partial^2 E(z)}{\partial z^2} + k_0^2 E(z) - \frac{4\pi i \omega}{c^2} \int_{-\infty}^{+\infty} \sigma_+(z-z',\omega) E(z') \, dz = ik_0 \frac{4\pi J_s(z)}{c},$$
(49.3)

where

$$4\pi J_s(z)/c = \pi^{-1} \sum_{n=-\infty}^{+\infty} [H(0_+) \delta(z - 2nL) + H(L_+) \delta[z - (2n+1)L]].$$
(49.4)

49. THE "PROBLEM" OF BOUNDARIES

The field $E(z)$ and the current $J_s(z)$ are conveniently represented by infinite Fourier series:

$$E(z) = \sum_{n=0}^{+\infty} a_n \cos k_n z, \qquad (49.5)$$

$$J_s(z) = \sum_{n=0}^{+\infty} b_n \cos k_n z, \qquad (49.6)$$

where

$$k_n = n\pi/L \qquad (49.7)$$

and

$$b_n = [H(0_+) + H(L_+) \cos k_n L]/L. \qquad (49.8)$$

$H(0_+)$ and $H(L_+)$ are the magnitudes of the rf magnetic fields at the planes $z = 0_+$ and $z = L_+$, respectively. Substituting Eqs. (49.4) into (49.3) and solving for the coefficient a_n, we find that

$$a_n = -\frac{i}{L} k_0 [H(0_+) + H(L_+) \cos k_n L] \frac{[2 - \delta_{n0}]}{k_n{}^2 - k_0{}^2 \epsilon_+(k_n, \omega)}, \qquad (49.9)$$

where $k_0 = \omega/c$ with δ_{n0} the Kronecker delta. The quantity $\epsilon_+(k, \omega)$ is the relative dielectric constant of the medium, i.e., it is given by Eq. (46.1).

The outgoing-wave boundary condition $E(L_-)/H(L_-) = 1$ permits us to express the field $E(z)$, Eq. (49.5) in terms of $H(0_+)$ alone. That is to say,

$$E(z) = H(0_+) \left\{ f(z, L) + \frac{1}{2} \frac{H(L_+)}{H(0_+)} [f(L - z, L) + f(L + z, L)] \right\}, \qquad (49.10)$$

where

$$f(z, L) = \frac{-2ik_0}{L} \sum_{n=0}^{\infty} \frac{[2 - \delta_{n0}] \cos k_n z}{k_n{}^2 - k_0{}^2 \epsilon_+(k_n, \omega)} \qquad (49.11)$$

and

$$H(L_+)/H(0_+) = -f(L, L)/[1 + f(0, L)]. \qquad (49.12)$$

The form of Eq. (49.10) is particularly revealing; $E(z)$ is generated by three terms. The first arises from the current at the plane $z = 0$; the other two terms result from the symmetrically placed image currents at the planes $\pm L$. In the limit $L \to \infty$, the last two terms vanish (i.e., the images recede to infinity). The summation over n then goes over into an integral

$$L^{-1} \sum_{n=0}^{\infty} \to \pi^{-1} \int_0^{\infty} dk \qquad (49.13)$$

and the expression for $E(z)$ becomes

$$E(z) = -\pi^{-1} H(0_+) k_0 \int_{-\infty}^{+\infty} \frac{e^{ikz}\, dk}{k^2 - k_0^2 \epsilon_+(k, \omega)}. \qquad (49.14)$$

Equations (49.10)–(49.14) are exact solutions of the boundary value problem when the field is normal to the surface, the Fermi surface spherical and the scattering specular. These are highly restrictive conditions which are seldom achieved in real experimental situations. However, Eqs. (49.10)–(49.14) seem to give qualitatively (and in some cases quantitatively) correct results even when these conditions are violated. We can understand this fact by means of a few simple arguments.

Equations (49.10)–(49.14) are equivalent to an infinite-medium problem excited by a fixed set of current sheets. The infinite-medium problem is, of course, independent of boundary scattering; however, in a real boundary value problem most of the field is indeed concentrated in a narrow-skin region very much like the delta function exciting currents in the infinite-medium problem. We expect, or at least it is plausible to think, that the fields deep within the specimen do not critically depend on the details of what goes on in the skin region. Thus any model with a localized excitation, e.g., a current sheet model, should give a reasonably accurate description of the long-range properties of the field.

With this point in mind we will go on to discuss Eq. (49.14) and see what it predicts about the fields in the medium. For the field normal geometry, it gives us back our helicon modes and predicts a whole set of single-particle-like excitations, (the so-called Gantmacher–Kaner oscillations) (*40, 41*) which have been observed. We will then compare the detailed predictions of Eqs. (49.10)–(49.14) with a more realistic calculation and convince ourselves that at least in this case these equations do indeed give us a reasonably accurate picture of the problem.

The denominator in the integrand in Eq. (49.14) is the Fourier transform of the wave equation [see Eq. (41.2)]. The behavior of the field $E(z)$ is determined by the singularities present in $\epsilon_+(k, \omega)$. The poles of the integrand are simply the zeros of the denominator, i.e., they are determined by the dispersion relation for the normal modes [Eq. (42.15)],

$$k^2/k_0^2 = \epsilon_+(k, \omega). \qquad (49.15)$$

These normal mode singularities give contributions to the field which are of the form

$$E_m(z) = a_m \exp(ik_m z). \qquad (49.16)$$

The a_m are the residues at the poles k_m. The position in the complex k plane of the root k_m is determined by the field, the frequency, and by the parameters characterizing the medium; e.g., τ. When modes exist, i.e., there are roots k_m which are almost real then the fields E_m have a very long range and they dominate the large z behavior of the fields. In the propagating region, then, the field at large z, or equivalently the transmission coefficient, is determined by the poles k_m and their residues a_m. Using what is essentially Eq. (49.10) several people have computed the transmission amplitude for finite slabs of metal. Figure VIII-29

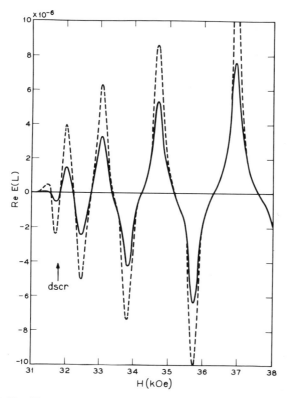

FIG. VIII-29. Transmission amplitude for diffuse (solid) and specular (dashed) boundary conditions. $f = 50$ MHz ; $L = 0.023$ cm; $l = 0.030$ cm.

shows such a plot made for a slab of thickness L by Baraff (42). The vertical axis is the real part of the electric field at L. The dotted curve assumes purely specular reflection, i.e., Eq. (49.10) and the solid curve a slightly more complicated diffuse boundary condition. The general features are quite similar. The oscillations are simply the oscillations in phase of the wave vector **k** associated with the propagating helicon mode. The results are as we have seen in quite good agreement with experiment (see Fig. VIII-3).

The other type of singularity present in the integrand of Eq. (49.10) is the logarithmic singularity with a branch point at

$$k_E = (\omega - \omega_c + i/\tau)/v_F. \tag{49.17}$$

Physically, k_E corresponds to a wave vector such that those electrons out at the tip of the Fermi surface are in Doppler-shifted cyclotron resonance (see p. 158) with the disturbance at wave vector k. It corresponds to the "onset" of single-particle-like excitations.

Schematically Fig. VIII-30 shows the singularities present in the integrand of Eq. (49.14) when $\omega_c \tau \to \infty$ and for a value of magnetic

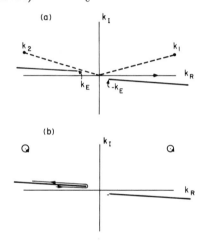

Fig. VIII-30. The singularities of the integrand in Eq. (49.14) for values of the experimental parameters where there is no helicon propagation.

field such that there are no real roots of Eq. (49.15). The two poles on the physical sheet which are present are labeled by $k_{1,2}$. The position of these poles is approximately given by the roots of Eq. (49.15) for $z \ll 1$, [see Eqs. (46.1), (46.2)], i.e.,

$$k_{1,2} = \exp(i\theta_{1,2})(3\pi/4)(\omega_p^2 \omega/c^2 v_F)^{1/3} \tag{49.18}$$

with $\theta_1 = \pi/6$ and $\theta_2 = 5\pi/6$].

The path of integration for Eq. (49.14) is along the real k axis. This contour may be distorted as shown in Fig. VIII-30. On the one hand, contributions from the two poles at $k_{1,2}$ give rise to the anomalous-skin-

49. THE "PROBLEM" OF BOUNDARIES

effect field. On the other hand, the asymptotic behavior of the field is determined by the singularity nearest the real k axis. In the non-propagating regime the closest real singularity is the end of the branch cut. It dominates the tail of the anomalous skin effect.

Utilizing Eq. (49.14), deforming the contour around the branch cut as shown in Fig. VIII-30, and neglecting the contributions from the pole, we see that

$$E(z) \underset{z\to\infty}{\sim} \frac{1}{\sigma_+^2(\omega, k = k_E)} \int (\sigma_+^+ - \sigma_+^-) e^{ikz} \, dk. \tag{49.19}$$

The integration in Eq. (49.19) is along the branch cut and $(\sigma_+^+ - \sigma_+^-)$ is the discontinuity across that cut. Since only k values near k_E contribute, we may expand $\sigma_+(k, \omega)$ about $k = k_E$: Utilizing Eq. (46.1) for σ_+, we see that near the branch point

$$(\sigma_+^+ - \sigma_+^-) \sim (u - 1)/(\omega - \omega_c + i/\tau) \tag{49.20}$$

with

$$u \equiv x^{-1} = kv_F/(\omega - \omega_c + i/\tau) \tag{49.21}$$

so that

$$E(z) \underset{z\to\infty}{\sim} \frac{1}{\sigma_+^2(\omega, k = k_E)} \oint_1^\infty u(u - 1) e^{iu\alpha} \, du \tag{49.22}$$

with

$$\alpha = k_E z. \tag{49.23}$$

The integral in Eq. (49.22) is easily evaluated and

$$E(z) \underset{z\to\infty}{\sim} \frac{1}{\sigma_+^2(\omega, k = k_E)} \frac{1}{z^2} \exp(ik_E z). \tag{49.24}$$

The z^{-2} in Eq. (49.24) has its origins in the fact that the conductivity vanish at the branch point, i.e., that there are no electrons with exactly the maximum velocity along the field. The oscillations in phase given by $\exp(ik_E z)$ are the so-called Gantmakher–Kaner oscillations (40).

Equation (49.24) has a simple physical origin. To derive it we use an argument given by Weisbuch and Libchaber (43). The skin-effect field, as we have mentioned several times, is the dominant field in the metal. It is well represented by a simple exponential, i.e.,

$$E \simeq E_0 e^{-qz}. \tag{49.25}$$

Here q^{-1} is of the order of the anomalous skin depth. We are interested in the field at distances z such that $qz \to \infty$. The electrons in the skin depth gain momentum from the field and carry it into the bulk of the sample. At a point z' the momentum gained by one electron in a time dt is

$$dp_{z'} = eE_0 e^{-qz'} dt. \qquad (49.26)$$

The electrons' momentum in the absence of the field **E** rotates in accordance with

$$\dot{\mathbf{p}} = (e/c)(\mathbf{v} \times \mathbf{H}_0). \qquad (49.27)$$

Each increment of the momentum dp_z gained from the electric field at z will rotate according to Eq. (49.27) so that at z, $dp_{z'}$ becomes

$$dp_{zz'} = eE_0 e^{-qz} e^{ik(z-z')} d(z-z')/v_z \qquad (49.28)$$

where

$$k = (\omega - \omega_{\text{c}})/v_z \qquad (49.29)$$

and $k(z - z')$ is the angle of rotation of the momentum between z and z'. The $d(z - z')/v_z$ is simply the length of time dt that an electron with velocity v_z takes to go from z to z'.

The current at z due to an electron with velocity v_z is the sum of all currents originating at different depths z', i.e.,

$$j(v_z, z) \cong \frac{e^2 E_0}{m v_z} \int_0^z e^{qz'} \exp[ik(z - z')] \, dz'. \qquad (49.30)$$

Since $q \sim (\delta_0)^{-1} \gg k$

$$j(v_z, z) \cong \frac{e^2 E_0}{m v_z} \frac{1}{q} \exp(ikz). \qquad (49.31)$$

The current given in Eq. (49.31) must be summed over the Fermi surface to obtain the total current at z. To do this we multiply (49.31) by the probability $P(v_z)$ of finding an electron with velocity v_z and integrate over v_z. Since

$$P(v_z) = \tfrac{3}{4}(n_0/v_{\text{F}})(1 - (v_z/v_{\text{F}})^2), \qquad (49.32)$$

the total current at z is simply

$$j^{\text{T}}(z) = \int_{-v_{\text{F}}}^{v_{\text{F}}} dv_z \, P(v_z) j(v_z, z). \qquad (49.33)$$

In the limit $z \to \infty$, Eq. (49.33) is equivalent to Eq. (49.22) and can be evaluated in a similar manner. We find that,

$$j^{\text{T}}(z) \sim z^{-2} \exp(ik_E z). \qquad (49.34)$$

49. THE "PROBLEM" OF BOUNDARIES

Since
$$j^T(z) \sim E(z) \tag{49.35}$$

when z is large, Eq. (49.34) is equivalent to Eq. (49.24).

Figure VIII-31 shows a typical field normal transmission spectrum taken by Schultz (25) and co-workers at microwave frequencies. The

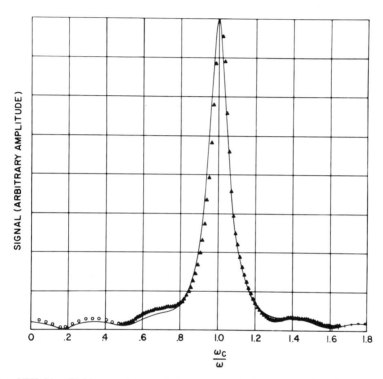

FIG. VIII-31. Field normal transmission spectrum in Na compared to the theoretical estimate [Eq. (49.10)]. $d = 0.16$ mm; $T = 1.20°$K; $\omega = 6.1 \times 10^{10}$ Hz.

points are the experiment. The solid curve is a theoretical fit to the data using a formula completely analogous to Eq. (49.24). The details are unimportant. The fit is impressive.

This brief and highly oversimplified discussion of boundaries is intended only as an introduction to the problem. It stresses some of the physics and leaves out much of the rather sophisticated mathematical techniques that have been employed in solving this type of boundary value problem. The reader who is interested in more of the mathematical details of the boundary value problem is referred to the excellent articles by Baraff (37, 42).

REFERENCES

1. W. P. Allis, S. J. Buchsbaum, and A. Bers, "Waves in Anisotropic Plasmas." MIT Press, Cambridge, Massachusetts, 1963.
2. M. H. Cohen, M. J. Harrison, and W. A. Harrison, *Phys. Rev.* **117**, 937 (1960); J. J. Quinn and S. Rodriquez, *ibid.* A **133**, 1589 (1964).
3. E. A. Stern, *Phys. Rev. Lett.* **10**, 91 (1963).
4. J. L. Stanford and E. A. Stern, *Phys. Rev.* **144**, 534 (1966).
5. A. Libchaber and C. C. Grimes, *Phys. Rev.* **178**, 1145 (1969).
6. A. W. Overhauser, *Phys. Rev. Lett.* **9**, 421 (1964).
7. A. W. Overhauser and S. Rodriquez, *Phys. Rev.* **141**, 431 (1966).
8. H. Mayer and B. Heitel, *in* "Optical Properties and Electronic Structure of Metals and Alloys" (F. Abeles, ed.), p. 47. North-Holland Publ., Amsterdam, 1966.
9. N. Smith, *Phys. Rev.* **183**, 634 (1969). J. Brabiskin and P. G. Siebenman, *Proc. Int. Conf. Electron Mean Free Paths in Metals, Sept. 1968*, in *Phys. of Condensed Matter* 113, (1969).
10. S. A. Werner, E. Gorman, and A. Arrott, *Phys. Rev.* **186**, 705 (1969).
11. L. D. Landau, *J. Phys. (USSR)* **10**, 25 (1946).
12. L. Spitzer, "Physics of Fully Ionized Gases." Wiley (Interscience), New York, 1956.
13. E. A. Kaner and V. I. Skobov, *Zh. Eksp. Teor. Fiz.* **45**, 610 (1963) [Engl. transl., *Sov. Phys.—JETP* **18**, 419 (1963)].
14. S. J. Buchsbaum and P. M. Platzman, *Phys. Rev.* **154**, 395 (1967).
15. J. R. Houck and R. Bowers, *Phys. Rev.* **166**, 397 (1968).
16. W. M. Walsh and P. M. Platzman, *Phys. Rev. Lett.* **15**, 784 (1965).
17. M. Ya Azbel and E. A. Kaner, *J. Phys. Chem. Solids* **6**, 113 (1958).
18. C. C. Grimes and A. F. Kip, *Phys. Rev.* **132**, 1991 (1963).
19. P. M. Platzman, W. M. Walsh, and E-Ni Foo, *Phys. Rev.* **172**, 689 (1968).
20. S. Schultz and G. Dunifer, unpublished data. This data and several others to appear in this manuscript may be found in G. Dunifer's Ph.D. Thesis, "Spin Waves in the Alkali Metals." Univ. of California at San Diego, San Diego, California, 1968.
21. W. M. Walsh, *in* "Electrons in Metals" (J. F. Cochran and R. R. Haering, eds.), pp. 127–252. Gordon & Breach, New York, 1968.
22. E. A. Kaner and V. G. Skobov, *Fiz. Tverd. Tela* **6**, 1104 (1964) [*Sov. Phys.—Solid State* **6**, 851 (1964)].
23. E. A. Kaner and V. G. Skobov, *Physics* **2**, 165 (1966).
24. E. A. Kaner and V. G. Skobov, "Electromagnetic Waves in Metals in a Magnetic Field." Taylor & Francis, London, 1968.
25. S. Schultz, unpublished work.
26. T. Stix, "Theory of Plasma Waves." McGraw-Hill, New York, 1962.
27. I. Bernstein, *Phys. Rev.* **109**, 10 (1958).
28. S. J. Buchsbaum and A. Hasegawa, *Phys. Rev.* **143**, 303 (1966).
29. P. Nozières, *in* "Polarisation, Matière et Rayonnement," Volume jubilaire à l'honneur de A. Kastler. Presses Universitaires de France, Paris, 1969.
30. G. E. Smith, L. C. Hebel, and S. J. Buchsbaum, *Phys. Rev.* **129**, 156 (1963).
31. W. M. Walsh, Jr., P. M. Platzman, P. S. Peercy, L. W. Rupp, Jr., and P. H. Schmidt, *Proc. 12th Int. Conf. Low Temp. Meeting, Kyoto, Japan Sept. 1970*, p. 619.
32. E-Ni Foo, *Phys. Rev.* **182**, 674 (1969).
33. J. F. Koch, R. A. Stradling, and A. F. Kip, *Phys. Rev.* A **133**, 240 (1964).
34. G. A. Baraff, *Phys. Rev.* **187**, 851 (1969).
35. G. H. Reuter and E. H. Sondheimer, *Proc. Roy. Soc. (London)* **195**, 130 (1948).

REFERENCES

36. M. Ya Azbel and E. A. Kaner, *J. Phys. Chem. Solids* **6**, 113 (1958).
37. G. A. Baraff, *Phys. Rev.* **167**, 625 (1968).
38. P. M. Platzman and S. J. Buchsbaum, *Phys. Rev.* **132**, 2 (1963).
39. R. E. Prange and T. W. Nee, *Phys. Rev.* **168**, 779 (1968).
40. V. F. Gantmakher and E. A. Kaner, *Zh. Eksp. Teor. Fiz.* **48**, 1572 (1965) [Engl. transl., *Soviet Phys.—JETP* **21**, 1053 (1965)].
41. P. S. Peercy, W. M. Walsh, Jr., L. W. Rupp, Jr., and P. H. Schmidt, *Phys. Rev.* **178**, 713 (1968).
42. G. A. Baraff, *Phys. Rev.* **178**, 115 (1969).
43. G. Weisbuch and A. Libchaber, *Phys. Rev. Lett.* **19**, 498 (1967).

IX. Conduction Electron Spin Resonance in an Electron Gas

50. Hydrodynamic or Long-Wavelength Treatment of CESR

To this point we have focused our attention on the orbital degrees of freedom of the conduction electrons and completely neglected the weak magnetism associated with their spin. We would now like to discuss spin dynamics. For the noninteracting electron gas, this is essentially the problem of conduction electron spin resonance (CESR) in a spatially inhomogeneous field. An understanding of this problem is a prerequisite to our later discussion of spin dynamics in interacting Fermi systems.

Experimentally, CESR involves the detection of magnetic dipole transitions between the Zeeman energy levels ($\Delta E = \hbar\omega_s$) of electrons in a static magnetic field. More specifically it involves a measurement of the change in surface impedance or transmission coefficient of a bulk slab of metal due to the resonant absorption of energy by the spins of the conduction electrons. Such transitions are caused by an oscillating magnetic field transverse to the steady magnetic field $\mathbf{H}_0$ at a frequency near ω_s.

Our discussion of spin resonance will deal almost exclusively with the so-called transmission electron spin resonance (TESR) experiments (1, 2). The important features of microwave transmission experiments have already been discussed in some detail in Section 42. The sample, in the shape of a thin parallel slab of material, forms the wall between two cavities. The rf field in the slab is essentially confined to the rather narrow skin region near the front surface. The electronic spins in this region can resonantly absorb rf energy. They may then transmit that energy to the receiver cavity either collectively, as in the wave-propagation problems we have been discussing, or as a single particle, merely reradiating the absorbed energy at the back surface as in the case of the Gantmakher–Kaner oscillations briefly described in the preceding chapter. For TESR in relatively impure specimens, we shall see that it is the latter picture which is the correct one.

CESR was discovered in Na by Griswold, Kip, and Kittel as early as 1952 (3). The essential features of the phenomenon had been explored experimentally by Feher and Kip (4), and a beautiful theoretical analysis of the problem presented by Dyson in 1955 (5), when the concept of

AKCR had barely been proposed. Dyson used a simple noninteracting model for the electrons in the metal. In this picture the idea of single-particle transmission, independent of temperature, magnetic field, and purity of the specimen, is always valid. Spin wave propagation, or collective aspects of the CESR experiment were not discovered until 1966 (6). The technical reason for the long time lapse separating the two sets of experiments will become apparent in the section on spin waves (Section 58).

In this chapter we will only consider the noninteracting or Dysonian picture of CESR in a metal. When electron–electron interaction effects are neglected we will see that there are no wavelike disturbances associated with this resonance effect. In contrast the cyclotron resonance has pronounced wavelike phenomena associated with it even in the noninteracting system. This is an important distinction between the two types of resonances.

As one sweeps the static field $\mathbf{H}_0$ through resonance, i.e.,

$$\omega = \omega_s \equiv g\mu_B H_0 \tag{50.1}$$

there is an increasing magnitude of the transmitted field accompanied by a continuous phase shift. A typical transmission electron spin resonance (TESR) spectrum for Na is shown in Fig. IX-1. The data were taken on very pure Na at a temperature of 30°K so that $\omega_c \tau \cong 3$. The short mean free time arises from electron–phonon scattering effects. The reference phase was arbitrarily adjusted so that the line was symmetric about spin resonance. The "Dysonian theory" of normal CESR attempts to describe the rather curious diffraction like shape of the line as a function of the parameters characterizing the material, such as temperature, impurity concentration, and sample thickness.

For a group of noninteracting electrons, we may think of each electron as moving through the lattice carrying its precessing spin with it. Its orbital motion as in the cyclotron problem, is occasionally interrupted (at a rate τ^{-1}) by scattering from an impurity. An important distinction that should be kept in mind when contrasting spin resonance with cyclotron resonance is the considerable degree of decoupling between the spin and orbital degrees of freedom in most metals with small Z. In such a situation the concept of a spin lifetime is a useful one. One can think of the orbital scattering time τ as the time it takes for a nonequilibrium *current* to relax back to equilibrium. Similarly, one can define a spin lifetime T_2 that characterizes the rate at which a nonequilibrium magnetization or equivalently spin polarization relaxes to its equilibrium value.[†]

[†] We make no distinction here between T_2 and T_1 since in the metals under consideration they are roughly equal.

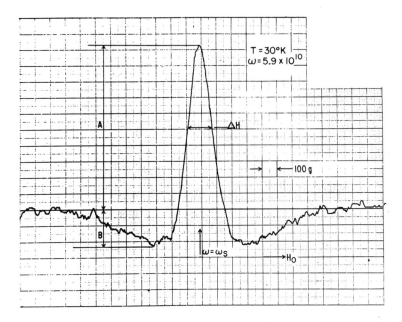

Fig. IX-1. The transmission spectrum (in the neighborhood of CESR) of a thin slab of pure Na at a temperature of 30°K; $\omega = 5.9 \times 10^{10}$.

The decoupling between spin and orbital degrees of freedom is directly reflected in the rather miniscule deviations of the observed CESR g value from the free electron value, for example, in potassium $g = 1.9997$. The observed value is to be contrasted with the free electron value of $g_0 \simeq 2.0013$. The quantity

$$\Delta g/g \equiv [(g - g_0)/g] \qquad (50.2)$$

is small, i.e., of order 10^{-3} for most of the alkalis. The smallness of $\Delta g/g$ directly reflects the smallness of the spin-orbit coupling. Due to this weak spin-orbit coupling, the spin lifetime T_2 is generally much longer than the orbital lifetime τ. It has been shown that the ratio of T_2 to τ may be crudely related to the g shift if we remember that Δg is proportional to some spin-orbit matrix element and that the inverse lifetime T_2^{-1} is proportional to a similar spin-orbit matrix element squared. This kind of reasoning implies that (7, 8)

$$T_2/\tau \sim (g/\Delta g)^2 \sim 10^6 \qquad (50.3)$$

in a low Z material like potassium. Thus, typically an electron makes many orbital deflecting collisions without losing its spin memory.

50. HYDRODYNAMIC TREATMENT OF CESR

Now consider the CESR experiment. The rf magnetic field in the narrow skin region induces a net rf nonequilibrium magnetization in the electron gas by causing transitions among the Zeeman levels for electrons which are in that region. As electrons move away from the skin region out into the bulk of the metal they carry the nonequilibrium magnetization with them. Since the spin lives for a very long time (T_2), it is intuitively clear that we can think of the electrons' motion as a kind of diffusion or random walk process. The limit $\omega_c \tau \ll 1$ is particularly simple. In this case the electrons' flight (between collisions) is, to a good approximation, a straight line, i.e., a small segment of a circle. The static magnetic field is unimportant in determining the *orbital* electronic trajectories.

The overall displacement of an electron in such a situation is accurately characterized as a random walk. The length of the step in the random walk is a mean free path $l \equiv v_F \tau$. The distance the electron diffuses in a time T_2 is given by

$$\delta_{\text{eff}}^0 = \frac{l}{\sqrt{3}} \left(\frac{T_2}{\tau}\right)^{1/2} \equiv (D_0 T_2)^{1/2}. \tag{50.4}$$

The quantity $D_0 = \frac{1}{3} v_F^2 \tau$ is the usual diffusion constant. In pure alkali metals the diffusion distance δ_{eff}^0 is of the order of one centimeter. The size of δ_{eff}^0 implies that the nonequilibrium magnetization, which was produced in the skin region, will eventually spread out into the sample and vary, in thick specimens, on a scale of distance comparable to δ_{eff}^0. Thus, one can hope to transmit spin information through quite thick samples. These ideas are central to the TESR technique.

When $\omega_c \tau \gg 1$, the situation regarding the electronic motion is slightly more complicated. In spite of this, we can still think of the electrons' motion as a type of "diffusion." Such a description implies that details of the electrons' motion are not important since the motion of the electrons, in a uniform field $\mathbf{H}_0$ such that $\omega_c \tau \gg 1$, is *not* diffusive. Each electron in fact traverses a circular orbit between collisions. However, it is true that the distance which an electron moves in a long time $T_2 \gg \tau$ is correctly described by a diffusion equation, which tells us how the orbit center diffuses (9). All distances which we compute in such a picture are only correct to something of the order of an orbit diameter.

The $\omega_c \tau \gg 1$ case differs from that with $\omega_c \tau \ll 1$ because the diffusion of the orbit center is anisotropic. To see this consider the electron ($\omega_c \tau \gg 1$) diffusing across the lines of force. The particle (see Fig. IX-2) goes around in a well-defined circular orbit and then skips, after scattering, to another orbit with a different center. The step in the

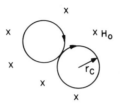

FIG. IX-2. Electrons' motion in the large $\omega_c\tau$ limit.

random walk process is no longer a mean free path l, but the cyclotron radius

$$R_c = v_F/\omega_c \equiv l/\omega_c\tau \ll l. \tag{50.5}$$

In effect the particles are tied to the lines of force and cannot easily diffuse across them.

The distance an electron diffuses across the field in a time T_2 is now given by

$$\delta_{\text{eff}}^{\pi/2} \equiv (D^{\pi/2} T_2)^{1/2} \tag{50.6}$$

where

$$D^{\pi/2} = D_0/[1 + (\omega_c\tau)^2]. \tag{50.7}$$

When $\omega_c\tau \gg 1$, Eq. (50.6) reduces to

$$\delta_{\text{eff}}^{\pi/2} = \frac{R_c}{\sqrt{3}} \left(\frac{T_2}{\tau}\right)^{1/2}. \tag{50.8}$$

For diffusion along the field, the step in the random walk process is still l and the diffusion distance is given by Eq. (50.4).

In general, the motion of the orbit center is described by an anisotropic diffusion constant (10)

$$D = D_0 \left[\frac{\sin^2 \Delta}{1 + (\omega_c\tau)^2} + \cos^2 \Delta\right]. \tag{50.9}$$

The quantity D determines the root mean square displacement of the orbit center in a direction Δ relative to $\mathbf{H}_0$. In a time T_2 the electron has diffused a distance,

$$\delta_{\text{eff}}^{\Delta} = (DT_2)^{1/2}. \tag{50.10}$$

Even at the highest magnetic fields in the best alkali metals, $\delta_{\text{eff}}^{\Delta}$ is still a fraction of a centimeter and the scale on which the magnetization vector varies is large compared to all other lengths in the problem. If we are concerned, as we will be, with the response of the magnetization to external fields which are varying rapidly in space on this scale, these arguments would indicate that the detailed variation of the skin-effect fields is unimportant. We will be able to show that the magnetic response of the medium, in the long spin relaxation time limit, depends only on certain average properties of the skin effect field. These properties, in turn, can be simply related to the surface impedance in the absence of magnetic susceptibility.

Now that we have a qualitative picture of the electrons' motion we can ask how broad we might expect a resonance line to be. The naive answer to this question is startling and discouraging. When $l \gg \delta_s$, the time an electron spends in the skin region in pure metals is roughly

$$t_s \sim \delta_s/v_F . \tag{50.11}$$

Here δ_s is the skin depth. For a $v_F \simeq 10^8$ cm sec^{-1} and $\delta_s \simeq 10^{-5}$ cm sec^{-1}, $t_s \simeq 10^{-13}$ sec. At first glance, the short time spent in the skin region might lead one to believe that the resonance line would be severely broadened. An uncertainty principle argument suggests that $\Delta\omega \simeq 1/t_s$. This large broadening would make it experimentally impossible to observe CESR in metals.

It was Dyson (5) who realized the fallacy of this naive argument. He pointed out that the CESR line would *not* be as broad as $1/t_s$, because an electron does not simply traverse the skin depth *once*. In the course of its migrations into the bulk and over to the other side it returns several times to the skin region. Thus, a typical electron senses a set of pulsed rf fields whose intervals are random but whose *phases are coherent*. In Fig. IX-3 we have sketched the time dependence of the pulse spectrum sensed by a such an electron. The width of the pulse is approximately t_s,

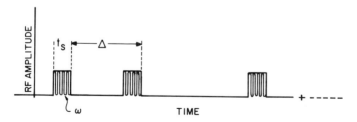

FIG. IX-3. The rf amplitude sensed by an electron repeatedly returning to the skin region.

the intervals are random and the Fourier transform of a single pulse is

$$F_\alpha(\Omega) \sim e^{-i\alpha(\Omega-\omega)} \left\{ \frac{\exp[i(\Omega-\omega)t_s] - 1}{\Omega - \omega} \right\}. \tag{50.12}$$

Here ω is the frequency of the rf field and α is the time specifying the onset of the pulse. Equation (50.12) has a diffraction like spectrum centered about ω with width $1/t_s$. Adding a second pulse at another time α will, in general, lead to destructive interference at all but the central frequency ω. The net effect of adding many such pulses is to produce a spectrum which consists of a sharp line superimposed on a broad background of width $1/t_s$. The limiting width of the sharp piece of the spectrum is ultimately determined by the time T_2, i.e., the length of time the electron effectively samples the field before losing its spin memory rather than t_s.

This situation is similar in many ways to the Ramsey two-hairpin experiments of molecular beam physics (*11*). In these experiments, a group of excited molecules passes through two small region of rf field (hairpins) which are separated by a relatively long field-free region. The two rf fields are driven with a fixed phase relative to one another. The molecules spend only a short time in each region of field but a relatively long time in going from one hairpin to the other. Because the two regions are driven coherently, it is not the time spent in each region which limits the resolution but the time it takes electrons to traverse the long field free region which determines the uncertainty principle smearing of the line.

Before going on to discuss, in some detail, the extremely simple problem of transmission through a slab it is worthwhile to examine the infinite medium response function which incorporates the physics we have just discussed. The linear combination

$$\chi_+(\mathbf{k}, \omega) \equiv \chi_{xx}(\mathbf{k}, \omega) + i\chi_{xy}(\mathbf{k}, \omega), \tag{50.13}$$

i.e.,

$$m^+(\mathbf{k}, \omega) \equiv m_x(\mathbf{k}, \omega) + im_y(\mathbf{k}, \omega) = \chi_+(\mathbf{k}, \omega) h^+(\mathbf{k}, \omega), \tag{50.14}$$

characterizes the response of the resonant transverse magnetization to a delta-function driving rf field in real space at a frequency ω [Fourier transform $h^+(\mathbf{k}, \omega) = 1$]. It incorporates the diffusion and resonant properties we have just described in words. For the noninteracting gas,

$$\chi_+(\mathbf{k}, \omega) = \frac{-\chi_0 \omega_s}{(\omega - \omega_s + i/T_2 + iDk^2)} \tag{50.15}$$

(we will derive this result later) where D is given by Eq. (50.9).

The $\chi_+(\mathbf{k}, \omega)$ given in Eq. (50.15) summarizes mathematically all of the physics contained in our qualitative discussion of the preceding few pages. When $k = 0$, χ_+ shows a resonance at $\omega = \omega_s$ with a width determined by T_2^{-1}. The diffusion term iDk^2 influences the shape of the line only. In a typical boundary value problem, as we shall see, k will be fixed by the geometry, i.e., it will be of the order of L^{-1} for a slab of thickness L. The fact that k is not determined by δ^{-1} is the mathematical restatement of the fact that the electrons reenter the skin region several times while diffusing a distance L. This point will become clearer when we consider the boundary value problem.

The form of χ_+ given in Eq. (50.15) implies that the space and time dependent macroscopic magnetization in the infinite medium obeys a Bloch-like equation of the form

$$\partial \mathbf{M}/\partial t = \gamma_0(\mathbf{H}) \times \mathbf{M} + D\,\nabla^2(\mathbf{M}) + (1/T_2)(\mathbf{M}_0 - \mathbf{M}) \qquad (50.16)$$

with $\gamma_0 \mathbf{H}_0 = \omega_s$ and $\mathbf{M}_0 \equiv \gamma_0 \mathbf{H}_0$. Here $\mathbf{M}(\mathbf{r}, t)$ is the magnetization density in the infinite medium. Equations (50.15) and (50.16) are completely equivalent. In the finite medium, it is more convenient to think of the Bloch equation (50.16) along with a boundary condition on $\mathbf{M}(\mathbf{r}, t)$ as the defining equations. We will come back and discuss the appropriate boundary conditions in the next section.

We have, on the basis of physical argument, merely written down Eq. (50.15) and its equivalent, Eq. (50.16), without any real justification. It may be derived from more fundamental considerations even though it is clear that Eq. (50.16) is correct; since any *microscopic* theory must yield, at long wavelengths, a set of macroscopic Bloch-like equations describing, if you wish, the hydrodynamics of the magnetization.

The time dependence of the nonequilibrium magnetization density must obey a continuity equation of a form analogous to the electrical continuity equation for charge, i.e.,

$$\partial \mathbf{M}/\partial t + \nabla \cdot \mathbf{J}_m = \gamma_0(\mathbf{H}) \times \mathbf{M} + (1/T_2)(\mathbf{M}_0 - \mathbf{M}) \qquad (50.17)$$

where $\mathbf{J}_m$ is the magnetization current (a tensor). The only difference between Eq. (50.17) and the usual continuity equation

$$\partial \rho/\partial t + \nabla \cdot \mathbf{j} = 0 \qquad (50.18)$$

is the free precession term $\gamma_0(\mathbf{H} \times \mathbf{M})$ and the phenomological spin relaxation term $(1/T_2)(\mathbf{M}_0 - \mathbf{M})$. The Bloch equation is equivalent to the assumption that

$$\nabla \cdot \mathbf{J}_m = D\,\nabla^2 \mathbf{M}, \qquad (50.19)$$

an assumption which is only valid for slowly varying quantities.

In many situations it is important to understand the derivation of Eq. (50.16). The derivation given in Chapter X depends on the solution of a transport equation analogous to the Boltzmann–Vlasov equation (45.1) (*12, 13*). We will return to an analysis and derivation of this type of transport equation for magnetization in Chapter X where we will extend the long-wavelength analysis presented here to shorter wavelength and at the same time include the effects of electron–electron interactions. In the noninteracting case, the long-wavelength susceptibility, i.e., the Bloch equation with an anistropic diffusion constant, is sufficient, as we shall see, to characterize the experiments.

51. CESR in a Bounded Medium

It is a simple matter to utilize the long-wavelength infinite-medium susceptibility, Eq. (50.15), or equivalently the Bloch equation, Eq. (50.16) along with a boundary condition on **M** to effect a solution of the finite-slab transmission problem. In the slab problem all of the fields are functions of z alone and the only vector component of interest is $\mathbf{M}_+ \equiv M_x \hat{u}_x + i M_y \hat{u}_y$. Since $(\partial/\partial z)\mathbf{M}_+$ is proportional to the magnetization current [see Eq. (50.19)], it is a good approximation to set

$$\frac{\partial}{\partial z} \mathbf{M}_+(z) = 0 \tag{51.1}$$

for $z = 0$ and $z = L$. Equation (51.1) simply reflects the fact that the spin current is conserved at the boundaries. In order to violate Eq. (51.1) the surface scattering must change the magnetization current. The only mechanism available for this is the existence of magnetic scatters on the surface, or equivalently a strong spin-orbit coupling which would lead to spin-flip scattering due to the rapid potential changes at the surface. All bulk magnetic scattering effects are incorporated into T_2. They do not affect the validity of Eq. (51.1).

We would like to compute the magnitude and phase of the magnetic field which is resonantly transmitted through a slab (lying in the x, y plane) (*9*). We begin by examining the magnetization in the finite medium. For the time-varying Fourier component at frequency ω,

$$\mathbf{M}_+(z) = \int_0^L \chi_+(z, z', \omega) \, \mathbf{H}_+(z') \, dz'. \tag{51.2}$$

The fixed $\mathbf{H}_+(z)$ is, to order χ_0, the unknown rf magnetic field which is determined by solving the skin effect problem. The quantity $\chi_+(z, z', \omega)$ is the finite-medium susceptibility, i.e., it is the solution of Eq. (50.16) with the boundary conditions Eq. (51.1) assuming that the driving term $\mathbf{H}_+$ is a spatial delta-function. Clearly,

$$\chi_+(z, z', \omega) = (1/L) \sum_{n=-\infty}^{+\infty} \cos(k_n z) \cos(k_n z') \chi_+(k_n \hat{\mathbf{u}}_z, \omega) \quad (51.3)$$

with $k_n = n\pi/L$ and $\chi_+(\mathbf{k}, \omega)$ given by Eq. (50.15).

In an infinite-medium problem we expect $\chi_+(z, z', \omega)$ to vary on a scale of distance determined by δ_{eff}. Typically δ_{eff} as we have discussed in Section 50 can be of the order of one centimeter. In a finite geometry problem when $\delta_{\mathrm{eff}} \gg L$ the susceptibility $\chi_+(z, z', \omega)$ varies on a scale of distance determined by the size of the sample itself. On the other hand, the rf magnetic field, whatever its detailed behavior, varies on a scale of distance determined by the skin depth δ_s. Since $\delta_s/L \ll 1$ we may set $z = 0$ in $\chi_+(z, z', \omega)$ and remove it from the integration in Eq. (51.2), i.e.,

$$\mathbf{M}_+(z) \cong \chi_+(z, 0) \int_0^L \mathbf{H}_+(z') \, dz'. \quad (51.4)$$

Maxwell's equations, more specifically the curl equation, tell us that

$$\nabla \times \mathbf{E} = (-1/c) \, \partial \mathbf{M}/\partial t \quad (51.5)$$

so that

$$\int_0^L \mathbf{H}_+(z') \, dz' \simeq \int_0^\infty \mathbf{H}_+(z') = (ic/\omega)(\mathbf{n} \times \mathbf{E}_+(0)). \quad (51.6)$$

Since

$$(\mathbf{n} \times \mathbf{E}_+(0)) = (c/4\pi) Z_0 \mathbf{H}_+(0) \quad (51.7)$$

where Z_0 is the surface impedance of the specimen,

$$\mathbf{M}_+(z) \cong -(c^2/4\pi\omega) \chi_+(z, 0) Z_0 \mathbf{H}_+(0). \quad (51.8)$$

Thus

$$\mathbf{M}_+(0) \cong -(c^2/4\pi\omega) \chi_+(0, 0) Z_0 \mathbf{H}_+(0), \quad (51.9)$$

$$\mathbf{M}_+(L) \cong -(c^2/4\pi\omega) \chi_+(L, 0) Z_0 \mathbf{H}_+(0). \quad (51.10)$$

Equations (51.9)–(51.10) simply express the fact that a predetermined spatially localized driving term ($\mathbf{H}_+$) acts to excite all modes, in this case purely diffusive, to produce a response. We will come back to these equations when we discuss spin waves.

The important point to make about the pair of Eqs. (51.9) and (51.10) is that all of the unknown detailed behavior of the skin-effect field has been lumped into the single unknown constant Z_0. The quantity Z_0

is somewhat dependent on the applied magnetic field. For example, it shows resonances at the AKCR cyclotron resonance described earlier. However, even in this case, significant variations in the surface impedance occur only when the magnetic field has gone through an appreciable fraction of a cyclotron line width, i.e., $\Delta H_0/H_0 \sim 1/(\omega_c \tau)$ and this happens only in the neighborhood of cyclotron resonance. Even for these large changes of field, Z_0 is only very mildly dependent on the magnetic field since very few of the electrons are taking part in cyclotron resonance. In fact, the spin resonance structure in the alkalis typically occurs away from cyclotron resonance $\omega_s \neq \omega_c$ and is confined to a great extent to a much smaller region in field, i.e., $\Delta H_0/H_0 \simeq 1/(\omega_s T_2)$. All of these considerations leads one to believe that it is a good approximation to neglect the magnetic field variation of Z_0 when analyzing spin resonance data.

We would now like to show that

$$\mathbf{M}_+(L_-) \sim Z_0 \mathbf{H}_+(L_+) \tag{51.11}$$

follows from very general considerations. Equation (51.11) simply states that the transmitted field at L_+, i.e., just outside the specimen, is proportional to the value of the magnetization just inside the specimen, i.e., at L_-. We will be able to show that Eq. (51.11) follows from rather general considerations independent of any of the complexities associated with the anomalous skin effect. It depends only on the fact that the skin-effect field is rapidly varying relative to the magnetization.

Consider the derivation of Eq. (51.11). The magnetization produced by the field in the skin region at $z = 0$ will carry along an rf electric field, albeit a very small electric field, with it. The continuity of the tangential components of $\mathbf{E}$ and $\mathbf{H}$ at the second boundary will violate the outgoing wave boundary condition (i.e., $\mathbf{E}_+(L_+) = \mathbf{H}_+(L_+)$) for the transmitted wave. The way to "fix up" the boundary condition is to add to the slowly varying field, associated with the magnetization, a rapidly varying skin-effect-like solution. This extra field will enable us to satisfy the outgoing wave boundary condition.

Formally, we can split our solutions up into a rapidly varying part (skin effect) plus a slowly varying driven part (magnetization wave). (In the analysis all polarization factors which introduce unnecessary and extraneous complications are neglected.) The rapidly varying part of the rf field satisfies equations of the form

$$\nabla \times \mathbf{H}^{(0)} = (4\pi/c)\boldsymbol{\sigma} \cdot \mathbf{E}^{(0)} \tag{51.12}$$

$$\nabla \times \mathbf{E}^{(0)} = (i\omega/c)\mathbf{H}^{(0)}, \tag{51.13}$$

where **σ** is the *nonlocal* conductivity operator. The slowly varying part satisfies the inhomogeneous equation

$$\nabla \times \mathbf{H}^{(1)} = (4\pi/c)\boldsymbol{\sigma} \cdot \mathbf{E}^{(1)}, \tag{51.14}$$

$$\nabla \times \mathbf{E}^{(1)} = (i\omega/c)[\mathbf{H}^{(1)} - 4\pi\mathbf{M}(z)], \tag{51.15}$$

where $\mathbf{M}(z)$ is a *prescribed* function given by Eq. (51.8). In Eq. (51.14), **σ** may be thought of as the local conductivity operator since the fields in Eq. (51.14) are forced to be slowly varying by the slow variation of $\mathbf{M}(z)$. The continuity conditions on **E** and **H** at the boundary $z = L$ are

$$\mathbf{E}^{(1)}(L_-) + \mathbf{E}^{(0)}(L_-) = \mathbf{E}(L_+), \tag{51.16}$$

$$\mathbf{H}^{(1)}(L_-) + \mathbf{H}^{(0)}(L_-) = \mathbf{H}(L_+). \tag{51.17}$$

The outgoing-wave boundary conditions implies that $\mathbf{E}(L_+) = \mathbf{H}(L_+)$ so that (within unimportant polarization factors),

$$\mathbf{E}^{(1)}(L_-) + \mathbf{E}^{(0)}(L_-) = \mathbf{H}^{(1)}(L_-) + \mathbf{H}^{(0)}(L_-). \tag{51.18}$$

For the slowly varying wave $\mathbf{H}^{(1)}$ it can be shown using Eq. (51.14) that

$$\mathbf{E}^{(1)}(L_-)/\mathbf{H}^{(1)}(L_-) \simeq Z_1^2(k_m/k_0), \tag{51.19}$$

where

$$Z_1^2 = -\omega(\omega + i/\tau)/\omega_p^2 \tag{51.20}$$

is the dimensionless impedance of the metallic medium in the local regime. The quantity k_m is a typical wave vector characterizing the spatial variation of the magnetization wave and $k_0 \equiv \omega/c$ is the free-space wavelength. The ratio k_m/k_0 is of order one and the quantity $Z_1^2 \ll 1$ at microwave frequencies. The actual magnitude of k_m/k_0 is, as we will show, unimportant.

For the skin-effect field,

$$\mathbf{E}^{(0)}(L_-)/\mathbf{H}^{(0)}(L_-) \equiv Z_0(4\pi/c) \equiv Z_0', \tag{51.21}$$

where Z_0 is unknown but certainly $Z_0' \ll 1$. Substituting Eq. (51.21) into (51.18) we find that (leaving out the arguments L)

$$\mathbf{H}^{(0)} = -\mathbf{H}^{(1)} \left[\frac{Z_1^2(k_0/k_m) - 1}{Z_0' - 1} \right], \tag{51.22}$$

so that

$$\mathbf{H}^{(0)} + \mathbf{H}^{(1)} = \mathbf{H}^{(1)} \left[1 - \frac{Z_1^2(k_0/k_m) - 1}{Z_0' - 1} \right],$$
$$\simeq -\mathbf{H}^{(1)} Z_0' = \mathbf{H}(L_+). \tag{51.23}$$

Since Eq. (51.19) tells us that $\mathbf{E}^{(1)} \ll \mathbf{H}^{(1)}$, Eq. (51.14) implies that

$$\mathbf{H}^{(1)} \cong 4\pi \mathbf{M}(L_-), \tag{51.24}$$

so that

$$\mathbf{H}(L_+) = -Z_0' 4\pi \mathbf{M}(L_-). \tag{51.25}$$

This completes the derivation of Eq. (51.11) in the anomalous as well as the classical regime. The fascinating thing about it is that it is really independent of complicated boundary value conditions characteristic of skin-effect problems. It depends solely on the fact that the magnetization field varies slowly compared to the skin-effect field.

Utilizing Eq. (51.10) and the approximate equation for $\chi_+(\mathbf{k}, \omega)$, Eq. (50.15), we may in fact obtain a closed form expression for the transmitted magnetic field, i.e.,

$$H_+(L_+) \sim M_+(L_-) \sim \sum_{n=-\infty}^{+\infty} \frac{(-1)^n}{(\omega - \omega_s + i/T_2 + i\,Dk_n^2)} \tag{51.26}$$

with

$$k_n^2 = n^2 \pi^2 / L^2. \tag{51.27}$$

Since

$$\sum_{n=-\infty}^{+\infty} \frac{(-1)^n}{A^2 - \pi^2 n^2} = \frac{\csc A}{A} \equiv \frac{\cot A/2 - \cot A}{A}, \tag{51.28}$$

Equation (51.28) implies that,

$$H_+(L_+) \sim [T_2 \Gamma^2 W \sin 2W]^{-1}. \tag{51.29}$$

Here

$$-4W^2 = [(1 - \omega_s/\omega)\,\omega T_2 + i]/T_2 \Gamma^2. \tag{51.30}$$

In this noninteracting case $T_2 \Gamma^2$ is a pure *imaginary* number simply related to the diffusion constant D, i.e.,

$$T_2 \Gamma^2 = -i D T_2 / L^2 \equiv -i(\delta_{\text{eff}}/L)^2. \tag{51.31}$$

Equation (51.29) describes quantitatively what we had already conjectured on the basis of physical arguments alone. The nature of the transmitted $\mathbf{H}$ field is completely characterized by the quantity $\delta_{\text{eff}}^d \equiv (DT_2)^{1/2}$, the "anisotropic" diffusion distance for the spins [see Eq. (50.10)]. By varying the ratio δ_{eff}/L and thus changing, in a nontrivial way, the number of times the electron returns to the skin depth before it finally gets to the opposite side of the specimen and reradiates its energy, we vary the so-called A/B ratio (see Fig. IX-1), i.e., change the shape of the line.

The width of the spin resonance line, or scale of frequency, is fixed by the "parameter" T_2. This quantity is, in a sense, adjustable since it characterizes the purity of the material and the magnitude of spin-orbit coupling effects. The "complex" proportionality constant in Eq. (51.29) is related to the surface impedance in the absence of magnetic effects and hence is unknown. In a real experiment the phase of this complex parameter (as pointed out briefly in Section 50) may be removed from the data by adjusting the phase of the reference signal to give a symmetric line.

Figures IX-4 and IX-5 show a series of experimental and theoretical

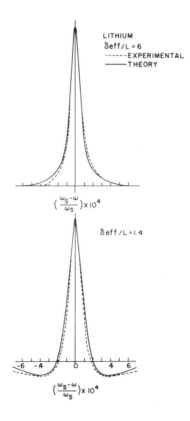

FIG. IX-4. Comparison of TESR data (dashed) in Li with the simple Dysonian theory (solid).

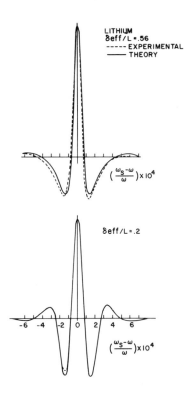

FIG. IX-5. Comparison of TESR data (dashed) in Li with the simple Dysonian theory (solid).

curves describing the transmission spectrum of slab of Li (a few tenths of a millimeter thick) in the neighborhood of CESR. The theoretical curves (solid) are generated using Eq. (51.29) and by fitting the two parameters T_2 and δ_{eff}/L. As the temperature is increased the main effect is a decrease in δ_{eff}/L. This in turn implies a line with bigger lobes on it, i.e., the electron returns fewer times to the skin depth before reradiating its energy. Figures IX-4a,b and IX-5a show how a temperature change makes τ smaller thus decreasing δ_{eff}/L. In Fig. IX-5b the thickness of the slab was increased giving rise to a still smaller δ_{eff}/L. These experimental results show that the Dyson theory is (in a simple metal like Li) in good quantitative agreement with the TESR results at elevated temperatures.

We deliberately picked such a poor material, i.e., high temperature, in order to illustrate the validity of Eq. (51.29). Too large an $\omega\tau$, i.e., too pure a specimen at low temperatures will show deviations from the simple Dysonian formula Eq. (51.29). It is this breakdown, characterized by the failure of the one electron approximation, which we will try to describe in the next chapter.

REFERENCES

1. R. B. Lewis and T. R. Carver, *Phys. Rev. Lett.* **12**, 693 (1964).
2. P. Monod and S. Schultz, *Phys. Rev.* **173**, 644 (1968).
3. T. W. Griswold, A. F. Kip, and C. Kittel, *Phys. Rev.* **88**, 951 (1952).
4. G. Feher and A. F. Kip, *Phys. Rev.* **98**, 337 (1955).
5. F. J. Dyson, *Phys. Rev.* **98**, 49 (1955).
6. S. Schultz and G. Dunifer, *Phys. Rev. Lett.* **18**, 283 (1967).
7. A. W. Overhauser, *Phys. Rev.* **89**, 689 (1953).
8. Y. Yafet, *Solid State Phys.* **14**, 2 (1963).
9. M. Lampe and P. M. Platzman, *Phys. Rev.* **150**, 340 (1966).
10. W. P. Allis, *in* "Handbuch der Physik" (S. Flugge, ed.), Vol. XXI, p. 395. Springer-Verlag, Berlin and New York, 1956.
11. N. F. Ramsey, *Phys. Rev.* **78**, 695 (1950).
12. M. Ya. Azbel, V. I. Gerasimenko, and I. M. Litshitz, *Zh. Eksp. Teor. Fiz.* **35**, 691 (1958) [Engl. transl., *Sov. Phys.—JETP* **8**, 480 (1959)].
13. E. P. Wigner, *Phys. Rev.* **40**, 749 (1932).

X. The Interacting Electron Liquid

52. The Phenomenological Formulation of Fermi-Liquid Theory

In Chapters VII–IX we have investigated the problem of wave propagation in metals within the framework of a noninteracting electron gas model. We would now like to include, in a quantitative manner, the effect of electron–electron interactions on the properties of waves in metals.

Since our noninteracting model (with an effective mass m^*) seems to work quite well for most of the electromagnetic properties of metals, we might be tempted to utilize some form of perturbation theory in an attempt to incorporate electron–electron interaction effects into the analysis. Unfortunately, such an approach is not possible in a real metal, since the Coulomb interaction between electrons is strong, not weak. The parameter which characterizes the relative strength of the Coulomb interaction is r_s (see Section 2, Chapter I). For the alkalis, r_s is larger than unity. It ranges from 3.22 for Li to 5.57 for Cs and is in the neighborhood of four for Na, K, and Rb. Thus the electrons in a metal are strongly interacting and we must look for a nonperturbative description of this strongly interacting system. Fortunately, such a description has been developed by Landau in 1956 and is called the Landau theory of Fermi liquids (1). Landau used this theory to describe the properties of He$_3$, a neutral Fermi liquid. Silin (2) later extended the Landau theory to include effects introduced by a long-range Coulomb force, i.e., to describe electrons in a metal. We will employ the Landau–Silin (LS) theory to discuss the interacting electron gas in a metal. We alluded to this theory and its nonperturbative character in Chapter III. Chapters X and XI will, largely, be devoted to a discussion of the LS theory and its application to a study of waves in metallic systems.

A normal Fermi liquid is defined as a system whose low-lying excited states are in one to one correspondence with the noninteracting system. A superconductor for example is not a normal Fermi liquid. The LS theory describes the low-lying excited states of a normal Fermi liquid, i.e., those states with excitation energies small compared to the Fermi energy E_F. It says nothing about the ground state of the system, nor phenomena which involve excited states at energies of the

52. FORMULATION OF FERMI-LIQUID THEORY

order of E_F. A description of the low-lying excited states enables one to calculate the response of such systems to slowly varying fields ($k \ll k_F$ and $\omega \ll E_F$) and to predict the thermodynamic behavior of such systems at low temperatures ($kT \ll E_F$). These are precisely the properties of the interacting Fermi system that are needed to describe wave propagation. For our purposes, the LS theory is an ideal one.

From the outset it is apparent that any such nonperturbative theory must be constructed along semiphenomenological lines. A complete, microscopic theory is clearly out of the question since we are dealing with a strongly interacting system of many electrons. However it has been shown, by Landau (3) and more completely by Luttinger and Nozières (4), that the Fermi-liquid theory is equivalent to a microscopic infinite-order perturbation (perturbation in r_s) treatment of the low-frequency response of the interacting electron system. The equivalence is true to all orders in r_s, and a priori it is not clear that any simple truncation of the perturbation series would lead to sensible results for "real" metals. However, calculations based on a selective summation of the terms in the perturbation series seem to give a rather consistent picture of the interaction effects. We will discuss such calculations in some detail in Chapter XI.

The LS theory is formulated in terms of a single unknown function which describes the scattering amplitude of two "quasi particles." This "Landau scattering function" may, in turn, be identified with a certain infinite sum of terms in the perturbation series. The spirit of the discussion of this chapter will be to assume that the Landau scattering function is an unknown but well-defined quantity. We will then use it to calculate properties of the electron fluid. The measured properties can then be used in reverse to extract information about the scattering function. This approach is analogous to the procedure one uses in analyzing the De Haas–Van Alphen effect to obtain Fermi-surface parameters (5).

Consider the noninteracting fermion gas; for the time being we ignore spin. At $T = 0$ all single-particle momentum states with $p < p_F$ are occupied and all others unoccupied. The low-lying excitations of the system are obtained by removing particles from $p < p_F$ and putting them into an empty state p' with $p' > p_F$. The excited states of this non-interacting system are completely specified by prescribing the particle distribution function $n(\mathbf{p}, \mathbf{r}, t)$. In the ground state $n(\mathbf{p}, \mathbf{r}, t) \equiv n^0(p)$ with

$$n^0(p) = \begin{cases} 1 & p < p_F, \\ 0 & p > p_F. \end{cases}$$

The quantity $\delta n \equiv n(\mathbf{p}, \mathbf{r}, t) - n^0(p)$ specifies the number of electrons excited above the Fermi surface and the number of holes (empty states) left behind.

In his original paper, Landau postulated that the low-lying excited states of an interacting Fermion system could also be described in terms of simple particle hole-like excitations about a sharp Fermi surface. These long-lived excitations he called "quasi particles." It is the "quasi-particle" distribution function which characterizes the excited states of the Fermi liquid. The fact that a single "quasi-particle" distribution function should specify the states of the interacting system is a completely nontrivial, nonobvious statement which we take, for the purpose of our discussions as a postulate. The reader is referred to a number of very good many-body books on the theory of Fermi liquids for further mathematical treatment of this point (6).

Such single-particle excitations can exist, even in strongly interacting Fermi systems because of the exclusion principle. If we excite a single electron with some small energy $(E_p^\circ - E_F)$ above the ground state, then it is clear that such a state is *not* an eigenstate of the many-electron system. This single-particle state decays into multiple-particle states via its Coulomb interaction with the Fermi sea. However, regardless of the details of this decay or the strength of the coupling, the *rate of decay* must be slow for a single-particle state close to the Fermi surface since the Fermi sea occupies almost all of the phase space into which the single-particle state can decay. This phase-space argument leads one to the conclusion that the lifetime τ of such a single-particle state must go to infinity as $(E_p^\circ - E_F)^{-2}$ i.e., $1/\tau \cong \omega_p[(E_p^\circ - E_F)/E_F]^2$. As a consequence, the low-lying single particle like states are *nearly* eigenstates of the interacting system. The Landau theory assumes that this is the case and deals with distributions of such "quasi particles." Note that, in some sense, it is the Pauli principle which turns a strongly interacting problem into what can be considered a weak-coupling theory.

As in the case of the noninteracting system we introduce the quasi-particle distribution function $n(\mathbf{p}, \mathbf{r}, t)$ (temporarily neglecting spin),

$$n(\mathbf{p}, \mathbf{r}, t) = n^0 + \delta n. \qquad (52.1)$$

The distribution function $n(\mathbf{p}, \mathbf{r}, t)$ has meaning only for $\mathbf{p}$ near p_F, i.e., it is δn in the neighborhood of p_F which tells us about the low-lying excited states. We choose $n^0(p)$ such that

$$\int (2\pi)^{-3} d^3p \, n^0(p) = n_0. \qquad (52.2)$$

The use of a semiclassical distribution function $n(\mathbf{p}, \mathbf{r}, t)$ implies that our quasi particles have position as well as momenta. A consequence of

52. FORMULATION OF FERMI-LIQUID THEORY

this assumption is that theory as developed here can only treat disturbances which are slowly varying on a scale set by p_F, i.e., $\Delta r \gg 1/p_F$ (7).

Formally the energy of the system is expressed as functional of $n(\mathbf{p}, \mathbf{r}, t)$

$$E = E\{n\}. \tag{52.3}$$

In particular, the change in energy due to some stationary excitation of quasi particles is given by

$$\delta E = \int (\delta E/\delta n)\, \delta n\, d^3r\, d^3p. \tag{52.4}$$

One conventionally writes this equation as

$$\delta E = \int E(\mathbf{p}, \mathbf{r}, n)\, \delta n\, d\tau \tag{52.5}$$

with

$$d\tau \equiv d^3r\, d^3p/(2\pi)^3. \tag{52.6}$$

The quasi-particle energy E, as in Hartree–Fock theory, is defined by

$$E(\mathbf{p}, \mathbf{r}, n) \equiv (2\pi)^3\, \delta E/\delta n(\mathbf{p}, \mathbf{r}, t). \tag{52.7}$$

It still depends functionally on n.

Since we consider small deviations from the equilibrium ground state we may expand $E(\mathbf{p}, \mathbf{r}, n)$ about $E(\mathbf{p}, \mathbf{r}, n^0)$. This expansion, to lowest order in δn, gives

$$E(\mathbf{p}, \mathbf{r}, n) \simeq E(\mathbf{p}, \mathbf{r}, n^0) + \int \left(\frac{\delta E[\mathbf{p}, \mathbf{r}, n(\mathbf{p}', \mathbf{r}', t)]}{\delta n(\mathbf{p}', \mathbf{r}', t)} \right)_{n=n^0} \delta n(\mathbf{p}', \mathbf{r}', t)(2\pi)^3\, d\tau'. \tag{52.8}$$

Thus one can rewrite Eq. (52.7) as

$$E(\mathbf{p}, \mathbf{r}, n) \simeq E(\mathbf{p}, \mathbf{r}, n^0) + \int F(\mathbf{p}, \mathbf{r}, \mathbf{p}', \mathbf{r}')\, \delta n(\mathbf{p}', \mathbf{r}', t)\, d\tau', \tag{52.9}$$

where

$$F(\mathbf{p}, \mathbf{r}, \mathbf{p}', \mathbf{r}') = (2\pi)^6 \left\{ \frac{\delta^2 E\{n\}}{\delta n(\mathbf{p}, \mathbf{r}, t)\, \delta n(\mathbf{p}', \mathbf{r}', t)} \right\}_{n=n^0}. \tag{52.10}$$

In a metallic system the effective forces between "quasi particles" has two parts, both coming from the Coulomb interaction;

(1) electrostatic interaction between average densities at the points $\mathbf{r}$ and $\mathbf{r}'$. This gives a term in the energy of the form (see Section 18),

$$\tfrac{1}{2} \int d^3\tau\, d^3\tau'\, \frac{e^2}{|\mathbf{r} - \mathbf{r}'|}\, \delta n(\mathbf{p}, \mathbf{r}, t)\, \delta n(\mathbf{p}', \mathbf{r}', t). \tag{52.11}$$

The potential in this expression has infinite range. It can be treated by introducing an electric field $\mathbf{E}_H$ which will be handled like an external field (2). Separating off the long-range part of the Coulomb field is precisely equivalent to the replacement of $\mathbf{E}_{\text{ext}}$ in the BV equation [see Section 22, Eq. (22.3)] by $\mathbf{E}_{\text{tot}}$, i.e.,

$$\mathbf{E}_{\text{tot}} = \mathbf{E}_{\text{ext}} + \mathbf{E}_H \tag{52.12}$$

(2) The remaining effective forces between particles are short range because of the Fermi–Thomas screening. As a consequence of this, the function F (once we have subtracted off $\mathbf{E}_H$) has a range in real space of order k_{FT}^{-1}. Since δn, due to external perturbations, varies slowly on this scale one can rewrite Eq. (52.9) as

$$E(\mathbf{p}, \mathbf{r}, n) = E_\mathbf{p}^0 + (2\pi)^{-3} \int d^3 p'\, f_{\mathbf{p}\mathbf{p}'}\, \delta n(\mathbf{p}', \mathbf{r}, t), \tag{52.13}$$

where

$$f_{\mathbf{p}\mathbf{p}'} \equiv \int d^3 r'\, F(\mathbf{p}, \mathbf{r}, \mathbf{p}', \mathbf{r}') \tag{52.14}$$

and we have assumed we are in a translationally invariant system. The function $f_{\mathbf{p}\mathbf{p}'}$ relates the change in energy of a quasi particle to the occupation number of other quasi particles which are present in the *same* spatial volume element.

Although we have not shown it to be true, the function $f_{\mathbf{p}\mathbf{p}'}$ is simply the quasi-particle–quasi-particle forward scattering amplitude. This connection should not be surprising. Changes in the energy momentum relation of an excitation are generally related to its forward scattering amplitude. For example, the dielectric constant of a medium, i.e., the dispersion relation (energy–momentum relation) of a light wave is directly proportional to the forward scattering amplitude for light (8).

The LS theory (neglecting spins) is completely characterized by the unknown functions $f_{\mathbf{p}\mathbf{p}'}$ and $E_\mathbf{p}^0$ in terms of which everything else will be expressed. For a rotationally invariant system and for small excitation energies, i.e., $p \simeq p_F$

$$E_\mathbf{p}^0 \simeq p_F(p - p_F)/m^*. \tag{52.15}$$

The quantity $f_{\mathbf{p}\mathbf{p}'}$ is the Landau scattering function.

For illustrative purposes, let us examine the form of $f_{\mathbf{p}\mathbf{p}'}$ in the Hartree–Fock theory. (For a discussion of transport theory in this approximation the reader is referred to the discussion in Section 18.) The ground state energy per particle in the Hartree–Fock approximation is given by (9)

$$E_T = (1/N) \left\{ \sum_{\mathbf{p},\alpha} (p^2/2m)\, n_{\mathbf{p},\alpha} - \tfrac{1}{2} \sum_{\mathbf{p},\mathbf{p}',\alpha}{}' V(|\mathbf{p} - \mathbf{p}'|)\, n_{\mathbf{p}',\alpha} n_{\mathbf{p},\alpha} \right\}. \tag{52.16}$$

Here

$$V(|\mathbf{p} - \mathbf{p}'|) = 4\pi e^2/|\mathbf{p} - \mathbf{p}'|^2 \qquad (52.17)$$

is the Fourier transform of the Coulomb repulsion between electrons. The prime on the summation in Eq. (52.16) indicates that the $\mathbf{p} = \mathbf{p}'$ terms are to be omitted. These terms are cancelled by the uniform background charge which must be present to preserve charge neutrality. The distribution function $n_{\mathbf{p},\alpha}$ is labeled with a spin index α. It takes on the values ± 1, i.e., up or down.

In order to obtain $f_{\mathbf{pp}'}$ we use our definition [Eq. (52.7)] of the quasi-particle energy

$$E_{\mathrm{HF}}(\mathbf{p}, \alpha) = (p^2/2m) - \sum_{\mathbf{p}'}{}' V(|\mathbf{p} - \mathbf{p}'|) n_{\mathbf{p}',\alpha}. \qquad (52.18)$$

This is just the usual one-electron energy in the Hartree–Fock approximation. Differentiating once again with respect to $n_{\mathbf{p},\alpha}$ we find that

$$f_{\mathbf{pp}'}^{\parallel} = -4\pi e^2/|\mathbf{p} - \mathbf{p}'|^2 \qquad \text{(parallel spins)} \qquad (52.19)$$

$$f_{\mathbf{pp}'}^{\perp} = 0 \qquad \text{(antiparallel spins)}. \qquad (52.20)$$

In general, of course, f will not be zero for antiparallel spins.

It is clear from this Hartree–Fock calculation that the inclusion of spin into the LS theory is essential. These effects are, at least phenomenologically, easily incorporated into the theory. We simply extend the definition of the semiclassical distribution function. It now must be thought of as a 2×2 matrix in spin space, i.e.,

$$n(\mathbf{p}, \mathbf{r}, t) \to \rho_{\alpha\beta}(\mathbf{p}, \mathbf{r}, t) \qquad (\alpha, \beta = 1, z) \qquad (52.21)$$

where

$$\rho_{\alpha\beta}(\mathbf{p}, \mathbf{r}, t) = n(\mathbf{p}, \mathbf{r}, t) \delta_{\alpha\beta} + \mathbf{\mu}(\mathbf{p}, \mathbf{r}, t) \cdot \mathbf{\sigma}_{\alpha\beta}. \qquad (52.22)$$

The Landau theory is assumed to go through as before using $\rho_{\alpha\beta}$ in place of n. Now we must, of course, allow the function f to be spin dependent, i.e., a 4×4 matrix in the spin of the two quasi particles. Thus the quasi-particle energy in a time independent problem is given by

$$E(\mathbf{p}, \mathbf{r}, \mathbf{\sigma}) = E_{\mathbf{p}}^{\circ} + (2\pi)^{-3} Tr_{\sigma'} \int d^3 p'\, f(\mathbf{p}, \mathbf{\sigma}, \mathbf{p}', \mathbf{\sigma}') \delta\rho(\mathbf{p}', \mathbf{r}, \mathbf{\sigma}'). \qquad (52.23)$$

The form of $f(\mathbf{p}, \mathbf{\sigma}, \mathbf{p}'\mathbf{\sigma}')$ is restricted by symmetry considerations. The most general form for f which is consistent with rotational invariance is

$$f(\mathbf{p}, \boldsymbol{\sigma}, \mathbf{p}'\boldsymbol{\sigma}') = f(\mathbf{p}, \mathbf{p}') + \boldsymbol{\sigma} \cdot \mathbf{g}(\mathbf{p}, \mathbf{p}') + \boldsymbol{\sigma}' \cdot \mathbf{g}'(\mathbf{p}, \mathbf{p}') + \sum_{ij} \sigma_i \sigma_j' \zeta_{ij}(\mathbf{p}, \mathbf{p}').$$
(52.24)

with $i, j = 1, 2, 3$. Since $f(\mathbf{p}, \boldsymbol{\sigma}, \mathbf{p}', \boldsymbol{\sigma}')$ is the second variational derivative of the energy with respect to the quasi-particle occupation numbers, it must be symmetric with respect to an interchange of primed and unprimed variables, i.e.,

$$f(\mathbf{p}, \boldsymbol{\sigma}, \mathbf{p}', \boldsymbol{\sigma}') = f(\mathbf{p}', \boldsymbol{\sigma}', \mathbf{p}, \boldsymbol{\sigma}). \tag{52.25}$$

Equation (52.25) in turn implies that

$$f(\mathbf{p}, \mathbf{p}') = f(\mathbf{p}', \mathbf{p}), \quad \mathbf{g}(\mathbf{p}, \mathbf{p}') = \mathbf{g}'(\mathbf{p}', \mathbf{p}), \quad \zeta_{ij}(\mathbf{p}, \mathbf{p}') = \zeta_{ij}(\mathbf{p}', \mathbf{p})$$
(52.26)

Since the only vectors one can form from $\mathbf{p}$ and $\mathbf{p}'$ are $\mathbf{p}, \mathbf{p}'$, and $\mathbf{p} \times \mathbf{p}'$ one sees that

$$\boldsymbol{\sigma} \cdot \mathbf{g}(\mathbf{p}, \mathbf{p}') = \boldsymbol{\sigma} \cdot [\mathbf{p}A + \mathbf{p}'B + (\mathbf{p} \times \mathbf{p}')C]. \tag{52.27}$$

Equation (52.27) is the most general form for the term linear in $\boldsymbol{\sigma}$. The terms proportional to $\boldsymbol{\sigma} \cdot \mathbf{p}$ and $\boldsymbol{\sigma} \cdot \mathbf{p}'$ are not allowed in metals with inversion symmetry since they change sign under spatial inversion. The term proportional to $\boldsymbol{\sigma} \cdot (\mathbf{p} \times \mathbf{p}')$ is in general allowed. Time reversal invariance $(\mathbf{p} \to -\mathbf{p}, \mathbf{p}' \to -\mathbf{p}', \boldsymbol{\sigma} \to -\boldsymbol{\sigma})$ of the scattering amplitude implies that $C = -C^*$, i.e., C is pure imaginary. However in a system with no spin-orbit coupling this linear term does not appear. The spin itself is a constant of the motion, so that the scattering amplitude must be invariant to rotations in spin space. Similar arguments apply to the term in ζ_{ij}. We find for nonmagnetic systems possessing a plane of reflection symmetry in the absence of spin orbit coupling that the most general form for f is

$$f(\mathbf{p}, \boldsymbol{\sigma}, \mathbf{p}', \boldsymbol{\sigma}') = f(\mathbf{p}, \mathbf{p}') + \zeta(\mathbf{p}, \mathbf{p}') \boldsymbol{\sigma} \cdot \boldsymbol{\sigma}'. \tag{52.28}$$

There are now two symmetric functions $f(\mathbf{p}, \mathbf{p}') = f(\mathbf{p}', \mathbf{p})$ and $\zeta(\mathbf{p}, \mathbf{p}') = \zeta(\mathbf{p}', \mathbf{p})$ replacing the single $f(\mathbf{p}, \mathbf{p}')$ which occurred for spinless particles. The two functions f and ζ are simply related to the $f^{\|}$ and $f^{\perp}$ [Eqs. (52.19), (52.20)] which we encountered in our brief Hartree–Fock discussion, i.e.,

$$f^{\|} = f + \zeta, \quad f^{\perp} = f - \zeta. \tag{52.29}$$

In Hartree–Fock theory we recall that $f^{\perp} = 0$ so that in this case $f = \zeta$.

We will restrict almost all of our discussion to systems with complete rotational invariance. In this uniform isotropic electron gas model f and ζ are only functions of $|\mathbf{p}|$, $|\mathbf{p}'|$ and $\cos\theta \equiv \hat{\mathbf{p}} \cdot \hat{\mathbf{p}}'$. Since the $|\mathbf{p}| = |\mathbf{p}'| = p_F$, f and ζ are really just functions of $\cos\theta$ which can, quite generally, be expanded in a series of Legendre polynomials, i.e.,

$$f(\mathbf{p}, \mathbf{p}') = \sum_n f_n P_n(\cos\theta) \qquad (52.30)$$

$$\zeta(\mathbf{p}, \mathbf{p}') = \sum_n \zeta_n P_n(\cos\theta). \qquad (52.31)$$

The set of dimensionless quantities,

$$A_n = m^* p_F f_n / (\pi^2 (2n+1)) \qquad (52.32)$$

$$B_n = m^* p_F \zeta_n / (\pi^2 (2n+1)) \qquad (52.33)$$

are, along with m^*/m_0, the set of *phenomenological* parameters that describe the transport properties of the interacting isotropic electron gas.

Real metals are clearly not completely rotationally symmetric. It is well known, however, that the Fermi surfaces of Na and K are very accurately spherical and that the lattice has little effect (small band gaps) on the free-electron-like band structure of these metals. Although the phonon spectrum of the alkalis is quite anisotropic, we will assume that the isotropic model for Na and K is a good one. We will come back and discuss this point again in Chapter XI. It is in fact possible to make a qualitative estimate of the accuracy of this picture as applied to all of the alkalis. The isotropic model turns out to be a good approximation for these systems.

53. Equilibrium Properties of the Interacting Electron Liquid

The analysis of wave phenomenon in an interacting electron gas involves an investigation of nonequilibrium transport properties. Before going on to a discussion of such transport properties we would like to begin our study of the Fermi liquid by considering two of its equilibrium properties.

The specific heat of a Fermi liquid is easily derived. In order to calculate thermodynamic properties we need an expression for the entropy $S\{n\}$ as well as the energy $E\{n\}$ [Eq. (52.3)]. A knowledge of the entropy is equivalent to specifying the equilibrium distribution of quasi-particles at a finite temperature T. Landau argued that the appropriate expression for the entropy is

$$S\{n\} = -k_B \int d\tau \, [n_p \ln n_p + (1 - n_p) \ln(1 - n_p)] \tag{53.1}$$

by analogy with the expression for a free Fermi gas. This equation is based upon the one-to-one correspondence of the quasi-particle states with those of the noninteracting Fermion system and the fact that the entropy can be obtained by a counting argument.

Maximizing the entropy subject to the usual constraints on total number of particles and total energy, as in the noninteracting case, leads to an equilibrium distribution of the form

$$n_p = \{\exp[\beta(E_p - \mu)] + 1\}^{-1}. \tag{53.2}$$

Here $\beta \equiv 1/k_B T$, μ is the chemical potential, and E_p is given by Eq. (52.13). Since n_p, at low temperature, deviates from zero or one only near the Fermi surface, these states alone contribute to S.

The heat capacity C_V is given by

$$C_V = T(\partial S/\partial T)_{V,N}. \tag{53.3}$$

Utilizing Eq. (53.2) our formula for S [Eq. (53.1)] gives

$$C_V = \int d\tau \, (E_p - \mu)(\partial n_p/\partial T). \tag{53.4}$$

The leading term (linear in T) in Eq. (53.4) may be evaluated by assuming that $E_p \equiv E_p{}^0$. Neglect of the interaction terms in E_p, i.e., terms proportional to δn, results in errors which can be shown to be of order T^3. In this low-temperature limit, Eq. (53.4) can be evaluated to give

$$C_V = (\pi^2/3) \, k_B{}^2 T (dv/dE). \tag{55.5}$$

The quantity dv/dE is the quasi-particle density of states, i.e.,

$$dv/dE = \pi^{-2} \, p_F{}^2 (dp/dE)_{p=p_F} = m^* p_F/\pi^2 \tag{53.6}$$

so that

$$C_V = \tfrac{1}{3}(k_B T) \, m^* p_F. \tag{53.7}$$

Thus the linear term in the specific heat measures m^*, the mass of the quasi particles and tells us nothing about $f(\mathbf{p}, \sigma, \mathbf{p}', \sigma')$.

Given an n_p of the form displayed in Eq. (53.2) we could have given a very qualitative argument suggesting that the interaction terms would

53. EQUILIBRIUM PROPERTIES

not enter. The quantity $\delta n = n(\mathbf{p}, T) - n(\mathbf{p}, 0)$ is of order T and *odd* about the Fermi momentum. Thus the term in the energy proportional to δn, is second order in the perturbation, i.e., of order T^2. The $(\delta n)^2$ terms, again because of the oddness of δn about the Fermi surface, are of order T^4. The T^2 term in the energy is the linear term in the specific heat and the T^4 terms do not contribute to this order.

The Pauli susceptibility χ is another important equilibrium property of interest. It is slightly more difficult to calculate than the specific heat. However, χ is somewhat more interesting in that (as we shall see) it is simply related to the exchange parameter B_0 in Eq. (52.33).

For a noninteracting Fermion system, each particle has a moment

$$\mathbf{m} = \gamma_0 \boldsymbol{\sigma} \tag{53.8}$$

with

$$\gamma_0 = ge\hbar/4m_0 c. \tag{53.9}$$

In equilibrium, in the absence of any external fields, the magnetization of the system is zero. As we turn on our uniform dc field $\mathbf{H}_0$, the density matrix is perturbed from its equilibrium value

$$n^0(\mathbf{p})\,\delta_{\alpha\beta} \to n^0(p)\,\delta_{\alpha\beta} + \delta\rho_{\alpha\beta}(\mathbf{p}, \boldsymbol{\sigma}). \tag{53.10}$$

The perturbation in the density matrix $\delta\boldsymbol{\rho}$ can be thought of as an addition of quasi particles and quasi holes. In the perturbed situation the magnetization is given by

$$\mathbf{M} = Tr_\sigma \int d\tau\, [\delta\boldsymbol{\rho}(\mathbf{p}, \boldsymbol{\sigma})\,\gamma_0\boldsymbol{\sigma}]. \tag{53.11}$$

In writing Eq. (53.11) we have *assumed* that the moment $\mathbf{m}$ per *quasi particle* is the same as that for a free particle, i.e., γ_0. Microscopic theories (6) have verified Eq. (53.11) when the forces are spin independent. The physics behind such a calculation is simple. Suppose we turn off the interactions and add a *single* extra particle to the system. The moment is $\gamma_0\boldsymbol{\sigma}$. If the forces are spin independent, i.e., the total magnetization of the system is a rigorously conserved quantity, then when the interactions are turned on the *total* moment is unchanged. Since the number of particles equals the number of quasi particles, Eq. (53.11) follows.

In order to calculate the susceptibility we need the change in the density matrix due to a uniform static field in the z direction. At finite temperatures in the presence of such a field the equilibrium density matrix is still given by

$$\rho(\mathbf{p}, \sigma) = \frac{1}{e^{\beta(E(\mathbf{p},\sigma)-\mu)} + 1}. \tag{53.12}$$

At $T = 0$

$$\rho(\mathbf{p}, \sigma) = \theta(E_F - E(\mathbf{p}, \sigma)), \tag{53.13}$$

where θ is the unit step function. Here

$$E(\mathbf{p}, \sigma) = E_p^0 + \delta E(\mathbf{p}, \sigma) \tag{53.14}$$

with

$$\delta E(\mathbf{p}, \sigma) = -\gamma(\mathbf{p})\sigma \cdot \mathbf{H}_0. \tag{53.15}$$

Equation (53.15) is nothing but the definition of $\gamma(\mathbf{p})$, the "clothed" gyromagnetic ratio of a "quasi particle." The quantity $\gamma(\mathbf{p})$ is, as yet, unknown and must be determined self-consistently. We may, however, simply express χ in terms of it.

Since the Fermi energy is independent of the magnetic field,

$$\delta\rho(\mathbf{p}, \sigma) = [(\partial n_0/\partial E)_{E=E_p^0}]\, \delta E(\mathbf{p}, \sigma)$$
$$= \delta(E_F - E_p^0)\, \gamma(\mathbf{p})\sigma \cdot \mathbf{H}_0. \tag{53.16}$$

Utilizing Eqs. (53.16) and (53.11), we see (taking $\mathbf{H}_0$ in the z direction) that

$$M_z \equiv \chi H_0 V = Tr_\sigma \int d\tau\, \gamma_0 \sigma_z\, \delta(E_F - E_p^0)\, \gamma(\mathbf{p})\, \sigma_z H_0 \tag{53.17}$$

In the isotropic case where $\gamma(\mathbf{p})$ is independent of the direction of $\mathbf{p}$, the integral over the delta function is easily performed, and Eq. (53.17) yields

$$\chi = \gamma_0 \gamma(p_F)\, m^* p_F/\pi^2. \tag{53.18}$$

We must now determine $\gamma(p_F)$. Utilizing our definition of quasi-particle energy[†] [Eq. (52.23)], we see that

$$E(\mathbf{p}, \sigma) = E_p^0 - \gamma_0 \sigma \cdot \mathbf{H}_0 + Tr_{\sigma'}(2\pi)^{-3} \int d^3p'\, f(\mathbf{p}, \sigma, \mathbf{p}', \sigma')\, \delta\rho(\mathbf{p}', \sigma') \tag{53.19}$$

[†] It is important to notice that the zeroth-order energy which we insert into Eq. (52.23) is the free electron energy in the presence of the uniform field H_0.

or

$$\delta E = -\gamma_0 \boldsymbol{\sigma} \cdot \mathbf{H}_0 + \gamma_0 Tr_{\sigma'}(2\pi)^{-3} \int d^3p' f(\mathbf{p}, \boldsymbol{\sigma}, \mathbf{p'}, \boldsymbol{\sigma'}) \boldsymbol{\sigma'} \cdot \mathbf{H}_0 \, \delta(E_F - E_{p'}^0). \tag{53.20}$$

The first term is the "bare result," while the second arises from the change in energy of the quasi particles due to the perturbed distribution function. It can be thought of as arising from the exchange field of the other electrons.

Equation (53.20) along with the definition of δE [Eq. (53.15)] immediately leads to a self-consistent linear integral equation for $\gamma(\mathbf{p})$, i.e.,

$$-\gamma(\mathbf{p})\, \sigma_z = -\gamma_0 \sigma_z + (2\pi)^{-3} \, Tr_{\sigma'} \int d^3p' \, \delta(E_F - E_{p'}^0)[f_{\mathbf{pp'}} + \zeta_{\mathbf{pp'}} \boldsymbol{\sigma} \cdot \boldsymbol{\sigma'}] \gamma(\mathbf{p'}) \sigma_z' \tag{53.21}$$

or

$$\gamma(\mathbf{p}) = \gamma_0 - (2\pi)^{-3} \int d^3p' \, \zeta(\mathbf{p}, \mathbf{p'}) \gamma(\mathbf{p'}) \delta(E_F - E_{p'}^0) \tag{53.22}$$

In the isotropic case where $\gamma(\mathbf{p})$ is independent of the direction of $\mathbf{p}$,

$$\gamma(p_F) = \gamma_0 - \gamma(p_F) \tfrac{1}{2} m^* p_F \pi^{-2} \int_{-1}^{+1} \zeta(\theta) \, d(\cos \theta). \tag{53.23}$$

Hence

$$\gamma(p_F) = \gamma_0/(1 + B_0). \tag{53.24}$$

It follows from Eqs. (53.24) and (53.18) that

$$\chi = \frac{m_0 p_F \gamma_0^2}{\pi^2} \left(\frac{m^*}{m_0}\right) \frac{1}{1 + B_0}. \tag{53.25}$$

For free fermions, the Pauli susceptibility is

$$\chi_0 = m_0 p_F \gamma_0^2 / \pi^2, \tag{53.26}$$

so that

$$\chi/\chi_0 = (m^*/m_0)(1 + B_0)^{-1}. \tag{53.27}$$

The quantity m^*/m_0 can be determined from a specific heat measurement [see Eq. (53.7)]. Thus, a measurement of the conduction electron susceptibility and of the electronic specific heat gives us a way of determining B_0. In Na two groups have made a direct measurement of the conduction electron susceptibility (*10, 11*), by measuring the *absolute* intensity of the CESR signal. This is accomplished by comparing the CESR signal with the nuclear magnetic resonance signal in the same sample. The most recent and best value, obtained by Schumacher and Vehse (*10*), is $B_0 = -0.28 \pm 0.03$. This one experiment is the only equilibrium measurement which gives unambiguous information about the scattering function $f(\mathbf{p}, \sigma, \mathbf{p}', \sigma')$.

The single parameter B_0 measured in a single element is rather meager information about the central function in the theory of strongly interacting fermions. In order to get more information about this function we must turn to a description of transport in metals in the presence of interaction effects. In essence we will be interested in setting up an equation, similar to the Boltzmann–Vlasov equation [Eq. (22.7)], which will enable us to compute the wave number and frequency-dependent conductivity $\boldsymbol{\sigma}(\mathbf{k}, \omega)$ and susceptibility $\boldsymbol{\chi}(\mathbf{k}, \omega)$ of the medium. Given expressions for $\boldsymbol{\sigma}(\mathbf{k}, \omega)$ and $\boldsymbol{\chi}(\mathbf{k}, \omega)$ in terms of $f(\mathbf{p}, \sigma, \mathbf{p}', \sigma')$, we may ask ourselves a number of relevant experimental questions.

(1) Does the presence of interactions influence the known wave propagation phenomenon in metals?
(2) Does the interacting theory predict a new set of disturbances?
(3) Can we use the experimental data to determine values for the numbers A_n and B_n?
(4) Are the results *consistent* with Fermi-liquid theory? That is, are there more experiments than parameters? (A_0, A_1,..., B_0, B_1,...).

We will see that the answer to all four of these questions is at least a partial yes.

54. Transport Equations for the Interacting Fermi Liquid

In order to compute the response of an interacting Fermi liquid to gentle perturbations, we require a transport equation similar to the BV equation [Eq. (22.7)]. The theory was outlined by Landau (*1, 3*) but the details were filled in by others (*4*). Since there are some minor conceptual difficulties involved in arriving at such a Boltzmann-like equation we proceed in two steps

1. *Free Fermions with Spin*

For free Fermions with spin we would like to treat the position $\mathbf{r}$ and momentum $\mathbf{p}$ of the electrons by semiclassical methods. This is not as simple as one might suppose since the spin of the electron must be treated fully quantum mechanically. Consider a *single* particle and the equation of motion of its quantum-mechanical density matrix $\boldsymbol{\rho}$,

$$\partial \boldsymbol{\rho}/\partial t = -(i/\hbar)[h, \boldsymbol{\rho}]. \tag{54.1}$$

54. TRANSPORT EQUATIONS

Here h is the single-particle Hamiltonian. For a system of *noninteracting electrons*, Eq. (54.1) is complete. The time dependence of any observable operator is obtained from the equation,

$$\bar{A} = \text{Tr}(\rho A). \tag{54.2}$$

We use, as a basis for describing the wave function of the system, a complete set of states for which the momentum $\mathbf{p}$ and the spin α of the electron are specified. In this case h and ρ are matrices in the spin and momentum variables. Hence,

$$\frac{\partial}{\partial t} \langle \alpha, \mathbf{p}_1 | \rho | \beta, \mathbf{p}_2 \rangle = -(i/\hbar) \sum_{\mathbf{p}_3, \gamma} [\langle \alpha, \mathbf{p}_1 | h | \gamma, \mathbf{p}_3 \rangle \langle \gamma, \mathbf{p}_3 | \rho | \beta, \mathbf{p}_2 \rangle$$
$$- \langle \alpha, \mathbf{p}_1 | \rho | \gamma, \mathbf{p}_3 \rangle \langle \gamma, \mathbf{p}_3 | h | \beta, \mathbf{p}_2 \rangle] \tag{54.3}$$

We can now introduce the Wigner (7) representation, which yields formulas for the density matrix much like their classical counterparts, i.e., the Boltzmann–Vlasov equation [Eq. (22.7)].

Defining

$$\rho_{\alpha\beta}(\mathbf{p}, \mathbf{r}) = \sum_\mathbf{q} \langle \alpha, \mathbf{p} + \mathbf{q}/2 | \rho | \beta, \mathbf{p} - \mathbf{q}/2 \rangle e^{i\mathbf{q}\cdot\mathbf{r}/\hbar} \tag{54.4}$$

and

$$h_{\alpha\beta}(\mathbf{p}, \mathbf{r}) = \sum_\mathbf{q} \langle \alpha, \mathbf{p} + \mathbf{q}/2 | h | \beta, \mathbf{p} - \mathbf{q}/2 \rangle e^{i\mathbf{q}\cdot\mathbf{r}/\hbar}, \tag{54.5}$$

Eq. (54.3) can be converted into an equation for $\rho(\mathbf{p}, \mathbf{r})$. The algebra is fairly tedious but straightforward. In the limit $\hbar \to 0$ the classical density matrix (matrix only in the spin variables) obeys an equation of the form

$$\frac{\partial \rho^c}{\partial t} + \frac{i}{\hbar} [h^c, \rho^c]_\sigma + \sum_\mu \frac{1}{2} \left(\frac{\partial h^c}{\partial p_\mu} \frac{\partial \rho^c}{\partial r_\mu} - \frac{\partial h^c}{\partial r_\mu} \frac{\partial \rho^c}{\partial p_\mu} \right)$$
$$+ \frac{1}{2} \left(\frac{\partial \rho^c}{\partial r_\mu} \frac{\partial h^c}{\partial p_\mu} - \frac{\partial \rho^c}{\partial p_\mu} \frac{\partial h^c}{\partial r_\mu} \right) = 0 \tag{54.6}$$

The superscript c on ρ and h denotes the classical density matrix and classical Hamiltonian respectively, i.e.,

$$\rho_{\alpha\beta}(\mathbf{p}, \mathbf{r}) \xrightarrow{\hbar \to 0} \rho^c_{\alpha\beta}(\mathbf{p}, \mathbf{r})[1 + \mathcal{O}(\hbar^2) + \cdots) \tag{54.7}$$

$$h_{\alpha\beta}(\mathbf{p}, \mathbf{r}) \xrightarrow{\hbar \to 0} h^c_{\alpha\beta}(\mathbf{p}, \mathbf{r})[1 + \mathcal{O}(\hbar^2) + \cdots) \tag{54.8}$$

with

$$h^c_{\alpha\beta}(\mathbf{p}, \mathbf{r}) \equiv p^2/2m^*\delta_{\alpha\beta} - \gamma_0\boldsymbol{\sigma}_{\alpha\beta} \cdot \mathbf{H}. \tag{54.9}$$

[From this point on we omit the superscript c.]

It is important to note that since $\gamma_0 H \equiv e\hbar H/2m_0 c$ we have *not* taken the limit $\hbar \to 0$ in the spin term. Only in the spatial or equivalently $\mathbf{r}$ dependence of $h_{\alpha\beta}$ have we let $\hbar = 0$. The second term in Eq. (54.6) is a commutator on the spin variables alone. Any dependence of h on spin must be proportional to $\hbar$ so that the "spin term" is actually of the same order as the orbital terms, i.e., $(\gamma_0 H_0)/\hbar \to \omega_s$ as $\hbar \to 0$. Our result Eq. (54.6) is the semiclassical result. In the absence of spin the second term is zero and the sum of the third and fourth terms in Eq. (54.6) is equal to the Poisson bracket of h with $\boldsymbol{\rho}$.

Collisions, as in the Vlasov case, are often included in Eq. (54.6) phenomenologically. We replace the zero on the right-hand side of Eq. (51.7) by $\partial \boldsymbol{\rho}/\partial t \vert_{\text{coll}}$. We will require that $\partial \boldsymbol{\rho}/\partial t \vert_{\text{coll}}$ force a nonequilibrium density matrix back to the equilibrium one. The equilibrium density matrix in the absence of an external field is the Fermi distribution $n^0(p)$:

$$(\rho_0)_{\alpha\beta} \equiv n^0 \delta_{\alpha\beta}. \tag{54.10}$$

The use of n^0 in the collision integral immediately transcribes our single-electron transport equation into an equation describing the transport properties of a set of *noninteracting* fermions.

In the absence of external fields or even in the presence of a static dc field, the transport equation is a trivial one. It tells us either that n^0 satisfies Eq. (54.6) identically or equivalently that a nonequilibrium $\boldsymbol{\rho}$ relaxes back to n^0 with some characteristic time constant.

Equation (54.6) is nontrivial when we apply space- and time-dependent external fields. To write the equation for $\boldsymbol{\rho}$ explicitly in the presence of these fields we replace $\mathbf{p}$ by $\mathbf{p} - e/c\mathbf{A}$ in h and add on to h the scalar potential φ, i.e.,

$$h = [\mathbf{p} - (e/c)\mathbf{A}]^2/2m^* - \gamma_0\boldsymbol{\sigma} \cdot \mathbf{H} + e\varphi. \tag{54.11}$$

The electric and magnetic fields are just

$$\mathbf{E} = -(1/c)\partial \mathbf{A}/\partial t - \boldsymbol{\nabla}\varphi \tag{54.12}$$

$$\mathbf{H} = \boldsymbol{\nabla} \times \mathbf{A} \tag{54.13}$$

The semiclassical distribution function $n(\mathbf{p}, \mathbf{r}, t)$ utilized in the

Boltzmann–Vlasov equation, and the magnetization density $\mu(\mathbf{p}, \mathbf{r}, t)$ are defined by [see Eq. (52.22)]

$$n(\bar{\mathbf{p}}, \mathbf{r}, t) = \tfrac{1}{2}\operatorname{tr}[\rho(\bar{\mathbf{p}}, \mathbf{r}, t)] \qquad (54.14)$$

$$\mu(\bar{\mathbf{p}}, \mathbf{r}, t) = \tfrac{1}{2}\operatorname{tr}[\sigma\rho(\bar{\mathbf{p}}, \mathbf{r}, t)]. \qquad (54.15)$$

Here

$$\bar{p}/m^* \equiv (1/m^*)(\mathbf{p} - (e/c)\mathbf{A}) = \mathbf{v} \qquad (54.16)$$

is the velocity of the particles. A knowledge of the single-particle distribution $n(\bar{\mathbf{p}}, \mathbf{r}, t)$ enables us to compute the current induced in the medium, i.e.,

$$\mathbf{j} \equiv e\langle \mathbf{v}\rangle = e(2\pi)^{-3}\int d^3\bar{p}\,(\bar{\mathbf{p}}/m^*)\,n(\bar{\mathbf{p}}, \mathbf{r}, t). \qquad (54.17)$$

In a linearized problem, where there may be a large static magnetic field but only a weak rf field, the distinction between $\bar{\mathbf{p}}$ and $\mathbf{p}$ is important for the static portion of the field. We must include $\mathbf{H}_0$ in our definition of the velocity. Writing the transport equation in terms of the momentum $\bar{\mathbf{p}}$ or equivalently $\mathbf{v}$ has the advantage that all quantities in the formulas remain gauge invariant. The distribution functions $n°(\bar{p})$ will not, for example, depend on the absolute value of the vector potential of the dc field. In the absence of spin,

$$n° = \{\exp[\beta(\bar{E}°_\mathbf{p} - \mu)] + 1\}^{-1} \qquad (54.18)$$

with

$$\bar{E}°_\mathbf{p} = \bar{p}^2/2m^*. \qquad (54.19)$$

In order to put Eq. (54.6) in a more conventional form we must evaluate the commutator and the Poisson bracket expressions in it. The Poisson bracket of any two functions M and N is defined by

$$\left\{M\left(\mathbf{p} + \frac{e}{c}\mathbf{A}, \mathbf{r}\right), N\left(\mathbf{p} + \frac{e}{c}\mathbf{A}, \mathbf{r}\right)\right\} \equiv \frac{\partial M}{\partial p_i}\frac{\partial N}{\partial r_i} - \frac{\partial M}{\partial r_i}\frac{\partial N}{\partial p_i}. \qquad (54.20)$$

The implicit dependence of M and N on $\mathbf{r}$ gives a term proportional to the $\nabla \times \mathbf{A}$, i.e.,

$$\{M, N\}_A = \{M, N\}_{A=0} + \left(\frac{\partial M}{\partial \mathbf{p}} \times \frac{\partial N}{\partial \mathbf{p}}\right)\cdot\left(\frac{e}{c}\nabla \times \mathbf{A}\right). \qquad (54.21)$$

In addition the commutator

$$[h, \boldsymbol{\rho}] = i\gamma_0(\mathbf{H} \times \boldsymbol{\mu})\cdot\boldsymbol{\sigma}. \qquad (54.22)$$

These two, Eqs. (54.21) and (54.22), rather simple algebraic properties of the quantities appearing in Eq. (54.6), enable us to perform the operation implicit in Eq. (54.6) and to rewrite it.

The set of four equations for $n(\bar{p}, r, t)$ and $\mu(\bar{p}, r, t)$ are obtained from the resulting matrix equation by taking the trace (one equation) and by multiplying by σ and then taking the trace (three equations). The algebraic manipulations involved are particularly straightforward, tedious, and not illuminating so we omit them. The single-particle semiclassical distribution function satisfies the BV equation [Eq. (22.7)]. The magnetization density satisfies similar set of vectorlike equations

$$\frac{\partial \mu}{\partial t} + \left[eE + \frac{e}{c}\left(\frac{\partial \epsilon_1}{\partial p} \times H\right)\right] \cdot \frac{\partial}{\partial p}\mu$$
$$+ \frac{\partial \epsilon_1}{\partial p} \cdot \frac{\partial \mu}{\partial r} - \frac{\partial \mu}{\partial p} \cdot \frac{\partial \epsilon_1}{\partial r} + 2(\epsilon_2 \times \mu)$$
$$+ \frac{\partial n}{\partial r} \cdot \frac{\partial \epsilon_2}{\partial p} - \frac{\partial n}{\partial p} \cdot \frac{\partial \epsilon_2}{\partial r} - \frac{e}{c}\left(\frac{\partial n}{\partial p} \times H\right) \cdot \frac{\partial \epsilon_2}{\partial p}$$
$$= \frac{\partial \mu}{\partial t}\bigg|_{\text{coll}}. \qquad (54.23)$$

Here

$$\epsilon_1 = E^\circ_p = p^2/2m^*, \qquad (54.24)$$

$$\epsilon_2 = -\gamma_0 H. \qquad (54.25)$$

We have written Eq. (54.23) in a form suitable for extension to the more complicated case of interacting fermions. In this simple case several of the terms in Eq. (54.23) are obviously zero, for example $\partial \epsilon_2/\partial p = 0$.

We are interested in Eq. (54.23) when a large uniform static field H_0 is applied in the z direction. A small space- and time-dependent rf field h is then applied in the x, y plane. The magnetization vector μ may then conveniently be written as the sum of a zeroth-order static part and a first-order time- and space-dependent piece, i.e.,

$$\mu \equiv M^0 + m(p, r, t) \qquad (54.26)$$

The quantity M^0 is the static magnetization induced in the gas by the uniform field H_0 and m is the "small" rf magnetization density induced by h.

M_0 is easily evaluated. Since $\omega_s/E_F \ll 1$, the change in the density matrix due to the application of a static magnetic field is small. To first order in ω_s/E_F

$$\delta \rho = (\partial n^\circ/\partial E^\circ)\,\delta \epsilon \qquad (54.27)$$

Since
$$\delta\epsilon = -\gamma_0 \boldsymbol{\sigma} \cdot \mathbf{H}_0 \tag{54.28}$$
$$\delta\rho = -\gamma_0 (\partial n^\circ / \partial E^\circ) \boldsymbol{\sigma} \cdot \mathbf{H}_0 . \tag{54.29}$$

Multiplying Eq. (54.29) by $\boldsymbol{\sigma}$ and taking one-half the trace gives
$$\mathbf{M}^0 = -(\partial n^\circ / \partial E^\circ) \gamma_0 \mathbf{H}_0 . \tag{54.30}$$

We are now in a position to write a simple transport equation for $m^\pm = m_x \pm im_y$. Inserting ϵ_1 and ϵ_2 into Eq. (54.23), utilizing Eq. (54.30) and linearizing in $\mathbf{m}$, we obtain the following BV-like equations for $m^\pm$,

$$\frac{\partial m^\pm}{\partial t} + \left[\mathbf{v} \cdot \boldsymbol{\nabla} + \frac{e}{c}(\mathbf{v} \times \mathbf{H}_0) \cdot \frac{\partial}{\partial \mathbf{p}} \pm i\omega_s \right] m^\pm$$
$$= \frac{\gamma_0}{2} \left[\mathbf{v} \cdot \boldsymbol{\nabla} + i\omega_s \right] \left(\frac{\partial n^\circ}{\partial \epsilon} \right) h^\pm + \frac{\partial m^\pm}{\partial t} \bigg|_\text{coll} \tag{54.31}$$

where
$$h^\pm \equiv h_x \pm ih_y . \tag{54.32}$$

The left-hand side of Eq. (54.31) is familiar from our work on the BV equation. It describes the free streaming of rf magnetization density $\mathbf{m}(\mathbf{p}, \mathbf{r}, t)$. The only formal difference between the left-hand side of Eq. (54.31) and the BV equation (22.7) is the additional $i\omega_s$ term describing the free precession of the magnetic moment of the electrons. The right-hand side of Eq. (54.31) consists of a driving term, i.e., the term proportional to $h^\pm$ (this term is similar to the $e\mathbf{E} \cdot \mathbf{v}$ in the BV equation), and a collision term. The $\partial m^\pm / \partial t |_\text{coll}$ in Eq. (54.32) is the phenomenological collision term, which in a noninteracting model, characterizes the relaxation of the nonequilibrium magnetization density to its equilibrium value. It is the analog of the phenomenological collision term introduced into the BV equation. We will treat it, as for the orbital problem, within the framework of a relaxation time approximation. We must keep in mind, however, the discussion in Section 50 where we argued that there were two distinct relaxation times, τ and T_2.

The collision integral must take into account the two physically different scattering mechanisms. We choose $\partial m^\pm / \partial t |_\text{coll}$ to be of the form,

$$\frac{\partial m^\pm}{\partial t} \bigg|_\text{coll} = \frac{\partial m^\pm}{\partial t} \bigg|^1 + \frac{\partial m^\pm}{\partial t} \bigg|^2 \tag{54.33}$$

Here
$$\frac{\partial m^\pm}{\partial t} \bigg|^1 = -\frac{1}{\tau} \left[m^\pm - \left(\frac{\partial n^\circ}{\partial E^\circ} \right) \frac{(2\pi)^{-3} \int d^3p \, m^\pm}{(2\pi)^{-3} \int d^3p \, (\partial n^\circ / \partial E^\circ)} \right] \tag{54.34}$$

and

$$\left.\frac{\partial m^{\pm}}{\partial t}\right|^{2} = -(1/T_2)\, m^{\pm}. \tag{54.35}$$

The form of Eq. (54.33) is only slightly more complicated than Eq. (22.7). It contains the two distinct scattering times τ and T_2. The $\partial m^{\pm}/\partial t\, |^1$ characterizes the orbital or nonmagnetic scattering. It is "cooked up" in a way such that the total rf magnetization density $(2\pi)^{-3} \int m^{\pm}\, d^3p$, is conserved for this type of collision. This scattering term randomizes the momentum but does not relax the magnetization density, i.e.,

$$(2\pi)^{-3} \int d^3p \left.\left(\frac{\partial m^{\pm}}{\partial t}\right)\right|^1 = 0. \tag{54.36}$$

The second term characterizes the magnetic scattering. It relaxes the nonequilibrium transverse magnetization to zero.

Fourier analyzing the magnetization density $m^{\pm}$, i.e.,

$$m^{\pm}(\mathbf{p},\mathbf{r},t) \equiv m^{\pm}(\mathbf{p})\, e^{i(\mathbf{k}\cdot\mathbf{r}-\omega t)}, \tag{54.37}$$

and defining the quantity g by

$$m^{\pm}(\mathbf{p}) \equiv (\partial n^{\circ}/\partial E^{\circ}) g^{\pm}, \tag{54.38}$$

the transport equation (54.31) is,

$$-i\omega g^{\pm} + [i\mathbf{k}\cdot\mathbf{v} + i\omega_c\, \partial/\partial\varphi \pm i\omega_s] g^{\pm}$$
$$= \tfrac{1}{2} i\gamma_0(\mathbf{k}\cdot\mathbf{v} + \omega_s)\, h^{\pm} - (1/\tau)\left[g^{\pm} - (4\pi)^{-1} \int g^{\pm}\, d\Omega\right] - (1/T_2) g^{\pm}. \tag{54.39}$$

Equation (54.39) is, for the zero-temperature noninteracting Fermion system, an equation in angle, i.e., the angle the velocity vector $\mathbf{v}$ makes with respect to the coordinate system defined in Fig. VIII-1. The quantity $d\Omega$ is the increment of solid angle. Except for the more complicated collision term, Eq. (54.39) is almost identical to the BV equation (45.1) for the particle density $n(\mathbf{p},\mathbf{r},t)$.

2. *Interacting Fermions with Spin*

Landau conjectured that, in the *interacting* system, the density matrix for quasi particles satisfies a transport equation which is, in form, identical with Eq. (54.6). In this case however the semiclassical hamiltonian h is simply equal to the energy $E(\mathbf{p},\mathbf{r},\sigma)$ of the quasi particles given by Eq. (52.23). In the presence of external fields he further suggested that

$$h = E(\mathbf{p} - (e/c)\mathbf{A}, \mathbf{r}, t) - \gamma_0 \boldsymbol{\sigma} \cdot \mathbf{H} + e\varphi \qquad (54.40)$$

The more complicated form of $E(\mathbf{p}, \mathbf{r}, t)$ distinguishes the noninteracting single-particle density matrix equation from the interacting one. $E(\mathbf{p}, \mathbf{r}, t)$ depends implicitly on the coordinate $\mathbf{r}$ and time t because the distribution function for the excited quasi particles is nonuniform in space and time due to the space- and time-dependent fields which are present.

The quantity $\partial \mathbf{p}/\partial t |_{\text{coll}}$ describes the collisions of quasi particles with lattice defects. As in the noninteracting case we will, in our subsequent discussion treat this term phenomenologically. It will be assigned a form consistent with all known conservation laws. Since we will be primarily interested in pure materials at low temperatures and high frequencies ($\omega \tau \gg 1$), $\partial \mathbf{p}/\partial t |_{\text{coll}}$ will be, in some sense, negligible. Thus, the approximate form used for it will not be very important.

As in the case of the noninteracting system, we can derive a set of linearized BV-like transport equations. To illustrate the procedure and indicate specifically how the LS transport equation differs from the non-interacting case, consider the derivation of such a transport equation in the absence of spin. In this case Eq. (54.6), in the classical limit, is simply

$$\frac{\partial n}{\partial t} + \frac{\partial n}{\partial \mathbf{r}} \frac{\partial E}{\partial \mathbf{p}} - \frac{\partial E}{\partial \mathbf{r}} \frac{\partial n}{\partial \mathbf{p}} = \frac{\partial n}{\partial t}\bigg|_{\text{coll}}. \qquad (54.41)$$

where

$$n = n^\circ(\mathbf{p} - (e/c)\mathbf{A}) + \delta n(\mathbf{p} - (e/c)\mathbf{A}, \mathbf{r}, t), \qquad (54.42)$$

and

$$E = E^0[\mathbf{p} - (e/c)\mathbf{A}] + \delta E(\mathbf{p} - (e/c)\mathbf{A}, \mathbf{r}, t) + e\varphi \qquad (54.43)$$

with

$$\delta E(\mathbf{p}) = (2\pi)^{-3} \int d^3 p' \, f_{\mathbf{p}\mathbf{p}'} \, \delta n(\mathbf{p}', \mathbf{r}, t). \qquad (54.44)$$

In order to obtain the final transport equation we need the relations,

$$\frac{\partial n^\circ}{\partial t} = -\frac{e}{c}\left(\frac{\partial n^\circ}{\partial E^\circ}\right)\left(\frac{\partial E^\circ}{\partial \mathbf{p}}\right) \cdot \left(\frac{\partial \mathbf{A}}{\partial t}\right) = -\frac{e}{c}\left(\frac{\partial n^\circ}{\partial E^\circ}\right)\left(\mathbf{v} \cdot \frac{\partial \mathbf{A}}{\partial t}\right) \qquad (54.45)$$

and, utilizing (54.21),

$$\{\delta n, E^\circ\}_\mathbf{A} = \{\delta n, E^\circ\}_{\mathbf{A}=0} + \frac{e}{c}(\mathbf{v} \times \mathbf{H}_0) \cdot \frac{\partial(\delta n)}{\partial \mathbf{p}} \qquad (54.46)$$

$$\{n^\circ, \delta E\}_\mathbf{A} = \{n^\circ, \delta E\}_{\mathbf{A}=0} + \frac{e}{c}(\mathbf{v} \times \mathbf{H}_0) \cdot \frac{\partial}{\partial \mathbf{p}}(\delta E)\left(\frac{\partial n^\circ}{\partial E^\circ}\right). \qquad (54.47)$$

Equations (54.45)–(54.47) when inserted into the density matrix equation (54.6) yield

$$\frac{\partial}{\partial t}(\delta n) + e\mathbf{E} \cdot \left[\frac{\partial n^\circ}{\partial \mathbf{p}}\right] + \left[\mathbf{v} \cdot \nabla + \frac{e}{c}(\mathbf{v} \times \mathbf{H}_0) \cdot \frac{\partial}{\partial \mathbf{p}}\right][\delta n + \delta E] = \frac{\partial n}{\partial t}\bigg|_{\text{coll}}.$$
(54.48)

The variable $\mathbf{v}$ in Eq. (54.48) is, as in our early discussion, the velocity, i.e., $[\mathbf{p} - (e/c)\mathbf{A}_0]/m^*$.

Equation (54.48) is the transport equation, for the interacting liquid. It plays the same role as the Boltzmann–Vlasov equation (45.1) did for the noninteracting system. It is in fact, in form, very similar to Eq. (45.1). The primary difference is the additional term δE which appears on the left-hand side of Eq. (54.48). We will shortly discuss this term in some detail.

A similar set of manipulations leads to the transport equations for the transverse components of the rf magnetization. These equations are analogous to Eqs. (54.31) and may be obtained by substituting into Eq. (54.23) the appropriate interaction dependent expressions for the energy, i.e.,

$$\epsilon_1 = E^\circ(p) + (2\pi)^{-3}\int d^3p'\, f(\mathbf{p}, \mathbf{p'})\, \delta n(\mathbf{p'}, \mathbf{r}, t)$$
$$\equiv E^\circ(p) + \delta E$$
(54.49)

and

$$\epsilon_2 = -\gamma_0 \mathbf{H} + (2\pi)^{-3}\int d^3p'\, \zeta(\mathbf{p}, \mathbf{p'})\, \mu(\mathbf{p'}, \mathbf{r}, t).$$
(54.50)

The result for small rf driving fields is

$$\frac{\partial m^\pm}{\partial t} + \left[\mathbf{v} \cdot \nabla + \frac{e}{c}(\mathbf{v} \times \mathbf{H}_0) \cdot \frac{\partial}{\partial \mathbf{p}} \pm i\Omega_0\right][m^\pm + \delta m^\pm]$$
$$= \frac{\gamma_0}{2}[\mathbf{v} \cdot \nabla \pm i\Omega_0]\left[\frac{\partial n^\circ}{\partial E^\circ}\right] h^\pm + \frac{\partial m^\pm}{\partial t}\bigg|_{\text{coll}}$$
(54.51)

where

$$\delta m^\pm = (2\pi)^{-3}\int d^3p'\, \zeta(\mathbf{p}, \mathbf{p'})\, m^\pm(\mathbf{p'}, \mathbf{r}, t)$$
(54.52)

and

$$\Omega_0 \equiv \omega_{\text{s}}/(1 + B_0).$$
(54.53)

Equations (54.48) and (54.51) are the complete set of transport equations relevant to our forthcoming discussion of wave phenomenon in the interacting Fermi liquid.

55. The Electrical Transport Equation for Fermi Liquids

Since the driving term $e\mathbf{E} \cdot \partial n°/\partial \mathbf{p}$ in Eq. (54.48) confines the deviations in the distribution function to the neighborhood of the Fermi surface, it is convenient to define the quantity g by

$$\delta n \equiv -(\partial n°/\partial E°)g. \tag{55.1}$$

The linearized transport equation for g becomes

$$\frac{\partial g}{\partial t} + \left[\mathbf{v} \cdot \nabla + \frac{e}{c}(\mathbf{v} \times \mathbf{H}_0) \cdot \nabla_\mathbf{p}\right][g + \delta\epsilon_1] = e\mathbf{E} \cdot \mathbf{v} + \frac{\partial g}{\partial t}\bigg|_{\text{coll}}. \tag{55.2}$$

Here,

$$\delta\epsilon_1 = 2(2\pi)^{-3} \int d^3p' \, f(\mathbf{p}, \mathbf{p}') g(\mathbf{p}', \mathbf{r}, t). \tag{55.3}$$

If we assume that collisions conserve the number of particles, we may integrate both sides of Eq. (55.2) to obtain the continuity equation

$$\partial \rho/\partial t + \nabla \cdot \mathbf{j} = 0 \tag{55.4}$$

where the charge density ρ is given by

$$\rho = e(2\pi)^{-3} \int d^3p \, g \, \delta(E_\mathbf{p}° - E_\mathrm{F}) \tag{55.5}$$

and the current density

$$j_\alpha = e(2\pi)^{-3} \int d^3p \, (p_\alpha/m^*)(g + \delta\epsilon_1) \, \delta(E_\mathbf{p}° - E_\mathrm{F}). \tag{55.6}$$

Equation (55.6) may be rewritten as

$$j_\alpha = e(2\pi)^{-3} \int d^3p \, v_\alpha g \, \delta(E_\mathbf{p}° - E_\mathrm{F}) \tag{55.7}$$

where the quasi-particle velocity

$$v_\alpha = \partial E(\mathbf{p}, \mathbf{r}, t)/\partial p_\alpha$$
$$\equiv (p_\alpha/m^*) + 2(2\pi)^{-3} \int d^3p' \, f(\mathbf{p}, \mathbf{p}')(p_\alpha'/m^*) g \, \delta(E_\mathbf{p}° - E_\mathrm{F}). \tag{55.8}$$

Since the integral over p_α in Eq. (55.8) picks out the first spherical harmonic, we can rewrite Eq. (55.7) as

$$j_\alpha = e(2\pi)^{-3} (m_0/m^*)(1 + A_1) \int d^3p \, g \, \delta(E_\mathbf{p}° - E_\mathrm{F})(p_\alpha/m_0). \tag{55.9}$$

In Landau's original paper, he utilized Eq. (55.9) to "prove" the relation

$$m^*/m_0 = 1 + A_1 . \tag{55.10}$$

He argued, as we did in the spin case, that if the interacting system is translationally invariant, then the total momentum, i.e., current induced by a uniform field, is conserved in quasi-particle collisions. It then follows that the current associated with a single excited quasi particle must also be conserved as the interactions are turned off, i.e., j_α must be given by

$$j_\alpha = e(2\pi)^{-3} \int d^3p \, g \, \delta(E_\mathbf{p}^\circ - E_\mathrm{F})(p_\alpha/m_0) \tag{55.11}$$

so that Eq. (55.10) follows.

We have presented the above argument to illustrate two important points:

(1) in a translationally invariant system m^*/m_0 and A_1 are not independent parameters;

(2) the relation between m^*/m_0 and A_1 is not valid in a metal where phonons and band structure effects are important.

The function $g + \delta\epsilon_1 \equiv \bar{g}$ plays a prominent role in the transport equation. It has a simple physical interpretation. The quantity $\delta n = n - n^\circ$ is the deviation of the distribution function from the equilibrium distribution function with no quasi particle excited, i.e., $n^\circ = \theta[E_\mathbf{p}^\circ - E_\mathrm{F}]$. Now define the quantity,

$$\overline{\delta n} \equiv n - \overline{n^0} \tag{55.12}$$

In Eq. (55.12) $\overline{n^0}$ is the local equilibrium distribution, i.e.,

$$\overline{n^0} = \theta(E - E_\mathrm{F}) \tag{55.13}$$

is the Fermi distribution with the local quasi-particle energy E as the argument. To leading order in δn,

$$\overline{\delta n} = \delta n - (\overline{n^\circ} - n^\circ) \cong \delta n - (\partial n^\circ/\partial E^\circ) \, \delta E \tag{55.14}$$

Thus,

$$\overline{\delta n} = \delta n - (\partial n^\circ/\partial E^\circ)(2\pi)^{-3} \int d^3p' \, f(\mathbf{p}, \mathbf{p}') \, \delta n(\mathbf{p}', \mathbf{r}, t) \equiv -(\partial n^\circ/\partial E^\circ)[g + \delta\epsilon_1]. \tag{55.15}$$

Equation (55.15) implies that $\bar{g} \equiv g + \delta\epsilon_1$ is simply proportional to the deviation of the distribution function from local equilibrium. This is a useful bit of information. Although we do not know the explicit form of the collision integral, we do know that whatever its detailed behavior, it must be such that it forces relaxation back to *local* equilibrium, i.e., it is a function of $\bar{g}$ alone. The so-called relaxation time approximation (p. 152) is equivalent, within the context of the LS theory to the assumption that

$$\partial g/\partial t \mid_{\text{coll}} = -\bar{g}/\tau. \qquad (55.16)$$

For disturbances of the form $e^{i(\mathbf{k}\cdot\mathbf{r}-\omega t)}$ and a coordinate system defined in Fig. VIII-1, Eq. (55.2), is

$$-i\omega g + (i\mathbf{k}\cdot\mathbf{v} + \omega_c\, \partial/\partial\varphi)(g + \delta\epsilon_1) = e\mathbf{E}\cdot\mathbf{v} - (1/\tau)(g + \delta\epsilon_1). \qquad (55.17)$$

Equation (55.17) is the Fermi-liquid analog of the BV equation (45.8). It determines all of the low-frequency, long-wavelength electron transport for a uniform interacting electron gas. It is an integro-differential equation which explicitly involves the spin independent part of the LS scattering function.

Before going into a rather detailed description of the properties of the CW in the interacting system it is useful to examine some rather general features of Eq. (55.17). Since both g or $\bar{g}$ are perfectly acceptable distribution functions [see Eqs. (55.5)–(55.6)], it is clear that in the limit of low frequencies ($\omega/kv_F \to 0$ and/or $\omega/\omega_c \to 0$) Eq. (55.17) reduces to the old BV equation with $\bar{g}$ replacing g. This result implies that in the low-frequency regime we cannot expect the Landau scattering function $f(\mathbf{p}, \mathbf{p}')$ to affect the conductivity tensor significantly. Qualitatively this is the reason why the dispersion properties of helicons do *not* show many-body effects.

Conversely, in the high-frequency regime ($\omega/kv_F \gg 1$ and/or $\omega/\omega_c \gg 1$) we can use g as our distribution function and aside from an overall scale factor arrive at the standard expression for the long-wavelength response of an electron gas. In the absence of a field $\mathbf{H}_0$ when $\omega\tau \gg 1$ this conductivity is

$$\sigma = (n_0 e^2/i m_0 \omega)(m_0/m^*)(1 + A_1). \qquad (55.18)$$

A measurement of the *local* high-frequency conductivity would enable one to determine experimentally the quantity A_1. However, it is difficult to envisage an experiment which in fact precisely measures the high-frequency *local* conductivity. The anomalous skin effect, the presence of phonons (to be discussed), and the existence of interband transitions complicates the interpretation of infrared reflectivity measurements in terms of a local conductivity of the form given in Eq. (55.18).

There is still another point worth making before investigating specific solutions of Eq. (55.17). This argument, first given by Azbel (12), implies that the mass measured in AKCR is m^* ($\omega_c = eH_0/m^*c$), i.e., the quasiparticle mass. The argument is qualitative and simple enough to describe even though we have avoided any real discussion of AKCR. The essential features of AKCR were shown in Fig. VIII-9 and described in Chapter VIII. With the magnetic field parallel to the surface, the electron moves in a highly nonuniform electric field. The rf field, as we have already pointed out, has momentum components $k \sim \delta^{-1}$ in it so that typically $kR_c \gg 1$. In order to characterize AKCR resonance mathematically we need the various components of the conductivity tensor for $\mathbf{k} \perp \mathbf{H}_0$. The important values of k are crudely speaking determined by the spatial variation of the electric field, i.e., $k \sim \delta_s^{-1}$. As in Chapter VIII we must solve the BV-like equation (55.17) for the distribution function g and then evaluate the conductivity by averaging the velocity over the Fermi surface. In order to do this we *assume* that the term $\delta\epsilon_1$ in Eq. (55.17) is small. In this case the solution for g is given by the solution of the noninteracting BV equation [see Eq. (45.11)].

To see if our assumption concerning the smallness of $\delta\epsilon_1$ is self-consistent, we must insert the distribution function g into Eq. (55.17), evaluate the integral over momenta $\mathbf{p}'$, and compare it to the original g. In the large kR_c limit the Bessel functions J_n in the expressions for the distribution function are simply proportional to $\cos(kR_c \sin\theta - n\pi/2 - \pi/4)$ as in Eq. (48.29). The essential point is that they are a rapidly oscillating function of $\sin\theta$, where θ is the angle that the momentum vector makes on the Fermi surface.[†] The method of stationary phase tells us that an integral over θ for any term in the conductivity tensor picks out a small band of electrons centered about $\theta = \pi/2$. The width of this band is crudely proportional to $(kR_c)^{-1}$, i.e.,

$$\Delta\theta \sim (kR_c)^{-1}. \tag{55.19}$$

This situation is analogous to the anomalous-skin-effect problem where only those "effective" electrons whose velocities are closely parallel to the surface of the specimen contribute. The fact that only a small number of electrons contribute to AKCR is enough to insure that the $\delta\epsilon_1$ term is small. Since the scattering function $f(\mathbf{p}, \mathbf{p}')$ is almost surely smoothly

---

[†] The reader is referred to Section 48 and a discussion of the asymptotic ($kR_c \to \infty$) behavior of the conductivity tensor.

varying in angle, it will average out the "sharply" peaked distribution function and give practically no contribution to the AKCR impedance. Thus the experiment measures the mass m^*.

In the various limiting regimes of wave vector and frequency space the effects of the LS interaction function are relatively simple. Unfortunately they give us only a minor amount of information regarding the behavior of the Landau scattering function $f(\mathbf{p}, \mathbf{p}')$. As a result of these general arguments regarding the solution of the Landau–Silin transport equation (55.17), we are forced to conclude that it would be desirable to examine those regions of frequency and wave vector space such that $\omega \sim kv_\text{F} \sim \omega_\text{c}$. In this regime we would a priori expect the Landau scattering function to play a nontrivial role. This is precisely the CW regime. In the next section we will investigate the behavior of solutions of Eq. (55.17) which are relevant to a discussion of CW propagation in the presence of electron correlation effects.

56. CW in the Presence of Interactions

In order to discuss the effects of interactions on the propagation characteristics of the CW we must, in principle, solve Eq. (55.17) for the distribution function, evaluate the various components of the conductivity tensor, and then as in Chapter VIII solve the wave equation for the appropriate dispersion relation. For an arbitrary scattering function, there is no general solution to Eq. (55.17). We are left with two alternatives.

(1) If we truncate the scattering function, i.e., assume that there are only a finite number of moments in its spherical harmonic expansion, the integral equation becomes a finite set of coupled linear equations in the moments. There is no real justification for this procedure except for the fact that it leads to tractable results. It is true, however, that approximate calculations of the scattering functions (*13*) (and, as we shall see, experiments) seem to support the idea that the Legendre series falls off rapidly with increasing order. This approach (truncation) has been used by several authors in fitting some of the experimental data taken in transmission (*14*). Unfortunately, for the CW case, the algebraic complication involved in such an approximate analysis obscures much of the physics that can be extracted. The reason for this complication will become clear as we continue our discussion. In essence, the interaction function $f(\mathbf{p}, \mathbf{p}')$ has no non-trivial effect on the conductivity unless we take a nonzero A_2. This leads to a rather complicated three-by-three determinantal equation which must be solved in order to obtain the

dispersion relationship of the normal mode solutions of Maxwell's equations. In the spin case, this situation does not arise and we may, in fact, get most of the interesting physics out by taking $B_0 \neq 0$ and $B_n = 0$, $n \geqslant 1$. We will make this kind of approximation in the discussion on spin waves (Section 59).

(2) The other possible way of solving Eq. (55.17), at least approximately, is to look in the region where $\mathbf{k} \cdot \mathbf{v}$ is a small perturbation. In this limit the Landau BV equation may be solved for an arbitrary scattering function. This is the procedure we will adopt in this section.

At $k = 0$ the eigenmodes or solutions of Eq. (55.17) (neglecting collisions) are spherical harmonics. To see this we expand g in a series of spherical harmonics,

$$g = \sum_{n,m} \alpha_{n,m} Y_{n,m}(\theta, \varphi) \tag{56.1}$$

so that

$$g + \delta\epsilon_1 = \sum_{n,m} \alpha_{n,m}(1 + A_n) Y_{n,m}(\theta, \varphi)$$

$$\equiv \sum_{n,m} \tilde{\alpha}_{n,m} Y_{n,m}(\theta, \varphi). \tag{56.2}$$

For a uniform field in the z direction, Eq. (55.17) may be written as *(15)*

$$\sum_{n,m} i(\omega - m\omega_c(1 + A_n)) \alpha_{n,m} Y_{n,m} \sim eEY_{1,0}. \tag{56.3}$$

The modes of the system, i.e., the solution to the transport equation in the *absence* of a driving term are clearly the $Y_{n,m}$. The eigenfrequencies in turn are given by *(16)*

$$\omega = \omega_{n,m} = m\omega_c(1 + A_n) = m\omega_c \gamma_n^A. \tag{56.4}$$

As $k \to 0$, the electric field in the z direction [right-hand side of Eq. (56.3)] only couples to the $n = 1$, $m = 0$ mode. Uniform fields in the x and y direction would couple to the $n = 1$, $m = \pm 1$ modes. As k increases, the $\mathbf{k} \cdot \mathbf{v}$ term mixes in the higher $\omega_{n,m}$ modes and shifts the position of the resonances.

For the moment, we forget the presence of the driving term and compute the effect of the $\mathbf{k} \cdot \mathbf{v}$ term on the dispersion characteristics of the modes $\omega_{n,m}$. For *zero* external field and infinite τ, the equation generated by taking the (n, m) moment of Eq. (55.17) is ($\mathbf{k} \perp \mathbf{H}_0$),

$$(\omega/\omega_c - m\gamma_n^A) \alpha_{n,m} - kv_F \gamma_n^A \sum_{n',m'} C_{n,m}^{n',m'} \alpha_{n',m'} = 0. \tag{56.5}$$

The quantities $C_{n,m}^{n',m'}$ are the Clebsch–Gordon coefficients and are given by

$$C_{n,m}^{n+1,m\pm 1} = \frac{1}{2}\left[\frac{(n\pm m+1)(n\pm m+2)}{(2n+1)(2n+3)}\right]^{1/2}, \tag{56.6}$$

$$C_{n,m}^{n-1,m\pm 1} = -\frac{1}{2}\left[\frac{(n\mp m)(n\mp m-1)}{(2n-1)(2n+1)}\right]^{1/2}. \tag{56.7}$$

To second order in kv_F the eigenvalues $\omega_{n,m}$ may be found using standard second-order perturbation theory, i.e. (17),

$$\omega = \omega_{n,m} = m\omega_c\gamma_n{}^A + \beta_{n,m}k^2, \tag{56.8}$$

where

$$\beta_{n,m} = \frac{v_F{}^2}{\omega_c}\sum_{n',m'}\frac{(\gamma_n{}^A)(\gamma_{n'}^A)\,|\,C_{n,m}^{n',m'}\,|^2}{(m\gamma_n{}^A - m'\gamma_{n'}^A)}. \tag{56.9}$$

More explicitly,

$$\beta_{n,m} = \frac{1}{4}\frac{v_F{}^2}{\omega_c}\frac{\gamma_n{}^A}{(2n+1)}$$

$$\times\left\{\frac{\gamma_{n-1}^A}{(2n-1)}\left[\frac{(n-1+m)(n+m)}{m\gamma_n{}^A - (m-1)\gamma_{n-1}^A} + \frac{(n-1-m)(n-m)}{m\gamma_n - (m+1)\gamma_{n-1}^A}\right]\right.$$

$$\left.+\frac{\gamma_{n+1}^A}{2n+3}\left[\frac{(n+1-m)(n+2-m)}{m\gamma_n{}^A - (m-1)\gamma_{n+1}^A} + \frac{(n+1+m)(n+2+m)}{m\gamma_n - (m-1)\gamma_{n+1}^A}\right]\right\}. \tag{56.10}$$

In the limit $\gamma_n{}^A \to 1$, i.e., no Fermi-liquid effects, *all* $\beta_{n,m} = 0$.

Equation (56.10) implies that the normal mode solutions of the transport equation are *unshifted* to order k^2 if there are no Fermi-liquid effects. However, this does *not* mean that the CW, i.e., the zeros of the conductivity tensor are unshifted to order k^2. We know from the discussion in Section 48 that for $\mathbf{J} \parallel \mathbf{H_0}$ the CW near the first cyclotron harmonic does in fact behave as

$$\omega_{2,1} = \omega_c + \tfrac{1}{10}k^2 v_F{}^2/\omega_c. \tag{56.11}$$

Despite this fact there is a rather simple connection between the modes $\omega_{n,m}$ and the zeros of the conductivity tensor at least to order k^2. To illustrate this connection we consider the $\mathbf{E} \parallel \mathbf{H_0}$ to geometry. At $k=0$ the distribution function driven by the electric field is proportional to $Y_{1,0}$. Successive powers of $kv_F \cos\theta \sin\varphi$ couple in the higher (n,m) spherical harmonics. It is clear from the structure of Eq. (56.5) that the parity of the modes coupled in is conserved. The $\cos\theta\sin\varphi$ changes both n and m by ± 1 so that the sum $n + m$ is either even or odd. For $\mathbf{E} \parallel \mathbf{H}_0$, $n + m$ is evidently odd, and only the odd modes $(1, 0)$, $(2, 1)$, $(2, -1)$, etc. enter. The (n, m) harmonic enters the chain of equations only after the application of $k^{(n-1)}$ terms in kv_F. The chain is displayed graphically below

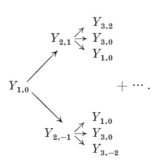

The first column is the local contribution of order $(k)^0$. The second column is of order k, the third of order k^2, etc. Since the conductivity σ_{zz} is proportional to $\int g Y_{10}\, d\Omega$ the "shortest" path back to $Y_{1,0}$ in our chain which touches on a particular $Y_{n,m}$ gives us the term which is leading order in k and resonant at $\omega_{n,m}$. Thus to order k^2 there is only one path, i.e., the one via $Y_{2,\pm 1}$. This term is resonant at $\omega_{2,1}$ and the fundamental singularity is shifted by A_2. In general the leading term (for a particular $\omega_{n,m}$) in σ_{zz} is of the form

$$\delta\sigma_{zz}^{n,m} \sim \frac{(kv_F/\omega_c)^{2(n-1)}}{\omega - \omega_{n,m}}. \tag{56.12}$$

The dispersion relation is given by the zero of σ_{zz}. For $\mathbf{E} \parallel \mathbf{H}_0$, it is clear that near a singularity the zero is essentially determined by $\omega_{n,m}$, i.e.,

$$\omega = \omega_{n,m} + \mathcal{O}(kv_F/\omega_c)^{2(n-1)}. \tag{56.13}$$

To order k^2, for $n > 2$, the zeros of σ_{zz} and its poles are identical. The $(2, 1)$ mode is clearly a special case since it enters into the conductivity to order k^2. The dispersion of this mode is readily found by setting $\alpha_{1,0} = 0$ and solving for the shifted frequency as in Eq. (56.8). This yields

$$\omega_{2,1} = \omega_c(1 + A_2) + \frac{(kv_F)^2 (\gamma_2{}^A)(\gamma_3{}^A)}{140\omega_c}\left[\frac{6}{\gamma_2{}^A} + \frac{20}{(\gamma_2{}^A - 2\gamma_3{}^A)}\right]. \tag{56.14}$$

The essential point here is that the intercepts of the CW dispersion relations near the harmonics are shifted by the A_n values. The fact that they have a k^2 dispersion away from it is interesting, but at present experimentally undetectable. The coefficient of the k^2 term for all but the (2, 1) mode is proportional to A_n, $n > 2$. In practice this means it is expected to be small, and a small k^2 term is difficult to pull out of the experiments.

The electron interaction effects at long wavelength split the degeneracy present in the noninteracting case where $\omega_{n,m} = \omega_{n',m}$ for all n and n'. In addition they introduce a finite k^2 dispersion to the higher modes. The reflection data of Walsh et al. have been reinterpreted in light of these results (18). In Fig. X-1 we examine more closely the experimental data

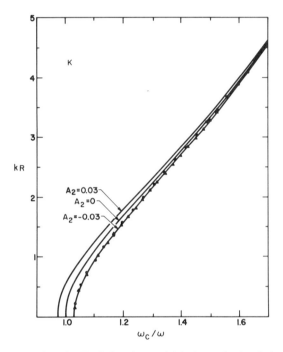

FIG. X-1. A comparison in K of the observed (circles and triangles) and predicted (solid) dispersion of the ordinary CW near the first harmonic of the cyclotron frequency ($\omega = \omega_c$). The different solid curves correspond to different values of the LS parameter A_2.

on potassium for CW propagation near the first harmonic in the $\mathbf{E} \parallel \mathbf{H_0}$ geometry (see Fig. VIII-17). There is no observed splitting of modes, presumably because the A_n and the experimental values of $\omega_c \tau$ are too small. There is, however, a definite shift in the intercept of the (2, 1) mode toward the high-field side of the first harmonic. The reflection data in potassium agrees well with an infinite τ dispersion curve shifted by an amount $A_2 \simeq -0.03$. The intercept on the second harmonic is consistent with an $|A_3| < 0.01$. In Na the data (see Fig. VIII-24) are too rudimentary to attempt a quantitative fit of the form shown in Fig. X-1. However a similar, very rough, fitting procedure gives an $A_2 \simeq -0.05 \pm 0.02$, with an $|A_3| < 0.02$.

The use of a shifted infinite τ dispersion relation to fit the data is definitely empirical but seems to work well. Different samples with different resistance ratios measured at different frequencies all fall nicely on the same universal curve. This is the justification for a simple fitting procedure. In principle infinitely pure specimens should enable one to read off the A_n for $n \geq 2$, as the intercept of the dispersion curves for $\mathbf{J} \parallel \mathbf{H}_0$.

Since the A_n seem to fall off rapidly with increasing n, it is particularly important to get information about A_0 and A_1. Unfortunately A_0 and A_1 do not enter into the final dispersion relations. This is, as we shall see, a general property of these *zero charge, zero current* modes. The parameter A_0 occurs in the transport equation multiplying $Y_{0,0}$, i.e., the net charge density produced by a disturbance. A_0 can, therefore, only influence waves possessing some degree of longitudinal character. This suggests that A_0 should enter explicitly into the extraordinary wave dispersion relation, Eq. (48.14).

Although we will not show it explicitly here, it is quite easy to demonstrate that each of the magnetoconductivity tensor components σ_{xx}, σ_{yy}, and σ_{xy} involves A_0 (*18*). However, the particular combination of the components given in Eq. (48.14) does not involve A_0. This situation arises from the fact that the combination of conductivities in question corresponds to a divergenceless current, hence zero net charge. To see this consider the equation for the x, y component of the current:

$$j_x = \sigma_{xx} E_x + \sigma_{xy} E_y, \tag{56.15}$$

$$j_y = \sigma_{yy} E_y + \sigma_{yx} E_x. \tag{56.16}$$

In the CW geometry, i.e., $\mathbf{k} \parallel \mathbf{x}$ the divergence of the current is proportional to j_x,

$$\nabla \cdot \mathbf{j} = i k_x j_x, \tag{56.17}$$

so that $\nabla \cdot \mathbf{j} = 0$ implies $j_x = 0$. If we arbitrarily set $j_x = 0$, then Eqs. (56.15) and (56.16) yield ($\sigma_{xy} = -\sigma_{yx}$)

$$j_y = \left[\frac{\sigma_{xx}\sigma_{yy} + \sigma_{xy}^2}{\sigma_{xx}} \right] E_y. \tag{56.18}$$

The effective conductivity tensor [square brackets in Eq. (56.18)] for the divergenceless current for any value of j_y cannot involve A_0. Since the zeros of this conductivity tensor approximate the roots of the CW dispersion relation, they, in turn, will not involve A_0.

The validity of the A_0 argument depends on having the correct combination of the various components of the conductivity tensor. It will begin to break down when the displacement current starts to become comparable to the conduction current. When this happens we must use the full ϵ_{ij}. The errors made using just the conduction current are of order $(\omega_c/\omega_p)^2$. Since $(\omega_c/\omega_p)^2$ is a very small quantity at microwave frequencies, it follows that A_0 does not, in any significant way, affect the CW dispersion relations.

The quantity A_1 enters the transport equation as a coefficient of $Y_{1,m}$, the current associated with the disturbance. Since the essence of the CW calculation is to seek zeros of the appropriate conductivities and therefore of their associated currents, it follows that A_1 cannot influence the CW dispersion as long as $\omega_p^2/\omega^2 \gg k^2/k_0^2$, i.e., we are in the long-wavelength limit. A_1 might, however, become significant at short but perhaps attainable wavelength as one passes into the Kaner–Skobov regime ($kR \gtrsim 20$). It should also be noted that, inasmuch as A_1 enters the problem via the current in the wave, it could also affect the degree of coupling, i.e., the actual amount of incident power which is transmitted through a slab versus ω_c/ω. However, in the absence of a solution of the boundary value problem, any such coupling information is probably uninterpretable. From this type of argument it is clear that it would be difficult to learn anything quantitative about A_0 and A_1 by examining the behavior of CW.

If one wishes to fit the experimental CW dispersion curves at finite but nonnegligible k values, it becomes necessary to truncate the spherical harmonic expansion of the scattering function after a few terms and solve the resulting set of coupled linear equations. The procedure is laborious but straightforward. We may write Eq. (55.17) ($\tau \to \infty$) as

$$(-i\omega + \omega_c \partial/\partial\varphi + i\mathbf{k}\cdot\mathbf{v})g \equiv \mathcal{O}g = e\mathbf{E}\cdot\mathbf{v} - [i\mathbf{k}\cdot\mathbf{v} + \omega_c \partial/\partial\varphi]\,\delta\epsilon_1(\theta,\varphi)$$
$$\equiv e\mathbf{E}\cdot\mathbf{v} + \Gamma(\theta,\varphi). \qquad (56.19)$$

Formally, then, g is given by

$$g = \mathcal{O}^{-1}[e\mathbf{E}\cdot\mathbf{v} + \Gamma(\theta,\varphi)]. \qquad (56.20)$$

Since $\Gamma(\theta,\varphi)$ is a functional of g, Eq. (56.20) is an expression for g in terms of itself. To solve it, take moments $\int g Y_{n,m}(\theta,\varphi)$ of both sides of this equation and obtain a set of coupled linear equations for the moments. The simplest case to handle is for $A_1 \neq 0$ and $A_n = 0, n \geq 2$. The result is *(15)*

$$\sigma_{zz} = \frac{\sigma_{zz}^0}{1 - [A_1/(1 + A_1)] \, 3a^2 F(a, kR_c)} \tag{56.21}$$

where

$$F(a, kR_c) = \sum_{n=0}^{\infty} \int_0^{\pi} \frac{J_n^2(b) \cos^2 \theta \, d\theta}{(1 + \delta_{n0})(a^2 - n^2)} \tag{56.22}$$

$a = \omega/\omega_c$ and $b = kR_c \sin$. Equation (56.21), while not very illuminating, is illustrative of our discussion concerning the importance of A_1 in the CW dispersion relations. The zeros of Eq. (56.22) are the zeros of σ_{zz}^0, i.e., they do not involve the A_n parameters in this A_1 approximation.

This concludes our brief discussion of the effect of interactions on CW propagation in the alkalis. The analysis of the data is not, at present, on very firm ground. One is left with the feeling that some of the observed deviations in the data may be due to an incorrect treatment of the boundary value problem. We do not know how good our approximations really are although it is true that all of the available data and the theory seem to be consistent. Empirically, the use of higher frequencies and purer samples (large $\omega_c \tau$) will tell us something about the validity of the approximations. It is clear however that the A_n, $n \geq 2$ can be obtained from experiment by observing the properties of CW. Their effects are felt as small corrections to a phenomenon which is present in the absence of interactions.

57. CW Propagating Parallel to the Magnetic Field

In the presence of interactions there are, also, a family of long-wavelength electromagnetic modes propagating parallel to a dc magnetic field in the vicinity of $\omega = \omega_c$ *(19)*. These modes are interesting because, in contrast to the CW, such modes do not exist in the absence of Fermi-liquid effects. In some sense they are the orbital analog of the paramagnetic spin wave modes which we will discuss in the next section. Unfortunately, as we shall see, their propagation characteristics are such that they are not readily observed under existing experimental conditions *(20, 21)*.

In the geometrical configuration where the wavevector **k** is parallel to the magnetic field, we have seen that the helicon mode will only propagate in a metal when its frequency $\omega \ll \omega_c$. To propagate a high-frequency long-wavelength mode in a metal we are restricted to situations where there is a possible zero of the conductivity. We refer the reader to our discussion in Section 48. In the absence of interactions in the helicon geometry, for a frequency $\omega \sim \omega_c$ no modes will propagate, i.e., there are no long-wavelength zeros of the conductivity $\sigma_+(\mathbf{k}, \omega)$ [Eq. (46.1)].

In the interacting case the situation is qualitatively different (19). We are faced with the task of solving the LS equation (55.17) in a new geometry, i.e.,

$$i\omega g + (ikv_F \cos\theta + \omega_c \, \partial/\partial\varphi)\bar{g} = e\mathbf{E}\cdot\mathbf{v} - (1/\tau)\bar{g}. \tag{57.1}$$

At $k = 0$, in the absence of a driving electric field, the modes of the system are the same modes discussed in Section 56. At finite k these modes, as in the CW case, take on some dispersion, i.e.,

$$\omega_{n,m} = m\omega_c \gamma_n{}^A + \beta'_{n,m} k^2. \tag{57.2}$$

In this case,

$$\beta'_{n,m} = \frac{v_F^2}{m\omega_c} \sum_{n'} \frac{\gamma_n{}^A \gamma_{n'}^A \, |\, D_{n,m}^{n',m}|^2}{\gamma_n{}^A - \gamma_{n'}^A}, \tag{57.3}$$

where the Clebsch–Gordon coefficients $D_{n,m}^{n',m}$ are defined by

$$\cos\theta\, Y_{n,m} = \sum_{n'} D_{n,m}^{n',m} Y_{n',m}. \tag{57.4}$$

They are given by

$$D_{n,m}^{n+1,m} = \{[(n+1)^2 - m^2]/[4(n+1)^2 - 1]\}^{1/2} \tag{57.5}$$

and

$$D_{n,m}^{n-1,m} = -[(n^2 - m^2)/(4n^2 - 1)]^{1/2}. \tag{57.6}$$

The coefficient of the quadratic term $\beta'_{n,m}$ diverges in the limit of vanishing Fermi-liquid interaction ($\gamma_n{}^A \to 1$). This is a reflection of the fact that at $\gamma_n{}^A = 1$, Eq. (57.1) has a continuous spectrum for $|\omega \pm m\omega_c| \leqslant kv_F$, i.e., there is cyclotron damping no matter how long the wavelength (see Section 46). In the perpendicular geometry the single-particle spectrum is discrete for arbitrary k. There is no continuum which begins at $\omega = m\omega_c$. Formally, we might recall that an expansion of the nonlocal helicon dispersion relation is an expansion in the parameter $kv_F/(\omega - \omega_c)$. At $\omega = \omega_c$ the expansion is invalid.

Given the $k = 0$ modes in the absence of a driving term, we can now go back, as in the CW case, and investigate the zeros of the conductivity $\sigma_+ \equiv \sigma_{xx} + i\sigma_{xy}$. The arguments closely parallel those given in Sections 48 and 56 and we will not repeat them in any detail. The circularly polarized driving field excites the $m = 1$ mode. The $\cos \theta$ does not mix in higher m values. The higher n modes $n > 2$ [given in Eq. (56.4)], because they enter the conductivity for powers of k greater than k^2, are also the locus of points where the conductivity vanishes.

The $n = 2$, $\omega_{2,1}$, mode is a special case just as it is in the perpendicular geometry. Its dispersion is readily obtained by setting, $\alpha_{1,1}$, the amplitude of the first spherical harmonic in an expansion of g [see Eq. (56.5)], equal to zero. This yields

$$\omega_{2,1} = \omega_c \gamma_2{}^A \left[1 + \left(\frac{k v_F}{\omega_c}\right)^2 \left(\frac{8}{35} \frac{\gamma_3{}^A}{(\gamma_2{}^A - \gamma_3{}^A)}\right)\right]. \tag{57.7}$$

The mode whose dispersion is given by Eq. (57.7) is the most important one in the limit of small A_n, $n \geqslant 2$, the experimentally interesting case. In fact, this mode is the only root when $A_2 \neq 0$ and $A_n = 0$ for $n > 2$. It arises because the electron interaction effects permit a zero in the current for a prescribed electric field. The current induced in the medium may be thought of as made up of a simple quasi-particle contribution along with a back flow piece [see Eq. (55.6)] due to the motion of the other charges in response to the motion of the quasi particle under consideration. The requirement of zero current in this instance is a requirement that the back flow cancel the direct portion of the current. Such a situation is obtained as long as $A_n \neq 0$ for $n \geqslant 2$ *(22)*.

Taking $A_2 \neq 0$ but $A_n = 0$ for $n > 2$, we can solve the LS transport equation in closed form. The dispersion, i.e., zero of the conductivity tensor is, in this case, given by *(21)*

$$-K_{\parallel} = (5\omega C_2 / k v_F)(1 - z K_{\parallel}) \tag{57.8}$$

where

$$C_2 \equiv A_2 / (1 + A_2) \tag{57.9}$$

$$z = (\omega - \omega_c + i/\tau)/k v_F \tag{57.10}$$

and

$$K_{\parallel}(z) = \tfrac{3}{2}\{z + \tfrac{1}{2}(1 - z^2) \ln[(z + 1)/(z + 1)]\}. \tag{57.11}$$

In Fig. X-2 we show a plot of the roots of Eq. (57.8) for an $A_2 = -0.03$ and infinite $\omega \tau$. The mode turns on at an $\omega_c / \omega = 1.03$ and runs into the continuum (the shaded region) at an $\omega_c / \omega = 1 - \tfrac{5}{3} C_2$. For small A_2, the mode propagates over a very limited range of field.

The effect of finite $\omega_c \tau$ is significant. A simple calculation utilizing Eq. (57.8) shows that

$$(\operatorname{Im} k)/(\operatorname{Re} k) \sim 1/|A_2 \omega_c \tau|. \tag{57.12}$$

57. CW PROPAGATING PARALLEL TO MAGNETIC FIELD

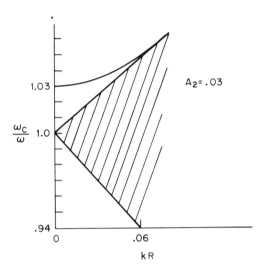

FIG. X-2. The dispersion curve for the CW propagating parallel to the magnetic field for a finite A_2. The shaded region is the continuum.

Equation (57.12) implies that very large $\omega_c \tau$ (for $A_2 \sim -0.03$) are necessary if one hopes to see a well-defined mode structure. To date experiments have not been performed under these idealized (large $\omega_c \tau$) conditions. However, an anomaly in the real part of the surface impedance $R(H_0)$ of a thick slab of pure potassium in a magnetic field oriented normal to the sample surface has been observed (20). This anomaly is shown in Fig. X-3. The data were taken at a frequency $\omega = 1.5 \times 10^{11}$ Hz at a temperature of 2.5°K. The sharp $R(H_0)$ peak occurs at a value of H_0 where we would expect (for $\mathbf{k} \parallel \mathbf{H}_0$) to propagate the $\omega_{2,1}$ mode. This peak has been interpreted as an experimental confirmation for the existence of this mode.

Detailed analysis of the boundary-value problem shows that the peak is consistent with mode propagation and $A_2 \cong -0.03$ (21). At mode turn on, $\omega = \omega_c(1 + A_2)$, we begin to feed additional energy into the specimen. This results in rapid increase in the surface resistance. The exact shape of the increase is primarily determined by the density of modes as a function of magnetic field. At mode cutoff (infinite $\omega_c \tau$) we expect the resistance to drop suddenly since we precipitously lose the degrees of freedom of the system into which we have been feeding electromagnetic energy. This accounts for the asymmetry, i.e., sharpness on the high-field side of the observed line (see Fig. X-3). The noninteracting theory (shown as the dotted line in Fig. X-3) shows no comparable structure (22, 23).

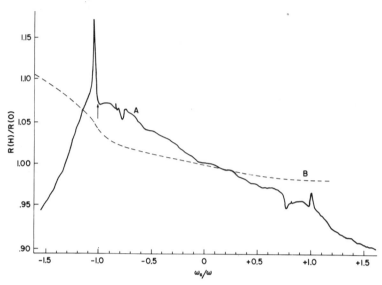

Fig. X-3. The dashed curve is the normalized surface resistance as a function of magnetic field predicted by the free electron model with diffuse surface scattering. The solid line denotes relative values of the surface resistance of K in the field normal geometry measured with predominantly circularly polarized radiation. $T = 2.5°K$; $\omega = 1.5 \times 10^{11}$ Hz.

While this anomaly in the surface impedance is striking and the shape of the line in rough agreement with our ideas on wave propagation, it would be more convincing to see these modes directly either in a transmission or in a two-sided reflection experiment. At very large $\omega_c \tau$, of the order of a few hundred, such experiments could be done. For $\omega_c \tau \sim 25$–30 (the value applicable to the experimental data in Fig. X-3) such experiments are not possible.

In the next section we will discuss another set of modes. The paramagnetic spin waves which, like the field parallel CW modes, do not exist in the absence of interactions. However, due to the peculiarities of excitation and the quantitative features of the interaction function, these modes, as we shall see, can be observed directly.

58. Spin Waves in the Interacting Electron Liquid

In order to discuss magnetic excitations in an interacting electron gas we must analyze the transport equation (54.51) for the magnetization density. Approximate solutions of the LS transport equation will give us expressions for the susceptibility $\chi_+(\mathbf{k}, \omega)$. We may then utilize this susceptibility to compute [see Eq. (51.26)], the power transmitted through a slab of metal in the vicinity of CESR. This procedure works because the correlations are local in space and hence do not affect the nature of the solutions to the boundary value problem.

The transport equation for the transverse magnetization vector $m^\pm$, is given by Eq. (54.51). Since the quantity m^+ is the linear combination which exhibits a resonance near the electron's precession frequency, it is sufficient to consider it alone. Defining the quantity g by

$$m^+ = -(\partial n^0/\partial E^\circ)g, \qquad (58.1)$$

the linearized transport equation for g becomes

$$-i\omega g + [i\mathbf{k}\cdot\mathbf{v} + \omega_c\,\partial/\partial\varphi + i\Omega_0]\bar{g}$$
$$= -(1/\tau)\left[\bar{g} - (4\pi)^{-1}\int \bar{g}\,d\Omega\right] - (1/T_2)\bar{g} + \tfrac{1}{2}i\gamma_0(\mathbf{k}\cdot\mathbf{v} + \Omega_0)\,h^+. \qquad (58.2)$$

The distribution function $\bar{g}$ is defined by

$$\bar{g} = g + \delta\epsilon_2, \qquad (58.3)$$

where

$$\epsilon_2 = 2(2\pi)^{-3}\int d^3p'\,\zeta(\mathbf{p},\mathbf{p}')\,g(\mathbf{p}')\,\delta(E^\circ_{\mathbf{p}'} - E_F). \qquad (58.4)$$

The quantity $\bar{g}$ is analogous to $\delta\bar{n}$ defined by Eq. (55.15). It specifies the deviation of the magnetization density from its local equilibrium value. The collision terms on the right-hand side of Eq. (58.2) are phenomenological in character. They have been inserted, as in the noninteracting case, to describe the two types of collisional process present, orbital and spin. Both collisional terms are constructed so as to produce relaxation to *local* equilibrium.

The frequency Ω_0 is defined in Eq. (54.53). It is simply the frequency (energy) required to flip the spin of an electron in the presence of an external field and an isotropic exchange field coming from the other electrons. Its precise value follows from arguments similar to those used in obtaining the dc susceptibility (see Section 55).

To solve Eq. (58.2) we are again forced to make some approximations. We can, keeping an arbitrary scattering function, examine the solutions of Eq. (58.2) in the long-wavelength limit. This is the procedure we will follow in analyzing the experimental data. However, for the case of spin disturbances it is also interesting to arbitrarily truncate the scattering function. If we assume that the scattering function is a constant independent of $\hat{\mathbf{p}}\cdot\hat{\mathbf{p}}'$, we can obtain an exact solution for the transverse susceptibility. In contrast to the CW case this simple assumption for the scattering function, $B_0 \neq 0$, $B_n = 0$, $n \geq 1$ leads to physically interesting results.

The presence of an isotropic term in the scattering function modifies the uniform susceptibility of the gas [see Eq. (53.23)]. This effect is, of course, analogous to a change in the "compressibility" for the case of density fluctuations. In a charged system, however, density fluctuations involve large energies $\hbar\omega_p$ and such distrubances are beyond the scope of the LS theory. In the spin problem, the presence of an isotropic exchange scattering function leads one to predict the existence of a spin wave mode. This mode is directly analogous to the zero-sound mode which is present in an uncharged Fermi liquid and may be analyzed within the framework of the LS theory since it has a low energy at long wavelengths.

The algebra is somewhat simplified if we turn off all of the collisional or absorptive processes, i.e., $\tau \to \infty$ and $T_2 \to \infty$. We will reintroduce these phenomenological scattering times later. In the "isotropic" scattering function approximation the transport equation (58.2) is

$$\partial g/\partial \varphi - i(a - b\cos\varphi)g = -i(b\cos\varphi + c)B_0 g_0 + \tfrac{1}{2}i\gamma_0(b\cos\varphi + c)h^+. \tag{58.5}$$

The coordinate system is specified in Fig. VIII-1. In Eq. (58.5)

$$a = (\omega - kv_\text{F}\cos\theta\cos\varDelta - \varOmega_0)/\omega_c \tag{58.6}$$

$$b = kv_\text{F}\sin\theta\sin\varDelta/\omega_c \tag{58.7}$$

$$c = (\varOmega_0 + kv_\text{F}\sin\varDelta\cos\theta)/\omega_c \tag{58.8}$$

and

$$g_0 = (4\pi)^{-1}\int g\, d\varOmega. \tag{58.9}$$

For arbitrary $\varDelta$, Eq. (58.5) may be solved exactly by the methods given in Section 56 [see Eqs. (56.19)–(56.21)]. However, by specializing our direction of propagation by assuming $\mathbf{k} \parallel \mathbf{H}_0$ ($\varDelta = 0$), we may immediately obtain an algebraic solution. In this case $b = 0$ and g is independent of φ, i.e., $\partial g/\partial \varphi = 0$. The solution for g is then

$$g = \left[-B_0 g_0 + \frac{\gamma_0}{2}h^+\right]\left[1 - \frac{\omega}{\omega - kv_\text{F}\cos\theta - \varOmega_0}\right]. \tag{58.10}$$

The zeroth moment of g is proportional to $\chi_+(\mathbf{k}, \omega)$, i.e.,

$$g_0 \sim \chi_+(\mathbf{k}, \omega)\, h^+. \tag{58.11}$$

Since a mode of the system is determined by requiring a finite response

58. SPIN WAVES

even in the absence of a driving field we may, by setting $h^+ = 0$ in Eq. (58.10) and assuming that g_0 is finite, χ_+ infinite, find the values of ω and k (dispersion relation) which are consistent with these assumptions.

Setting $h^+ = 0$ and integrating both sides of Eq. (58.10) over solid angle, we find that

$$-\frac{1}{B_0} = 1 - \frac{s}{2} \ln\left(\frac{s - s_0 + 1}{s - s_0 - 1}\right), \tag{58.12}$$

where $s = \omega/kv_F$ and $s_0 = \Omega_0/kv_F$. For values of ω and k lying outside of the region bounded by the curves (see Fig. X-4)

$$\omega = \Omega_0 \pm kv_F, \tag{58.13}$$

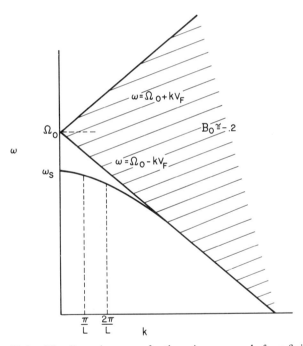

FIG. X-4. The dispersion curve for the spin wave mode for a finite B_0.

the dispersion relation is purely real, i.e., Eq. (58.12) has solutions with real ω and k. Within the region in the ω, k plane bounded by Eq. (58.13) the logarithmic function in this equation has an imaginary part and there is no solution of real ω and k.

The boundary curves, Eq. (58.13), are the spin analog of Eq. (45.28). They simply express the fact that a collective excitation, involving a coherent superposition of single-electron spin flips, may decay into an electron–hole pair with opposite spin orientation. This equation is nothing more than statement of energy–momentum conservation in this process. If we flip the spin of an electron near the Fermi surface and at the same time give the electron–hole pair a momentum **k** in the z direction, the pair, depending on the initial velocity of the particle, has a spread in energy which for $k/k_F \ll 1$ is given by Eq. (58.13). When **k** is arbitrarily oriented relative to $\mathbf{H}_0$, the continuum is still determined by Eq. (58.13); but with **k** replaced by k_z, i.e., our fan in Fig. X-4 has an angular width determined by $\cos \varDelta$, as in Eq. (45.28).

For small k (large s) we may expand the logarithm in Eq. (58.12) and obtain the approximate dispersion relation

$$\omega = \Omega_0(1 + B_0)\left\{1 + \frac{k^2 v_F^2}{3 B_0 \Omega_0^2} + \cdots\right\}. \tag{58.14}$$

The expansion is valid whenever

$$k^2 v_F^2 / \Omega_0^2 B_0 < 1. \tag{58.15}$$

The spin wave mode Eq. (58.14) starts, at $k = 0$, at an $\omega = \Omega_0(1 + B_0) \equiv \omega_s$, the unmodified conduction electron spin resonance. The spectrum is quadratic with a negative curvature for $B_0 < 0$ ferromagnetic exchange. A typical dispersion relation, i.e., the locus of points along which $\chi_+(\mathbf{k}, \omega)$ is infinite (no collisions) is shown in Fig. X-4 for $B_0 = -0.2$, $B_n = 0$ for $n \geqslant 1$, and $\varDelta = 0$. The shape and curvature, as we shall see in more detail, are functions of $\varDelta$.

The presence of this additional line of singularities in $\chi_+(\mathbf{k}, \omega)$ leads to an important modification of the transmission or reflection coefficients calculated utilizing Eqs. (51.10) and (51.11). It is clear from Eq. (51.3) that the finite geometry of the slab selects a set of discrete k_n values ($k_n = n\pi/L$) which are excited. Each k_n value resonates at a slightly different frequency, or equivalently, magnetic field. This state of affairs is also shown in Fig. X-4. We might expect, without actually evaluating Eq. (51.10), a whole series of peaks or resonances corresponding to the infinities of $\chi_+(\mathbf{k}, \omega)$ at $k = n\pi/L$. The $n = 0$ mode is simply the old conduction electron spin resonance. In the absence of exchange, $B_n = 0$, there are no roots (other than the CESR at $k = 0$) corresponding to infinities of $\chi_+(\mathbf{k}, \omega)$. An analysis including collisions, as we shall see, would have led us back, in the small k regime, to our purely diffusive $\chi_+(\mathbf{k}, \omega)$ [see Eq. (50.15)].

Our neglect of the higher B_n is not, a priori, a good assumption. It is much more useful and correct procedure to evaluate the susceptibility directly in the long-wavelength limit. Before considering this analysis let us examine the $k = 0$ limit of Eq. (58.2) to determine the spectrum of normal spin-wave modes (16). At $k = 0$ a spin resonance experiment will only couple to the mode with $Y_{0,0}$ symmetry since the susceptibility is directly proportional

$$\chi_+(\mathbf{k}, \omega) \sim \int g \, d\Omega. \tag{58.16}$$

Equation (58.16) is to be compared with the conductivity tensor which is relevant to an analysis of the CW experiments. In this instance, the current ($e\mathbf{E} \cdot \mathbf{v}$) type of coupling only allows one to couple, at $k = 0$, to the modes with $Y_{1,m}$ symmetry [see Eq. (56.3)].

At $k = 0$, the normal modes or solutions of Eq. (58.2) (neglecting collisions) are spherical harmonics. Expanding g in a series of spherical harmonics, i.e.,

$$g = \sum_{n,m} \alpha_{n,m} Y_{n,m}(\theta, \varphi) \tag{58.17}$$

and substituting it into Eq. (58.2) (in the absence of a driving term) yields,

$$\sum_{n,m} -i[\omega - (m\omega_c + \Omega_0) \gamma_n{}^B] \alpha_{n,m} Y_{n,m} = 0. \tag{58.18}$$

Equation (58.18) implies that the magnetic normal-mode frequencies of the system are

$$\omega_{n,m} = (m\omega_c + \Omega_0) \gamma_n{}^B \tag{58.19}$$

where

$$\gamma_n{}^B = (1 + B_n) \tag{58.20}$$

The (0, 0) mode is

$$\omega_{0,0} = \Omega_0(1 + B_0) \equiv \omega_s, \tag{58.21}$$

the unmodified conduction electron spin resonance. This simple relation is to be expected since the total spin magnetization, in the absence of spin-orbit coupling, is a constant of the motion, i.e., it is unaffected by interactions. The higher-order modes as in the CW case mix with the (0, 0) mode and produce dispersion. In addition higher-order modes could couple directly, at $k = 0$, to spatially dependent rf magnetic fields.

Because of the fact that the conduction electron susceptibility is so small, the intensity, in the spin wave modes at finite k is weak and very difficult to observe. The higher modes $\omega_{n,m}$, $n \neq 0$, have even weaker oscillator strengths and to date have not been observed (14). These higher modes have been called combination resonances. They involve a magnetic or spin-flip transition (Ω_0) plus one or more orbital (ω_c) transitions. We might remind the reader that in the CW problem the modes which influence the conductivity tensor were purely orbital. The reason for this apparent lack of symmetry is evident. In the absence of spin-orbit coupling we need an rf magnetic field to flip a spin. The electric field will not do the trick. Beyond the dipole approximation (finite k), both electric and magnetic fields lead to an arbitrary number of orbital or Landau-level transitions.

It would be extremely interesting to observe these combination resonances in metals. It is not a priori clear what the mechanism for coupling to them will be. We know that there is some spin-orbit coupling in the alkalis since $g - g_{\text{free}} \neq 0$. Thus, the coupling to the combination resonances may occur via electric-type matrix elements which operate through the small spin-orbit coupling or they may occur via the higher-order spin wave modes which are excited via magnetic-type matrix elements.

Let us now return to Eq. (58.2) and calculate the long-wavelength behavior of $\chi_+(\mathbf{k}, \omega)$. Inserting the spherical harmonic expansion for g [Eq. (58.17)] into Eq. (58.2), multiplying by $Y_{0,0}$, and integrating over solid angle, we obtain

$$-i\left(\omega - \omega_s + \frac{\gamma_0^B}{T_2}\right)\alpha_{0,0} + ikv_F \frac{\gamma_1^B}{3^{1/2}}\left\{\cos\varDelta\,\alpha_{1,0} + \frac{\sin\varDelta}{2^{1/2}}(\alpha_{1,1} - \alpha_{1,-1})\right\}$$
$$= i(2\pi)^{1/2}\gamma_0\Omega_0 h^+. \quad (58.22)$$

In vector notation this equation is simply

$$-i(\omega - \omega_s + i\gamma_0^B/T_2)\,m^+ + i\mathbf{k}\cdot\mathbf{J}_{m^+} = i\gamma_0 M_0 h^+ \quad (58.23)$$

since

$$m^+ \sim \alpha_{0,0} \quad (58.24)$$

and

$$\mathbf{J}_{m^+} = \int (\mathbf{p}/m^*)\bar{g}\,d\Omega_p = \gamma_1^B \int (\mathbf{p}/m^*)g\,d\Omega_p\,. \quad (58.25)$$

Equation (58.23) is the continuity equation relating the time derivative of m^+ to the divergence of a magnetization current $\mathbf{J}_{m^+}$ [see Eq. (50.17)].

Two points are worth noting at this point:

(1) the kv_F term in Eq. (58.22) may not be considered small since $\omega \simeq \omega_s$ at resonance; and

(2) the "effective" magnetization relaxation time is modified by the exchange, i.e. (24),

$$T_2^* = T_2/\gamma_0^B. \quad (58.26)$$

Equation (58.26) implies that as the system become more ferro-magnetic ($\gamma_0{}^B \to 0$), the magnetic relaxation time for a given scatterer becomes longer and longer (25).

In order to solve Eq. (58.22) for $\alpha_{0,0}$ (the magnetization) we need an equation relating $\alpha_{1,m}$ back to $\alpha_{0,0}$. Multiplying Eq. (58.2) by $Y_{1,m}$, we obtain the set of three equations

$$-i(\bar{\omega} - \Omega_0\gamma_1{}^B - m\omega_c\gamma_1{}^B)\alpha_{1,m} + ikv_F \sum_{n',m'} C_{n,m}^{n',m'} \gamma_n^B \alpha_{n',m'}$$
$$= i(2\pi)^{1/2} \gamma_0(kv_F) C_{0,0}^{1,m} h^+, \quad (58.27)$$

where

$$\bar{\omega} \equiv \omega + i\gamma_1{}^B/\tau. \quad (58.28)$$

and $C_{n,m}^{n',m'}$ are the Clebsch–Gordon coefficients defined in Eqs. (56.6) and (56.7).

Unlike the coefficient of $\alpha_{0,0}$ in Eq. (58.22), the factor multiplying $\alpha_{1,m}$ is not small when $\omega \sim \omega_s$ no matter how small τ^{-1} is. This is true a posteriori because $\gamma_1{}^B \neq \gamma_0{}^B$ and $\omega_c \neq \omega_s$. Hence, we may treat the kv_F term in Eq. (58.27) as a small perturbation and neglect the $\alpha_{2,m}$ terms which appear in it since they will be higher order in kv_F, i.e., they will contribute terms of order $(kv_F)^4$ in $\alpha_{0,0}$. This procedure breaks down if $\gamma_1{}^B$ coincidentally is equal to $\gamma_0{}^B$. The term proportional to $(kv_F) h^+$ may also be neglected since this type of term will lead to corrections to the zeroth-order driving term in Eq. (58.22) which are unimportant in the long-wavelength limit.

In vector notation the set of three equations [Eq. (58.27)] in the long-wavelength limit becomes

$$i(\bar{\omega} - \Omega_0\gamma_1{}^B)\mathbf{J}_{m^+} - i\mathbf{J}_{m^+} \times \bar{\omega}_c{}^B - \mathbf{k}\gamma_0{}^B\gamma_1{}^B \tfrac{1}{3}v_F^2 \, m^+ = 0, \quad (58.29)$$

where

$$\bar{\omega}_c{}^B \equiv (eH_0/m^*c)\, \gamma_1{}^B. \quad (58.30)$$

Since

$$\chi_+(\mathbf{k}, \omega) \equiv m^+(\mathbf{k}, \omega)/h^+(\mathbf{k}, \omega) \quad (58.31)$$

we see that

$$\chi_+(\mathbf{k}, \omega) = \frac{\chi_0 \omega_s}{[\omega - \omega_s + i\gamma_0{}^B/T_2 + \beta k^2]}, \quad (58.32)$$

with

$$\beta = \frac{1}{3} v_F^2 \gamma_0{}^B \gamma_1{}^B (\bar{\omega}_s - \bar{\Omega}_0) \left[\frac{\sin^2 \Delta}{(\omega_c{}^B)^2 - (\bar{\omega}_s - \bar{\Omega}_0)^2} - \frac{\cos^2 \Delta}{(\bar{\omega}_s - \bar{\Omega}_0)^2} \right] \quad (58.33)$$

and

$$\bar{\Omega}_0 = \Omega_0 \gamma_1{}^B \tag{58.34}$$

$$\bar{\omega}_{\mathrm{s}} = (\omega_{\mathrm{s}} + i\gamma_1{}^B/\tau) \tag{58.35}$$

In the limit where interaction effects vanish, i.e., $\gamma_0{}^B = \gamma_1{}^B = 1$,

$$\beta = \frac{i}{3} v_{\mathrm{F}}^2 \tau \left[\cos^2 \varDelta + \frac{\sin^2 \varDelta}{1 + (\omega_{\mathrm{c}}\tau)^2} \right] \tag{58.36}$$

and we recover the anistropic diffusionlike susceptibility characteristic of Dysonian spin resonance [see Eq. (50.15)]. If the interaction effects are present and τ is large enough, then β is almost real and there is a new branch of singularities present in the susceptibility tensor.

If **k** is along the static field, then the requirement of long τ is equivalent to the condition

$$\omega_{\mathrm{s}}\tau \, |(B_0 - B_1)|/(1 + B_0) \gg 1. \tag{58.37}$$

For the alkali metals, the susceptibility measurements in Na (26, 27) indicates that B_0 is typically of order 0.1. This fact implies that $\omega_{\mathrm{s}}\tau$ must be somewhat larger than twenty in order to satisfy Eq. (58.37) and keep the spin wave excitations well defined. This explains why spin waves are not easily observed. It is difficult to purify the alkalis to the point where $\omega_{\mathrm{c}}\tau > 10$.

The infinities of $\chi_+(\mathbf{k}, \omega)$ [Eq. (58.32)] in the limit $\omega\tau \gg 1$ define a spin wave dispersion relation. Figure X-5 is a plot, with typical values of the Fermi-liquid parameters, of this long-wavelength dispersion relation, i.e., the zeros of the denominator in Eq. (58.32). It is the analog of the "simplified" long-wavelength dispersion curve given by Eq. (58.14). The main differences are that there is an angular dependence to the branch of singularities and that Eq. (58.32) is valid for an arbitrary Fermi-liquid scattering function as long as $k \to 0$. The quantities B_0 and B_1 are the only Fermi-liquid parameters entering into the long-wavelength limit of $\chi_+(\mathbf{k}, \omega)$. If we had retained terms to order k^4, then B_0, B_1, and B_2 would have entered. Thus a solution of Eq. (58.2) which is valid for arbitrary k but assumes that B_0, B_1, and B_2 are finite and $B_n = 0$ for $n \geqslant 3$ is *exact* to order k^4. This type of argument can, of course, be carried through to arbitrary order in k.

Figure X-6 shows a typical trace taken in transmission with an $H_0 \simeq 3.25$ kG parallel to the sample ($\varDelta = 90°$). The sample is a pure Na slab approximately 0.15-mm thick at a temperature of 1.3°K. The upper portion of the trace shows the "large" CESR signal with obvious sidebands on the low-field side. The lower trace has the vertical scale

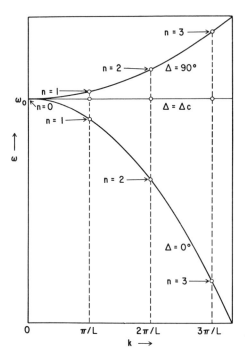

FIG. X-5. Plot showing the qualitative behavior of the spin wave branch as a function of the angle that **k** makes with respect to H_0.

expanded by about a factor of ten. The horizontal scale has been reduced by a factor of five. The CESR line is now off scale and the first, second, third, and possibly fourth sidebands are clearly visible in the experimental trace. These sidebands are the higher-order ($k = n\pi/L$) spin waves. They correspond to the resonances (infinities in χ_+) at different frequencies (magnetic fields) shown in Fig. X-5. As the magnetic field is rotated towards the normal, the spin wave structure moves in toward the main CESR line with all modes converging and then reappearing on the high-field side of CESR. Some typical experimental traces detailing this phenomenon are shown in Fig. X-7. The long-wavelength dispersion relation (see Fig. X-5), at least qualitatively, predicts this type of behavior.

In order to extract the Fermi-liquid parameters B_n from the data and test the predictions of the theory, we must make a detailed fit to the data for varying B_n, ultimately arriving at the best set of B_n values. At long wavelengths only B_0, B_1, and known quantities such as m^*/m_0, p_F, and L enter the dispersion relation, Eqs. (58.32) and

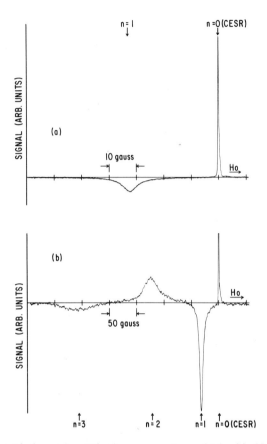

FIG. X-6. A typical experimental spin wave spectrum obtained in Na. $d = 0.15$ mm; $T = 1.3°K$; $\omega = 5.9 \times 10^{10}$ Hz; $\Delta = \pi/2$.

(58.33). For samples about 0.1-mm thick, it is easy to verify that the higher-order spin waves are not accurately described by a k^2 term alone, although the first spin-wave sideband ($k = \pi/L$) is moderately well described by such a long-wavelength approximation. The procedure actually used to fit the data has been tailored to these considerations. It consists of

(1) comparing the first spin wave mode with the k^2 theory in order to extract B_0 and B_1;

(2) using a truncation procedure which neglects B_n, $n \geqslant 3$, to determine the best values for B_2.

In the limit of infinitely long orbital collision times and in the long-

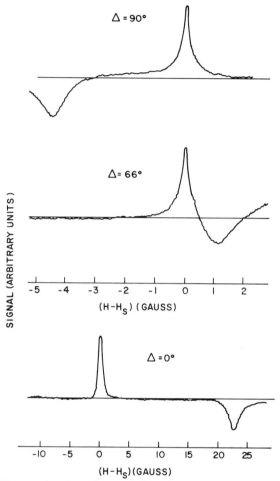

FIG. X-7. The angular dependence of the spin wave spectrum in K. $T = 1.4°K$; $\omega = 5.9 \times 10^{10}$ Hz.

wavelength limit, we can express the ratio of the location of the first spin-wave sideband at arbitrary Δ to that at $\Delta = \pi/2$ by

$$\alpha_\Delta/\alpha_{\pi/2} = (1 - A \cos^2 \Delta), \qquad (58.38)$$

where

$$\alpha \equiv (\omega - \omega_S) T_2 \qquad (58.39)$$

and the constant A is given by

$$A = [(m_0/m^*)(1 + B_0)(1 + B_1)/(B_0 - B_1)]^2. \qquad (58.40)$$

In the long-wavelength limit, the relation given by Eq. (58.38) holds for the particular value of the magnetic field which corresponds to the peak of any fixed spin-wave sideband. It describes the angular dependence of the position, relative to CESR of a spin-wave sideband. The fit to the angular dependence of the first spin-wave sideband is shown in Fig. X-8. The LS theory gives an excellent quantitative description of the

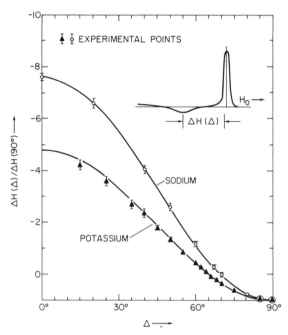

FIG. X-8. Comparison of the observed (circles and triangles) with the theoretical solid position of the first spin wave sideband as a function of magnetic field orientation Δ.

position of the lowest spin-wave. The existence of a spin-wave sideband and the experimental verification of the predicted angular variation are the best proof of the validity of the LS theory. Dunifer and Schultz (26, 14) have used these results to determine the value of A in Eq. (58.38) for Na and K. They find that

$$A = 8.8 \pm 0.1 \quad \text{(Na)} \qquad (58.41)$$

$$A = 5.95 \pm 0.05 \quad \text{(K)}. \qquad (58.42)$$

The constant A is an intrinsic property of the isotropic interacting electron gas in a metal. It does not depend on such things as sample thickness or relaxation time but only on B_0, B_1, and m^*. From the value of A and the values (27)

$$m^*/m_0 = 1.24 \pm 0.02 \quad \text{(Na)}$$
$$m^*/m_0 = 1.21 \pm 0.02 \quad \text{(K)}. \tag{58.43}$$

we can obtain a relationship between B_0 and B_1. This relationship is displayed graphically in Fig. X-9.

To determine B_0 and B_1 uniquely, we need one more equation linking the two parameters. This is found by measuring the position of the first spin-wave sideband at any fixed angle, e.g., $\varDelta = 0$. For infinite $\omega_c \tau$

$$\alpha_0 = \frac{\pi^2 v_F^2 \tilde{T}_2}{3L^2 \omega_s} \frac{(1+B_1)(1+B_0)^2}{(B_0 - B_1)} \tag{58.44}$$

or using the definition of A, Eq. (58.40), we see that

$$1 + B_0 = \frac{3L^2}{\pi^2} \frac{\omega_s \alpha_0}{v_F^2(m^*/m_0)\,\tilde{T}_2} \frac{1}{A^{1/2}}. \tag{58.45}$$

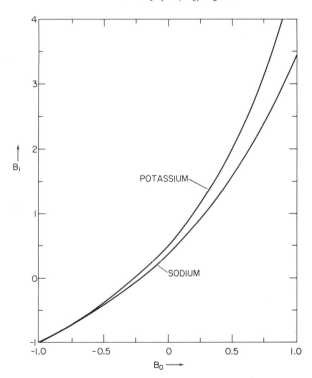

FIG. X-9. The theoretical relation between B_0 and B_1 as obtained from Eq. (58.40) and an experimental determination of A for Na and K.

Equation (58.45) allows one to determine B_0 from an absolute measurement of the sample thickness and the spacing of the first spin wave from the CESR. Using the values (28)

$$p_F = 0.925 \times 10^8 \text{ cm}^{-1} \quad \text{(Na)}$$
$$p_F = 0.7446 \times 10^8 \text{ cm}^{-1} \quad \text{(K)}. \tag{58.46}$$

Dunifer and Schultz (29) obtain the following set of values for B_0 and B_1

	Na	K
B_0	$-.215 \pm .03$	$-.285 \pm .02$
B_1	$-.005 \pm .04$	$-0.06 \pm .03$

(58.47)

In order to obtain additional information on the higher B_n, $n \geqslant 2$, Dunifer and Schultz (29) have examined the behavior of the spin waves with higher k values by using a "truncated" version of the susceptibility tensor (14). This procedure has been outlined in several places in the text and is merely an extension of our simple calculation for an isotropic scattering function of the spin-wave spectrum [see Eqs. (58.5)–(58.13)]. Figure X-10 shows a comparison of the theory and experimental line shape for a 300 G sweep through the first four modes of a sodium sample.

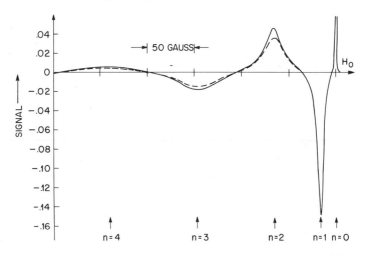

FIG. X-10. Comparison of experimental data (solid) and theoretical (dashed) curves for a complete spin wave spectrum in Na; $T = 1.3°K$; $\omega = 5.9 \times 10^{10}$ Hz, $\Delta = \pi/2$. The theoretical curve was computed using $B_0 = -0.18$, $B_1 = +0.05$, $B_2 = 0.0$, $T_2 = 2.0 \times 10^{-7}$ sec and $\tau = 3.95 \times 10^{-10}$ sec.

The fit is excellent. The final result of this careful fitting procedure to all the spin wave modes and their angular distribution is that the numbers B_0 and B_1 for Na and K as given above are correct and that the higher B_n are small.

Spin-wave structure has also been observed in Rb and Cs by Schultz and co-workers, but at the time of this writing the experiments are too preliminary to quote actual numbers. However, it does seem to be true that the numbers are, crudely speaking, similar to those for Na and K, i.e., $B_0 \simeq -0.2$, $|B_1/B_0| \ll 1$ and $B_n \simeq 0$, $n \geqslant 2$.

Having determined some of the B_n coefficients from the spin-wave dispersion, it would be important to compare these values to B_n values obtained from other distinct measurements. Such a comparison would provide us with a better check on the internal consistency of the LS theory. Unfortunately, the only other measurement that is available is the dc susceptibility measurement already mentioned. For Na this dc measurement (10) ($\chi = (1.13 \pm 0.05) \times 10^{-6}$ cgs) gives a $B_0 = -0.28 \pm 0.03$. This value is in substantial disagreement with the spin wave value given in Eq. (58.47). The value of $\chi = (0.95 \pm 0.1) \times 10^{-6}$ first reported by Schumacher and Slichter gives much closer agreement, i.e., $B_0 = -0.15$. B_0 has not been determined in this way for K.

REFERENCES

1. L. D. Landau, *Zh. Eksp. Teor. Fiz.* **30**, 1058 (1956) [Engl. transl., *Soviet. Phys.—JETP* **3**, 920 (1956)].
2. V. P. Silin, *Zh. Eksp. Teor. Fiz.* **33**, 495 (1957) [Engl. transl., *Soviet Phys.—JETP* **6**, 945 (1958)].
3. L. Landau, *Sov. Phys.—JETP* **8**, 70 (1959).
4. J. M. Luttinger and P. Nozières, *Phys. Rev.* **127**, 1431 (1962).
5. A. V. Gold, *in* "Solid State Physics," Vol. 1: Electrons in Metals (J. F. Cochran and R. R. Haering, eds.). Gordon & Breach, New York, 1968.
6. P. Nozières, "Interacting Fermi Systems." Benjamin, New York, 1964; D. Pines and P. Nozières, "The Theory of Quantum Liquids." Benjamin, New York, 1966.
7. E. P. Wigner, *Phys. Rev* **40**, 749 (1932).
8. J. M. Jauch and F. Rohrlich, "The Theory of Photons and Electrons," Appendix A-7. Addison-Wesley, Reading, Massachusetts, 1955.
9. C. Herring, "Exchange Interactions Among Itinerant Electrons in Magnetism" (G. Rado and H. Suhl, eds.). Academic Press, New York, 1966.
10. R. T. Schumacher and W. L. Vehse, *J. Phys. Chem. Solids* **24**, 297 (1963).
11. R. T. Schumacher and C. P. Slichter, *Phys. Rev.* **101**, 58 (1956).
12. M. Ya Azbel, *Zh. Eksp. Teor. Fiz.* **34**, 766 (1958) [*Soviet Phys.—JETP* **7**, 527 (1958)].
13. T. M. Rice, *Phys. Rev.* **175**, 858 (1968).
14. A. R. Wilson and D. R. Fredkin, *Phys. Rev. B* **2**, 4656 (1970).
15. P. M. Platzman, W. M. Walsh, and E-Ni Foo, *Phys. Rev.* **172**, 689 (1968).

16. V. P. Silin, *Zh. Eksp. Teor. Fiz.* **35**, 1243 (1958) [Engl. transl., *Soviet Phys.—JETP* **8**, 870 (1959)].
17. N. D. Mermin and Y. C. Cheng, *Phys. Rev. Lett.* **20**, 839 (1968).
18. P. M. Platzman and W. M. Walsh, Jr., *Phys. Rev. Lett.* **19**, 514 (1967); **20**, 89E (1968).
19. Y. C. Cheng, J. S. Clarke, and N. D. Mermin, *Phys. Rev. Lett.* **20**, 1486 (1968).
20. G. A. Baraff, C. C. Grimes, and P. M. Platzman, *Phys. Rev. Lett.* **22**, 590 (1969).
21. G. A. Baraff, *Phys. Rev. B* **1**, 4307 (1970); *B* **2**, 637 (1970).
22. P. M. Platzman and K. C. Jacobs, *Phys. Rev. A* **134**, 974 (1964).
23. R. G. Chambers, *Phil. Mag.* **1**, 459 (1956).
24. W. F. Brinkman and S. Engelsberg, *Phys. Rev. Lett.* **21**, 1187 (1968).
25. W. M. Walsh, Jr., G. S. Knapp, L. W. Rupp, Jr., and P. H. Schmidt, *J. Appl. Phys.* **41**, 1081 (1970).
26. S. Schultz and G. Dunifer, *Phys. Rev. Lett.* **18**, 283 (1967).
27. W. M. Walsh, *in* "Solid State Physics," Vol. 1: Electrons in Metals (J. F. Cochran and R. R. Haering, eds.), pp. 127–252. Gordon & Breach, New York, 1968.
28. D. Schoenberg and P. J. Stiles, *Proc. Roy. Soc. Ser. A* **281**, 62 (1964).
29. G. Dunifer, "Spin Waves in Na and K," Thesis. Univ. of California at San Diego, San Diego, California, 1969; G. Dunifer and S. Schultz, to be published.

XI. The Microscopic Theory of Fermi Liquids

59. Introduction to the Microscopic Theory of Fermi Liquids

In Chapter X, we used the Landau Fermi-liquid theory to discuss the properties of transverse electromagnetic waves in metals. We have concentrated on the properties of simple waves in the simplest metals, and have stressed the importance of these waves in probing the properties of the interacting electron fluid in a metal. Throughout we have consciously avoided potentially interesting but complicated, problems, such as the effect of complex Fermi surfaces on wave propagation, and the characteristics of waves propagating at arbitrary angles to the field $\mathbf{H}_0$. These effects will no doubt prove to be of interest in the future development of the field. However one must keep in mind that they greatly complicate the formalism, thus obscuring the essential physics underlying the collective behavior of an interacting electron gas.

Within the framework of the analysis presented in Chapters VIII–X, we have seen that the observation and quantitative description of PSW has led to a rather precise determination of the first few magnetic B_n parameters in the LS theory. Similarly the observation and analysis of CW in several metals has enabled one to measure a few of the orbital A_n parameters of the LS theory. Both types of disturbances have been studied in detail in Na and K and been observed in the other alkalis, Cs and Rb. Thus for the alkalis, the parameters of the Fermi-liquid theory (except for A_0 and A_1) have been fairly well characterized. However, even for these simple metals there is still an unfinished portion of the story. The theory we have presented is phenomenological. It was developed from several plausible, rather general assumptions, but it has certainly not been derived in a rigorous manner. A complete theory requires a microscopic basis for the phenomenology. Such a basis allows one to investigate the limits of validity of the theory, i.e., to understand the conditions under which it is valid.

A microscopic treatment of the interacting electron fluid starts from the many-body Hamiltonian, including Coulomb interactions between electrons and their interaction with the vibrations of the lattice, the phonons. It should also, in principle, include the effects of band structure, a complication which to date we have largely ignored. Given a

Hamiltonian, one then proceeds to compute, usually by some form of perturbation theory, the important physical quantities; e.g., the wave number and frequency-dependent conductivity. A detailed analysis of the diagramatic, many-body perturbation theory required for this type of calculation is definitely beyond the scope of this book. The machinery required for such calculations has not even been presented. The reader is encouraged to look at several excellent books on the subject (*1, 2*).

It is, however, appropriate and possible for us to discuss qualitatively the microscopic basis for the Landau theory and to indicate how a more complete discussion might proceed. The Landau theory in its barest form consists of two sets of conjectures. It assumes that the excited states of the system can be characterized by a set of quasi particles whose energy is a functional of the number of quasi particles which are excited (conjecture number one). Given these single-particle energies, it then assumes that the transport properties, i.e., response functions of the system may be calculated utilizing a single quasi-particle Boltzmann–Vlasov equation. The form of the transport equation is fixed by assuming that the quasi particles' one-particle Hamiltonian is equal to the quasi-particle energy (assumption number two). We will say nothing about the derivation of assumption number two but, for reasons of simplicity, talk only about the excitation spectrum of the interacting fluid. Such a discussion, although limited, will enable us to make plausible the microscopic identification of the quantities $E°(\mathbf{p})$ and $f(\mathbf{p}, \sigma, \mathbf{p}', \sigma')$ of the Landau theory (see Chapter X). In addition we will be able to outline a number of approximate theoretical calculations of these quantities (*3–5*). These calculations predict the various LS parameters in terms of the known properties of metals, such as their phonon spectra, band structure, electron density, etc.

60. The General Formulation of the Microscopic Theory

While the Landau theory is formulated in terms of the energies of quasi particles, the microscopic theory is generally formulated in terms of Green's functions. The reason for this type of formulation is simply that the infinite set of time-dependent Green's functions (one particle, two particle, etc.) are a complete description and that they are convenient quantities to calculate using the well-developed machinery of quantum field theory.

The one-particle Green's function gives us all the information concerning the single-particle excited states of the many-body system. It will be the only such function we will discuss. The two-particle

60. GENERAL FORMULATION

Green's function contains all of the response function information. In a direct calculation of the conductivity, for example, we must evaluate the two-particle Green's function. The second conjecture that the form of the transport equation is fixed by the properties of the single-particle excited states as a functional of the density enables us to identify all of the important functions in the Landau theory without actually deriving a transport equation or equivalently evaluating the two-particle Green's function.

The single-particle Green's functions we will study, in order to discuss the microscopic theory, are analogous to the quantities $\langle c_p^+(t), c_{p+k}\rangle$ introduced in Section 18. In fact, the term $k=0$, the diagonal terms in this density matrix element is most closely related to the one particle Green's function [defined, e.g., in reference (1), p. 59]. More precisely the Green's function is defined to be

$$G_\sigma(\mathbf{p}, \omega) = i \int_{-\infty}^{+\infty} e^{i\omega t}\langle 0 \mid T\{c_p^\sigma(t)[c_p^\sigma(0)]^+\}\mid 0\rangle \, dt. \qquad (60.1)$$

Here $|0\rangle$ is the exact ground state of the interacting system expressed in the Schrödinger picture and the annihilation (c_p^σ) and creation $(c_p^\sigma)^+$ operators are expressed in the Heisenberg picture, i.e., their time dependence is determined by the full Hamiltonian of the many-body system,

$$c_p^\sigma(t) = e^{iHt} c_p^\sigma e^{-iHt}. \qquad (60.2)$$

The time-ordering symbol T is defined so that the operators with earlier times are placed to the right (with appropriate change of sign).

The Green's function G describes the propagation of a single particle or hole with spin σ which has been added to the system (1). It can be expressed in terms of its spectral weight function $A_\pm^\sigma(\mathbf{p}, \omega)$, i.e.,

$$G_\sigma(\mathbf{p}, \omega) = \int_0^\infty \left[\frac{A_+^\sigma(\mathbf{p}, \omega')}{(\omega' - \omega + E_F - i\delta)} - \frac{A_-^\sigma(\mathbf{p}, \omega')}{(\omega' + \omega - E_F - i\delta)}\right] d\omega'. \qquad (60.3)$$

Here δ is a positive infinitesimal. The spectral functions $A_\pm$ are given by

$$A_+^\sigma(\mathbf{p}, \omega) = \sum_n |\langle\varphi_n |(c_p^\sigma)^+| \varphi_0\rangle|^2 \, \delta(\omega - E_n)$$

$$A_-^\sigma(\mathbf{p}, \omega) = \sum_m |\langle\varphi_m | c_p^\sigma | \varphi_0\rangle|^2 \, \delta(\omega - E_m). \qquad (60.4)$$

For a free Fermi gas of particles with mass m_0 the $A_\pm^\sigma$'s are delta functions, i.e.,

$$A_+^\sigma(\mathbf{p}, \omega) = [1 - n_\sigma^\circ(p)] \, \delta(\omega - p^2/2m_0 + E_F) \qquad (60.5)$$

$$A_-^\sigma(\mathbf{p}, \omega) = n_\sigma^\circ(p) \, \delta(\omega + p^2/2m_0 - E_F) \qquad (60.6)$$

so that

$$G_\sigma^\circ(\mathbf{p}, \omega) = \frac{n_\sigma^\circ(p)}{(p^2/2m_0 - \omega) + i\delta} + \frac{[1 - n_\sigma^\circ(p)]}{(p^2/2m_0 - \omega) - i\delta}. \quad (60.7)$$

In general, $A^\sigma(\mathbf{p}, \omega)$ is not such a simple function of ω. However, in normal Fermi liquids, i.e., fermion systems with a Fermi surface and no spin order, it can be shown that for small ω and $p \sim p_F$ that $A^\sigma(\mathbf{p}, \omega)$ is approximately a Lorentzian function of the form

$$A_\sigma(\mathbf{p}, \omega) = Z_\mathbf{p} \left[\frac{\Gamma_\mathbf{p}}{\pi}\right]\left[\frac{n_\sigma^\circ(\mathbf{p})}{(\omega - E_\mathbf{p}^\circ + E_F)^2 + \Gamma_\mathbf{p}^2} + \frac{1 - n_\sigma^\circ(\mathbf{p})}{(\omega - E_\mathbf{p}^\circ - E_F)^2 + \Gamma_\mathbf{p}^2}\right]. \quad (60.8)$$

The delta function in Eq. (60.5) has been smeared out ($\Gamma_\mathbf{p}$) and the residue ($Z_\mathbf{p}$) at the pole is no longer unity. The peak in the Lorentzian at $\omega = E_\mathbf{p}^\circ - \mu$ [in the second term in Eq. (60.8)] is the quasi-particle peak. The width $\Gamma_\mathbf{p}$ in the limit $p \to p_F$ (because of the available phase space) goes to zero as $(p - p_F)^2$. Thus our Lorentzian is really quite like a delta function. The smallness of $\Gamma_\mathbf{p}$ is the mathematical statement of a well-defined quasi-particle-like excitation spectrum for the Landau–Fermi liquid.

In the general case G (leaving off spin indices) satisfies an equation of the form

$$G(\mathbf{p}, \omega) = \left[\frac{p^2}{2m_0} - \omega - \Sigma(\mathbf{p}, \omega)\right]^{-1}. \quad (60.9)$$

Here the self-energy $\Sigma(\mathbf{p}, \omega)$ contains all the effects of the interaction. The assertion that the spectral weight function $A_\sigma(\mathbf{p}, \omega)$ [Eq. (60.8)] has a Lorentzian form with a vanishing $\Gamma_\mathbf{p}$ implies that the Im $\Sigma(\mathbf{p}, \omega)$ is small for p near p_F. This in turn implies that G has "resonances" at

$$\text{Re}\left[\frac{p^2}{2m_0} - E_\mathbf{p}^\circ - \Sigma(\mathbf{p}, E_\mathbf{p}^\circ)\right] = 0. \quad (60.10)$$

These resonances define the quasi-particle energy $E_\mathbf{p}^\circ$.

Within the framework of perturbation theory the self-energy $\Sigma(\mathbf{p}, \omega)$ may be expanded in a series of diagrams involving successively higher powers of the interaction (1). For pure Coulomb interactions, the two lowest-order contributions to $\Sigma(\mathbf{p}, \omega)$ are shown in Fig. XI-1. Figure XI-1a represents the lowest-order "direct" contribution while Fig. XI-1b the lowest-order "exchange" contribution. These two diagrams yield the Hartree–Fock approximation (see Section 29). In a translation-invariant system, the contribution from the first diagram is

60. GENERAL FORMULATION

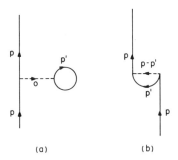

FIG. XI-1. Lowest-order Hartree–Fock self-energy diagrams for the electron gas problem.

cancelled by the presence of a uniform background of positive charge, while the second diagram contributes a term of the form

$$\Sigma_{\text{HF}} \sim \int d^3p'\, d\omega' \, \frac{e^2}{|\mathbf{p} - \mathbf{p}'|^2} \, G_0(\mathbf{p}', \omega') \tag{60.11}$$

In general, $\Sigma(\mathbf{p}, \omega)$ is the sum of all "irreducible" bubbles no matter how complicated; for example, see Fig. XI-2. The word "irreducible"

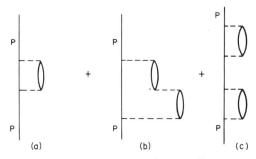

FIG. XI-2. Higher-order self-energy diagrams.

here means a diagram which cannot be divided into two equivalent parts by simply cutting or breaking a single propagator G_0. The diagram in Fig. XI-2c is not irreducible. In evalvating such diagrams the solid lines all contribute a factor G_0 (free electron propagators) and the dotted lines factors $V(q) = 4\pi e^2/q^2$. Using momentum and energy conservation at each vertex one finally integrates the resultant expression over all free internal momenta and energies.

The energy $E^\circ(\mathbf{p}) \equiv E[\mathbf{p}, n_\sigma^\circ(p)]$ in Eq. (60.10) is the quasi-particle energy in the ground state specified by $n_\sigma^\circ(p)$. To obtain the $f(\mathbf{p}, \sigma, \mathbf{p}', \sigma')$

function of Fermi-liquid theory we must consider the change in quasi-particle energy, $E(\mathbf{p}, n)$, when a quasi particle is added to the system so that [see Eq. (52.10)]

$$f(\mathbf{p}, \sigma, \mathbf{p}', \sigma') = \delta E(\mathbf{p}, \sigma)/\delta n(\mathbf{p}', \sigma')|_{n=n^0} \tag{60.12}$$

The functional dependence of $\Sigma(\mathbf{p}, \omega)$ on n, hence of the quasi-particle energy on n, is contained in the many G_0's which appear in each term of the perturbation series for $\Sigma(\mathbf{p}, \omega)$.

Equation (60.12) implies that we may find $f(\mathbf{p}, \sigma, \mathbf{p}' \sigma')$ by differentiating E or equivalently Σ with respect to n, i.e., with respect to all the G_0's which appear. The functional derivative of G_0 is,

$$\frac{\delta G_0}{\delta n} = \frac{1}{(p^2/2m_0 - \omega + i\delta)} - \frac{1}{(p^2/2m_0 - \omega - i\delta)} \equiv -2\pi i \delta\left(\omega - \frac{p^2}{2m_0}\right). \tag{60.13}$$

Equation (60.13) implies that we may think of the process of differentiation as the breaking of any one of the internal lines G_0, i.e., omitting the integration which is customarily associated with the energy variable ω carried by one of the internal propagators G_0 and setting its energy equal to $p^2/2m_0$. As an example the differentiation of the lowest-order Hartree–Fock diagrams (XI-1) is shown in Fig. XI-3.

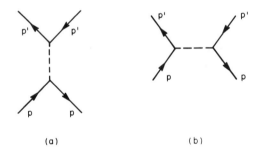

(a)   (b)

FIG. XI-3. Hartree–Fock contributions to the Landau scattering function.

This type of reasoning implies that $f(\mathbf{p}, \sigma, \mathbf{p}', \sigma')$ is simply related to the sum of all diagrams having two incoming and two outgoing lines, in fact one can show that (1) ($\epsilon_p \equiv p^2/2m_0$)

$$f(\mathbf{p}, \sigma, \mathbf{p}', \sigma') = 2\pi i Z_p Z_{p'} \Gamma(\mathbf{p}, \sigma, \epsilon_p, \mathbf{p}', \sigma', \epsilon_{p'}; \mathbf{p}', \sigma', \epsilon_{p'}, \mathbf{p}, \sigma, \epsilon_p), \tag{60.14}$$

where the four-point function Γ is graphically represented in Fig. XI-4.

60. GENERAL FORMULATION

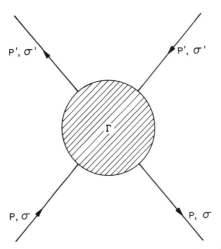

FIG. XI-4. Schematic representation of the forward particle–hole scattering amplitude.

It is the general forward scattering amplitude for an electron–hole pair. In the Hartree–Fock approximation Σ is given by Eq. (60.11), so that

$$f \sim \frac{\delta\Sigma}{\delta n(\mathbf{p}', \sigma')} \sim \frac{e^2}{|\mathbf{p}' - \mathbf{p}|^2}, \qquad \sigma = \sigma'$$
$$f = 0, \qquad \sigma \neq \sigma'. \tag{60.15}$$

In this discussion we have overlooked two important points which will be relevant to our discussion of the relationship between the approximate microscopic calculations and experiment. In general the four-point function is itself a function not only of the energy, momenta and spin of the incoming particles as in Eq. (60.14), but of the momentum and energy transfer

$$\bar{\omega} \equiv (\mathbf{q}, \epsilon) \tag{60.16}$$

as well. The general four-point function is displayed schematically in Fig. XI-4 with (leaving off spin indices temporarily)

$$\Gamma = \Gamma(\not{p}, \not{p}', \bar{\omega}) \tag{60.17}$$

and

$$\not{p} \equiv (\mathbf{p}, \omega) \tag{60.18}$$

It can be seen from Eq. (60.14) that the function f is related to the limit $\bar{\omega} \to 0$ of the general four-point function, i.e., it is the forward particle–hole scattering amplitude. Intuitively this identification is a

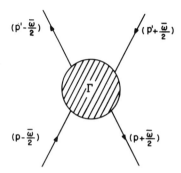

Fig. XI-5. Schematic representation of the four-point (particle–hole) scattering amplitude.

sensible one. We know, for example, that the dielectric constant for light propagating in a medium is directly proportional to the forward scattering amplitude. Since the dielectric constant determines the ω versus $\mathbf{k}$ relationship for the light, it is the quantity which yields the energy (ω) momentum ($\mathbf{k}$) relationship of the electromagnetic wave. In the same way the quasi-particle energy–momentum relationship is simply related to the forward quasi-particle scattering amplitude.

Unfortunately there are two technical difficulties associated with Eq. (60.14) which we can only briefly discuss. The first is related to the fact that the limit $\bar{\omega} \to 0$ is not well defined. The origin of the nonanalyticity of Γ is discussed in detail in Nozières' book (1). In essence the limiting form of the four-point function depends on the ratio

$$\eta \equiv q v_F / \epsilon. \tag{60.19}$$

All of the results of the theory may be expressed in terms of the two limiting values of Γ, i.e., Γ^0 ($\eta \equiv 0$) and Γ^∞ ($\eta = \infty$) which are in turn related to one another by the integral equation

$$\Gamma^\infty(\not{p},\not{p}') = \Gamma^0(\not{p},\not{p}') + \sum_{\not{p}''} \Gamma^0(\not{p},\not{p}'')[-2\pi i Z_{\mathbf{p}''}^2 \, \delta(\omega'' - E_F) \, \delta(\epsilon_{\mathbf{p}''} - E_F) \, \Gamma^\infty(\not{p}'',\not{p})]. \tag{60.20}$$

An analysis of the infinite set of perturbation diagrams shows that the function f is proportional to the function Γ^0. This is true because of the way in which we have described "quasi particles." While the momenta of the quasi particles are well defined, their energies are smeared by something of the order of $\Delta E_{\mathbf{p}} \simeq (\mathbf{p} - \mathbf{p}_F)^2/2m$. This implies that $q \equiv 0$ while ϵ is finite, i.e., $\eta = 0$.

60. GENERAL FORMULATION

The second "technical" difficulty associated with Eq. (60.14) is that we have is some sense ignored the problem of long-range interactions among the quasi particles. In writing the transport equation for the interacting electron liquid [Eq. (55.2)] a clear separation is made between long-range effects and short-range correlation effects. The electric field **E** in the LS equation contains all of the long-range interactions between quasi particles. It is determined from a self-consistent solution of Maxwell's equations and the LS transport equation. The function $f(\mathbf{p}, \sigma, \mathbf{p}', \sigma')$ characterizes the remaining short-range interactions between "quasi particles." This separation implies that we must also separate Γ into a short-range part $(\tilde{\Gamma})$ plus a part which is included in the self-consistent field (Γ_{Long}). Schematically then

$$\Gamma(\mathbf{\not p}, \mathbf{\not p}', \bar{\omega}) = \tilde{\Gamma}(\mathbf{\not p}, \mathbf{\not p}', \bar{\omega}) + \Gamma_{\text{Long}}. \tag{60.21}$$

The function f (dropping spin indices) is proportional to $\tilde{\Gamma}(\mathbf{\not p}, \mathbf{\not p}', \bar{\omega})$, i.e.,

$$f(\mathbf{p}, \mathbf{p}') = 2\pi i Z_p Z_{p'} \tilde{\Gamma}^0(\mathbf{p}, \mathbf{p}'). \tag{60.22}$$

Within the spirit of the Hartree–Fock approximation, Eq. (60.22) implies that we only include the exchange diagram Fig. XI-3b. The contribution from Fig. XI-3a is included in Γ_{Long}, i.e., the self-consistent field. In general, the diagrams contributing to Γ_{Long} are shown schematically in Fig. XI-6.

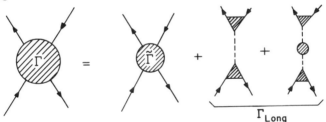

FIG. XI-6. The decomposition of the four-point function Γ into a short-range part $\tilde{\Gamma}$ plus an effective field piece Γ_{Long} (the last two diagrams).

It is convenient to define a function $r(\mathbf{p}, \mathbf{p}')$ by (leaving off spin indices) the relation

$$r(\mathbf{p}, \mathbf{p}') = 2\pi i Z_p Z_{p'} \tilde{\Gamma}^\infty(\mathbf{p}, \mathbf{p}'). \tag{60.23}$$

The functions $f(\mathbf{p}, \mathbf{p}')$ and $r(\mathbf{p}, \mathbf{p}')$ are dual functions in the many-body theory. The Landau theory may be completely formulated in terms of either function. The choice between $f(\mathbf{p}, \mathbf{p}')$ and $r(\mathbf{p}, \mathbf{p}')$ is entirely equivalent to our choice of distribution function. If we choose to use g, the deviation of the distribution function from its equilibrium value, then the theory is naturally formulated in terms of $f(\mathbf{p}, \mathbf{p}')$. If we choose to work with $\bar{g}$, then the transport equation naturally involves $r(\mathbf{p}, \mathbf{p}')$.

Equation (60.20) gives us a simple relation, for isotropic systems, between the Legendre coefficients of f and r. Defining Legendre coefficients r_l^s and r_l^a similar to the A_l and B_l, so that

$$\frac{m^* p_F}{2\pi^2} [r(\mathbf{p}, \uparrow, \mathbf{p}', \uparrow) \pm r(\mathbf{p}\uparrow, \mathbf{p}', \downarrow)] = \sum_l (2l+1) r_l^{s,a} P_l(\cos\theta_{\mathbf{p},\mathbf{p}'}) \quad (60.24)$$

we find from Eq. (60.20) that

$$r_l^s = A_l/(1 + A_l); \qquad r_l^a = B_l/(1 + B_l). \quad (60.25)$$

The Pauli exclusion principle imposes a condition in the form of a sum rule on the Landau parameters (6). The full four-point vertex function Γ because of the Fermi statistics is antisymmetric so that

$$\Gamma(\not{p}, \not{p}', \bar{\omega}) = -\Gamma[(\not{p} + \not{p}' - \bar{\omega})/2, (\not{p} + \not{p}' + \bar{\omega})/2, \not{p}' - \not{p}]. \quad (60.26)$$

By setting $\bar{\omega} = \not{p}' - \not{p}$, we see that both sides of Eq. (60.26) are identical (aside from the minus sign) and therefore zero. Putting $\not{p}$ and $\not{p}'$ on the energy shell and on the Fermi surface and taking the limit $\not{p}' \to \not{p}$, we arrive at the result

$$\Gamma^\infty(\mathbf{p}, \sigma, \mathbf{p}, \sigma) = 0. \quad (60.27)$$

Equation (60.27) (taking into account the fact that it is a condition on the full Γ) immediately leads to the important exact sum rule for the Landau parameters

$$\sum_{l=0}^{\infty} (2l+1) \left[\frac{A_l}{1 + A_l} + \frac{B_l}{1 + B_l} \right] = -\frac{1}{1 + A_0}. \quad (60.28)$$

The right-hand side of Eq. (60.28) comes from the limiting value of Γ_{long}. This quantity can be related, by means of a Ward identity, [see Ref. (1), Eq. (6-138)] to the compressibility (κ) of the interacting Fermi liquid,

$$\frac{\kappa}{\kappa_0} = \frac{m^*/m_0}{1 + A_0} \quad (60.29)$$

where κ_0 is the compressibility of the noninteracting (uncharged) Fermion gas. Transposing the right-hand side of Eq. (60.28) indicates that, in fact, A_0 does not enter the sum rule. For an uncharged system, interacting via short-range forces, there would be no long-range piece and thus no difference between $\tilde{\Gamma}$ and Γ. The right-hand side of Eq. (60.28) would be zero. The effect of the long-range force is to replace $A_0/(1 + A_0)$ by unity. In fact, if we had treated the Coulomb potential totally as a short-range force, then we would compute an infinite A_0 so that $A_0/(1 + A_0)$ would indeed be unity.

The sum rule Eq. (60.28) is only valid for a system of electrons interacting exclusively through their Coulomb forces. The exclusion principle of course is more generally applicable. When phonons are present, the sum rule is changed. The electron gas compressibility, the right-hand side of Eq. (60.28), is replaced by a simple average over the long-wavelength portion of the phonon spectrum, i.e., it depends on the combined lattice and electron gas compressibility. We will return to the exact form of this sum rule in Section 62 when electron phonon effects are discussed.

61. Coulomb Interactions and the Landau Scattering Function

For an electron gas, the dimensionless parameter which expresses the ratio of potential to kinetic energies is r_s (see Section 2) the interparticle separation in units of the Bohr radius. Small r_s corresponds to high density. In this limit, the kinetic energy becomes large compared to the potential energy and one can expand in powers of r_s and $r_s \ln r_s$. Such expansions however only converge in the region $r_s < 1$ (7). In the alkalis, r_s takes on the values $1.8 < r_s < 5.6$, which are far outside the range of high-density expansions. The electron gas at metallic densities is in an intermediate coupling regime, so that one cannot reliably calculate its properties from first principles. One is forced to proceed by calculating a limited set of graphs, using physical arguments to pick out what one hopes are the important ones. One must then compare the results with experiment to see if the answers are reasonable. A comprehensive review of the problem will be found in the review article by Hedin and Lundquist (8). We cannot in any sensible way discuss the details of these calculations. However, we can discuss them qualitatively and try to give the reader a qualitative feel for the content of the computations.

There have been two detailed calculations of the Landau f function for an electron gas at metallic densities (4, 5). Both calculations are essentially second-order perturbation calculations in an "effective" interaction $\tilde{V}(\mathbf{q}, \omega)$ between electrons. Going to second order in $\tilde{V}(\mathbf{q}, \omega)$ means summing the four diagrams shown in Fig. XI-7. The double wavy line in the diagram corresponds to the "effective" Coulomb interaction $[\tilde{V}(\mathbf{q}, \omega)]$ between electrons. Graph (a) is the Hartree–Fock (i.e., lowest-order exchange) result which we have already discussed.

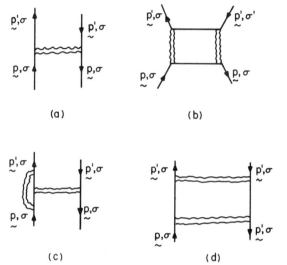

FIG. XI-7. Second-order contributions to the scattering function.

Within normalization factors its value is given by Eq. (60.11), with $4\pi e^2/q^2$ replaced by $\tilde{V}(\mathbf{q}, 0)$. Graphs (c) and (d) are the second-order exchange diagrams and graph (b) is the leading order "direct" contribution.

For non-spin-dependent forces, the exchange diagrams only contribute for parallel spins. The primed and unprimed variables are on the same solid line. At each vertex the spin indepent interaction can change momentum ($\mathbf{p} \neq \mathbf{p}'$) but it does not change the spin, i.e., $\sigma = \sigma'$. In the direct diagram the indices σ and σ' are independent of one another. The prime variables are on one solid line and unprimed on a separate solid line. It is the direct diagrams that lead to differences between the A_n and B_n coefficients. The fact that antiparallel spins $\sigma \neq \sigma'$ give zero contribution is equivalent [see Eq. (52.29)] to the statement $A_n = B_n$.

Although both calculations evaluate the same four graphs, they give different results insofar as they use different forms for the effective interaction. Hedin's (4) calculation is the easiest to describe. He uses a $\tilde{V}(\mathbf{q}, \omega)$ which is screened with a *static*, Lindhard dielectric function, i.e.,

$$\tilde{V}(\mathbf{q}, \omega) = \frac{4\pi e^2}{q^2 \epsilon(\mathbf{q}, 0)} \tag{61.1}$$

The dielectric function $\epsilon(\mathbf{q}, 0)$ is given by Eq. (20.2). This form for $\tilde{V}$ is simple enough so that one can evaluate the four graphs in Fig. XI-7 exactly. It contains the physics of "static" screening but leaves out effects due to dynamical screening, i.e., the ω dependence of $\epsilon(\mathbf{q}, \omega)$. Rice (3) on the other hand, uses a dynamically screened interaction of the form

$$\tilde{V}(\mathbf{q}, \omega) = 4\pi e^2/q^2 \epsilon(\mathbf{q}, \omega). \tag{61.2}$$

This choice incorporates more of the physics of the screening process. It includes into the wavy line all diagrams of the form shown in Fig. XI-8; i.e., $\epsilon(\mathbf{q}, \omega)$ is given by Eq. (19.5). Unfortunately, this choice

FIG. XI-8. The RPA approximation to the effective (double wiggly line) interaction.

makes the evaluation of diagrams (c) and (d) in Fig. XI-7 almost impossible. The procedure Rice followed was to evaluate graphs (a) and (b) exactly using $\tilde{V}(\mathbf{q}, \omega)$ given in Eq. (61.2) and to approximate graphs (c) and (d) using the so-called Hubbard (9) interpolation formula, i.e., an approximation for $\epsilon(\mathbf{q}, \omega)$.

In Table I we list the values of the Landau parameters obtained from these two calculations. They are denoted by A_l^{ee}. Rice finds A_1^{ee} positive for $r_s > 2$ corresponding to an enhanced effective mass, whereas Hedin finds A_1^{ee} negative, corresponding to a reduced effective mass. Silverstein (10) using a somewhat different approximation also found an enhanced mass. We shall see shortly that an enhanced mass is in better agreement with the experimental results for Na and K. There are also significant differences in the magnitudes of some of the other coefficients.

TABLE I

CALCULATED VALUES FOR LANDAU PARAMETERS OF AN ELECTRON GAS WITH $r_s = 4$

	Hubbard approximation	Hedin	Screened exchange
A_0^{ee}	-0.69	-0.55	-0.32
A_1^{ee}	$+0.06$	-0.03	-0.04
A_2^{ee}	-0.05	-0.025	-0.005
A_3^{ee}	$+0.004$		
B_0^{ee}	-0.28	-0.31	-0.32
B_1^{ee}	-0.06	-0.09	-0.04
B_2^{ee}	-0.02	-0.001	-0.005
B_3^{ee}	-0.001		
χ/χ_0	1.47	1.41	1.42
m^*_{ee}/m_0	1.06	0.97	0.96
Sum rule	$+0.23$	$+0.03$	$+0.26$

For comparison, the last column in Table I gives the results of a calculation keeping only the lowest-order exchange graph (Fig. XI-7a) using an $\epsilon(\mathbf{q}, \omega)$ given in Eq. (19.5) and evaluating these terms at $r_s = 4$. In this approximation, $A_n = B_n$. The results are closer to Hedin's, but there are significant differences in magnitude.

This discussion points up the inadequacies of such calculations. It is, in fact very difficult to go beyond certain simple lower-order graphs. Some higher-order effects are taken into account by the artifice of an effective interaction. However, there are many classes of graphs whose magnitude have not even been estimated. It could very well be that we make a serious error in neglecting such graphs. The essential difficulty is that there is no small parameter in the problem, hence no real justification for ignoring such contributions. Moreover, we know that there are many cancellations among higher-order graphs, so that great care must be exercised in going beyond the simpler approximations. It is possible that merely including more diagrams may not lead to more accurate results, especially if sum rules or conservation laws are violated (7).

It is interesting to see how well the various approximations satisfy the exclusion principal sum rule, given in Eq. (60.28). In the last row of Table I, we have tabulated that sum of Landau coefficients which should, in an exact theory, as per Eq. (60.28) be equal to zero. We see that the sum rule is reasonably well approximated by both calculations. Hedin's values give rather better numerical agreement than those calculated with the Hubbard approximation by Rice (5). This is not surprising since in Hedin's calculations the second-order exchange and the direct graphs were treated on the same footing, whereas in the Hubbard approximation the direct terms are included exactly while the exchange terms are only approximated.

We now wish to compare the electron gas calculations to experiment. In general, measured quantities depend upon both the electron–electron interactions and on electron–phonon effects which we will discuss shortly. There are however a few properties that depend only on the electron–electron interaction. These provide a direct check of the accuracy of this part of the calculation. For example, the Pauli spin susceptibility χ can be shown to be independent of the electron–phonon interaction (11). We recall that [Eq. (53.25)]

$$\frac{\chi}{\chi_0} = \frac{m^*/m_0}{1 + B_0} \tag{61.3}$$

The parameters m^* and B_0 are the observed values and contain electron–phonon effects which can be shown to cancel in the ratio.

The Pauli spin susceptibility has been measured directly for Na [see Section 53 (12)]. In addition, the spin wave measurements (see p. 270) combined with the effective mass values, $m^*/m_0 = 1.24$ for Na and $m^*/m_0 = 1.21$ for K measured by Grimes and Kip (11b) yield an indirect value for χ/χ_0. In Table II we tabulate these results along with two theoretical estimates.

TABLE II

EXPERIMENTAL AND THEORETICAL ESTIMATES OF χ/χ_0

	Na	K
Experiment[a] (1)	1.74 ± 0.1	1.58 ± 0.1
(2)	1.59 ± 0.07	1.69 ± 0.2
Theory[b] (1)	1.47	1.55
(2)	1.41	1.45

[a] Experiment (1) Na, from Schumacher and Vehse (11a), K, from Keck (12). Experiment (2) from spin wave data and effective mass measurements (11b).

[b] Theory (1) using the Hubbard approximation (3). Theory (2) from Hedin's and Lundquist's results (8).

The spin wave value (Experiment 2) is some 8 % lower than the directly measured value (Experiment 1). In view of the difficulties in making the direct measurement the overall agreement between the two experimental numbers appears quite satisfactory. The spin-wave value is in better agreement with the theoretical values from Rice's calculation (Theory 1) utilizing the Hubbard approximation to the dielectric constant than from Hedin's (Theory 2) static second-order results. In K there are no direct spin susceptibility measurements and the spin-wave data are less accurate at present. Keck (12) has estimated the spin susceptibility of K by studying the Knight shift in a series of Na–K alloys. He takes the directly measured value in Na as a reference point and extrapolates to pure K. In Table II we show this value and the theoretical estimates for K. The calculated value for K in the Hubbard approximation has been found by extrapolation from the results quoted in Ref. (3). Again the authors find reasonable agreement between theory and experiment.

62. The Electron–Phonon Interaction and the Landau Scattering Function

In metals we know that there are two sources of interaction between electrons. One is the Coulomb repulsion which we have just discussed. The second is due to scattering of the electrons from ions, i.e., electron–

phonon effects. These electron–phonon interaction effects are comparable in size to the electron–electron interaction effects. However, there is a significant difference from the computational point of view. In studying Coulomb interaction effects there is no small parameter in the problem. This meant that a priori all diagrams were important. We could not justify the neglect of any particular set of diagrams relative to the ones we retained.

For the electron–phonon problem, there is a small parameter to expand in. This weak coupling parameter is simply the square root of the ratio of electronic to ionic mass $[(m_0/M)^{1/2} \simeq 10^{-2}]$. The core of the argument first set forth by Migdal (13) is that, to leading order in $(m_0/M)^{1/2}$, only the lowest-order diagram contributes to the self-energy. This diagram, along with several higher-order ones, is shown in Fig. XI-9. In these

FIG. XI-9. Phonon contribution to the self-energy. Graph (a) is the lowest order diagram will (b) and (c) are higher order in the electron–phonon coupling. Graph (d) is a columbic renormalization of the lowest order diagram.

diagrams, wiggly lines represent phonon propagators while the solid lines are the electron propagators G_0 defined by Eq. (60.7). The graphs (b) and (c) are all of order $(m_0/M)^{1/2}$ compared to graph (a).

There is, however, one major difficulty with this type of argument. The vertices, black dots at the intersection of electron phonon lines in Fig. XI-9, characterize the coupling of the phonons to the electrons. A bare electron ion coupling is not sufficient. Migdal's argument does not allow us to neglect diagrams such as graph (d) in Fig. XI-9. In graph (d), the dashed lines are Coulomb interactions. The renormalization of the vertex due to Coulomb interactions (screened electron–phonon interaction) is important and we must have some way of characterizing the effective interaction at the vertex if we are to evaluate the graphs displayed in Fig. XI-9. Fortunately the pseudopotential theory provides us with a phenomenological way of evaluating this coupling.

62. ELECTRON–PHONON INTERACTION

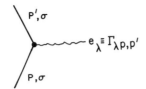

FIG. XI-10. The effective electron phonon coupling $\Gamma_{\lambda p, p'}$.

The vertex $\Gamma_{\lambda pp'}$ shown in Fig. XI-10 is nothing more than the pseudopotential matrix element

$$\Gamma_{\lambda p, p''} = \epsilon_\lambda \cdot (\mathbf{p} - \mathbf{p}'') V_{ps}(\mathbf{p} - \mathbf{p}''). \tag{62.1}$$

The magical pseudopotential V_{ps} includes the effect of the Coulomb renormalization of the vertex, i.e., all diagrams topologically equivalent to Fig. XI-9d.

A comprehensive review of how one goes about determining V_{ps} will be found in the review article by Cohen and Heine (14). Essentially, the shape of the Fermi surface, determined by the De Haas–Van Alphen effect, is used to determine the value of V_{ps} at reciprocal lattice vectors. The long-wavelength ($|\mathbf{p} - \mathbf{p}'| \to 0$) limit is known exactly since in this limit we can evaluate all vertex corrections using general conservation law arguments. Given these points, one can then interpolate smoothly between these values to obtain the Fourier transform, i.e., V_{ps} for all values of $\mathbf{p} - \mathbf{p}'$.

Given V_{ps} and the measured phonon spectrum, one can numerically compute graph (a) of Fig. XI-9. The first realistic calculation of the self-energy effects due to the electron–phonon interaction was carried out in this way by Ashcroft and Wilkens (15). Since then many authors (16) have applied the technique to a variety of simple metals. In general, it is fair to say that most of the results depend only weakly on the detailed shape of the pseudopotential used.

In order to make complete contact with the Landau theory we are still faced with the problem of computing the scattering function f and not simply the self-energy. For the moment, assume that only electron–phonon effects exist. The quantity f is related to the derivative of the self-energy by

$$f \sim \delta \Sigma / \delta n. \tag{62.2}$$

The process of differentiation breaks the solid line in Fig. XI-9 and yields Fig. XI-11a. Unfortunately, we have been slightly sloppy. It turns out that a careful analysis of the problem shows that the differentiation process really gives all the ladder-type diagrams of the type shown in

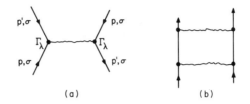

FIG. XI-11. The dominant phononic contribution to the scattering function.

Fig. XI-11b. However, if we had considered the function $r(\mathbf{p}, \sigma, \mathbf{p}', \sigma')$ instead of the function $f(\mathbf{p}, \sigma, \mathbf{p}', \sigma')$, then we would have found that graph (a) was the dominant one.

Our arguments suggest the essential result. It has been discussed by Prange and Sachs (11) and by Rice (5). The conclusion reached by considering the Migdal argument in detail is that one may calculate the electron–phonon contributions to the Landau $r(\mathbf{p}, \sigma, \mathbf{p}', \sigma')$ function by evaluating Fig. XI-11a, using an experimentally determined phonon spectrum and pseudopotential, i.e.,

$$r^{(ep)}(\mathbf{p}, \sigma, \mathbf{p}', \sigma') = 2\pi i Z_\mathbf{p} Z_{\mathbf{p}'} \Gamma_1^{ep}(\mathbf{p}, \mathbf{p}')(1 + \boldsymbol{\sigma} \cdot \boldsymbol{\sigma}'), \qquad (62.3)$$

where

$$\Gamma_1^{ep}(\mathbf{p}, \mathbf{p}') = \frac{1}{2NM} \sum_\lambda \left| \frac{\boldsymbol{\epsilon}_\lambda(\mathbf{p} - \mathbf{p}') \cdot (\mathbf{p} - \mathbf{p}')}{\omega_\lambda(\mathbf{p} - \mathbf{p}')} \right|^2 V_{ps}^2(\mathbf{p} - \mathbf{p}'). \qquad (62.4)$$

Equation (62.4) has been evaluated for Na and K by Rice (5) using the experimental phonon spectrum (17) along with several different pseudopotentials. We will present these numerical results in the next section where we tabulate much of the available data on the Landau parameters.

63. COMPARISON OF NUMERICAL ESTIMATES AND EXPERIMENTAL RESULTS

So far we have discussed the electron–electron and electron–phonon effects separately. To compare with the experimental values these two types of contributions must be combined. This is easier to do in the scattering limit, i.e., the $\epsilon/qv_F \to \infty$ limit. In this case the electron–phonon terms are particularly simple since Fig. XI-11a is the only contribution. To obtain the composite $r(\mathbf{p}, \sigma, \mathbf{p}', \sigma')$ function, we simply add the contributions to the four point vertex Γ and multiply by Z^2, so that

$$r(\mathbf{p}, \sigma, \mathbf{p}', \sigma') = 2\pi i Z_\mathbf{p} Z_{\mathbf{p}'} [\Gamma_{ep}^{(1)}(\mathbf{p}, \mathbf{p}')(1 + \boldsymbol{\sigma} \cdot \boldsymbol{\sigma}') + \Gamma_{ee}^\infty(\mathbf{p}, \sigma, \mathbf{p}', \sigma')]. \qquad (63.1)$$

With this prescription we can combine the second-order electron–electron calculation of f with the lowest-order electron–phonon result for r to obtain the composite A_n and B_n.

In the calculations on the electron gas, both Hedin and Rice (for no particular reason) computed the second-order contributions to $f(\mathbf{p}, \sigma, \mathbf{p}', \sigma')$ not $r(\mathbf{p}, \sigma, \mathbf{p}', \sigma')$. Since there is no small parameter in the electron gas problem, there is no way to determine which quantity one should in fact be calculating. Historically it has been f. The Migdal arguments relating to the electron–phonon part of the problem would suggest that r is the more natural object to compute. It sums more diagrams (not a priori a good thing to do) and it satisfies (assuming we go to a fixed order in perturbation theory) a sum rule like Eq. (60.28) exactly. Unfortunately, we have no way of knowing how good such a simple calculation of r would be.

When the moments A_n are small, there is no distinction between $f(\mathbf{p}, \sigma, \mathbf{p}', \sigma')$ and $r(\mathbf{p}, \sigma, \mathbf{p}', \sigma')$ [see Eq. (61.25)]. We will see that for the alkalis all of the Landau parameters A_n, B_n except A_0 are small, so that the difference between f and r will not bother us.

In Tables III and IV we list the numerical estimates and experimental values of the Landau parameters A_n and B_n for Na and K, respectively.

TABLE III

THE EXPERIMENTAL VALUES AND THEORETICAL ESTIMATES FOR THE LANDAU PARAMETERS IN Na

	Experiment (Na)	Theory[a]
A_0		−0.64
A_1		+0.11
A_2	−0.05 ± 0.01	−0.04
A_3	0.0 + 0.005	+0.005
B_0	−0.21 ± 0.05	−0.17
B_1	−0.005 ± 0.04	−0.005
B_2	0.0 ± 0.03	−0.02
B_3		+0.001
m^*/m_0	1.24 + 0.02	1.21

[a] Electron–electron using Hubbard approximation, electron–phonon in one plane-wave and Ashcroft pseudopotential.

The very good qualitative and even quantitative agreement between the two sets of numbers is striking. As we stressed earlier, there are no small parameters in the theory, and thus one could reasonably expect values of A_n and $B_n \simeq 1$ for small n. Such values are observed for example in He^3.

In fact the observed values in Na and K are much smaller, and the theoretical estimates are in agreement with this observation. The overall agreement is much better than one would have expected a priori. The question as to why the agreement is so good is an open one. Perhaps

TABLE IV

THE EXPERIMENTAL VALUES AND THEORETICAL ESTIMATES FOR THE LANDAU PARAMETERS IN K

	Experiment	Theory I[a]	Theory II[b]
A_0			−0.58
A_1			+0.04
A_2	−0.03 ± 0.005	−0.21	−0.02
B_0	−0.285 ± 0.02	−0.21	−0.24
B_1	−.06 ± 0.03		−0.04
B_2	0.0 ± 0.05		+0.003
m^*/m_0	1.2 ± 0.001	1.23	1.11

[a] Theory I: Electron–electron calculated using Hubbard and electron–phonon using Lee–Falicov potential.

[b] Theory II: Electron–electron from Hedin and same electron–phonon.

the approximations can be justified more rigorously. So far there has been no such justification.

The one Landau parameter which is predicted to be sizeable is A_0. Unfortunately A_0 and A_1 (see Section 56) have proved elusive. As we have already pointed out, A_0 does affect a uniform compression, i.e., change in density of the electron gas. In a metal the sound velocity is indeed proportional to the square root of the *total* compressibility, which contains other terms involving the ion–ion forces. In a strong magnetic field the effective compressibility of the electron gas, to high-frequency sound waves propagating hear cyclotron resonance, is magnetic field dependent. A measurement of the magnetic field dependence of the sound velocity in this case would give us a measure of A_0 (*18*).

In principle it is also possible to measure the electron electron contribution to A_1. The mass m^* measured in AKCR contains both electron–electron and electron–phonon interactions. Scher and Holstein (*19*) have shown that as one increases the frequency in a cyclotron resonance experiment to a significant fraction of a characteristic phonon frequency (ω_D), the mass enhancement due to phonons first increases, then at high frequencies $\omega_c \gg \omega_D$ disappears. In essence the lattice is unable to follow the cyclotron motion of the carriers. When electron phonon effects are negligible ($\omega_c \ll E_F$), then (neglecting some minor band structure corrections)

$$m^*_{\text{HF}}/m_0 = 1 + A_1^{ee}. \tag{63.2}$$

Equation (63.2) implies that an AKCR measurement in the infrared could possibly determine A_1^{ee}. In fact the entire Landau theory should be frequency dependent in the range of frequencies near ω_D. These considerations imply that such measurements might help unravel the electron–electron part of the interaction function, having a scale of energy of the order of E_F, from the electron–phonon part having a scale of energy of order ω_D. To date such experiments have not been successful although there has been a beginning (20).

To this point we have assumed $f(\mathbf{p}, \sigma, \mathbf{p}', \sigma')$ and $r(\mathbf{p}, \sigma, \mathbf{p}', \sigma')$ are isotropic. While this is true of the electron–electron part (at least in the alkalis), it is not strictly true of the electron–phonon contribution. The anisotropy does not arise from Fermi-surface anisotropies, since De Haas–Van Alphen measurements show that the Na and K Fermi surfaces are isotropic. Rather, it is caused by the rather large anisotropy of the phonon spectrum. The velocity of sound in metals, and in particular in Na and K, is highly anisotropic (21). The anisotropy due to the phonons is calculable since it arises from a phonon spectrum which is known quite accurately from neutron measurements. Calculations show that while there is a sizeable anisotropy in the phonon spectrum, the anisotropic coefficients of the $f(\mathbf{p}, \sigma, \mathbf{p}', \sigma')$ function are quite small (6). The prospects for measuring these anisotropies by doing single-crystal experiments are not bright. However, there is an important modification of the Landau sum rule [Eq. (60.28)] which is brought about by the anisotropy of the phonon spectrum (6). When one includes the interaction with phonons there is an additional contribution to the self-consistent field. This is obtained by adding to the right-hand side of Eq. (60.28) the contribution from one-phonon exchange. We can again evaluate this contribution exactly in the $\bar{\omega} \to 0$ limit. It depends on the direction of $\mathbf{q}$ relative to the crystal axes. Averaging the final result over all directions yields the result

$$1 + \frac{B_0}{1+B_0} + \sum_{l=1}^{\infty}(2l+1)\left[\frac{A_l}{1+A_l} + \frac{B_l}{1+B_l}\right]$$
$$= \frac{m^*}{3M}v_F^2 \sum_\lambda \left\langle \left(\frac{\epsilon_\lambda(\mathbf{q})\cdot\mathbf{q}}{v_\lambda(\mathbf{q})}\right)^2 \right\rangle_{\text{Av}}. \tag{63.3}$$

In Table V are the values obtained for the right side of Eq. (63.3) using the experimental values of m^* and the known elastic constants of Na and K (21). To obtain values for the left we need a value of A_1. This

TABLE V

TEST OF SUM RULE

	Na	K
A_1	0.10	0.15
Left-hand side of (6.3)	0.87 ± 0.3	0.67 ± 0.6
Right-hand side of (6.3)	0.65	0.47

cannot be determined directly from experiment but an estimate of its size can be made as follows. The Coulomb mass enhancement is obtained by combining the observed effective mass from Azbel–Kaner resonance with a calculated phonon enhancement (5). The Coulomb A_1 is then determined by the Landau equation for the effective mass [Eq. (63.2)]. The agreement between theory and experiment is very encouraging. Experimentally the moments seem to be falling off very rapidly. The fact that the first few moments of the Landau function, as determined from the experiments, reproduce the exact sum rule result gives an additional element of consistency to this description of a normal metal.

REFERENCES

1. P. Nozières, "Theory of Interacting Fermi Systems." Benjamin, New York, 1964.
2. A. A. Abrikosov, L. P. Gorkov, and I. E. Dzyaloshinski, "Methods of Quantum Field Theory in Statistical Physics." Prentice-Hall, New York, 1963.
3. T. M. Rice, *Ann. Phys. (New York)* **31**, 100 (1965).
4. L. Hedin, *Phys. Rev.* **139**, A796 (1965).
5. T. M. Rice, *Phys. Rev.* **175**, 858 (1968).
6. W. F. Brinkman, P. M. Platzman, and T. M. Rice, *Phys. Rev.* **174**, 495 (1968).
7. D. F. Dubois, *Ann. Phys. (Paris)* **8**, 24 (1959).
8. L. Hedin and S. Lundqvist, *Solid State Phys.* **23**, 2 (1969).
9. J. Hubbard, *Proc. Roy. Soc. Ser. A* **243**, 136 (1957).
10. S. Silverstein, *Phys. Rev.* **128**, 631 (1963); **130**, 912 (1963).
11. R. E. Prange and A. Sachs, *Phys. Rev.* **158**, 672 (1967).
11a. R. T. Schumacher and W. E. Vehse, *J. Phys. Chem. Solids* **24**, 297 (1963).
11b. C. C. Grimes and A. F. Kip, *Phys. Rev.* **132**, 1991 (1963).
12. J. A. Keck, *Phys. Rev.* **175**, 897 (1968).
13. A. B. Migdal, *Zh. Eksp. Teor. Fiz.* **34**, 1438 (1958) [Engl. transl., *Sov. Phys.—JETP* **7**, 996 (1958)].
14. M. L. Cohen and V. Heine, *Solid State Phys.* **24**, (1969).
15. N. Ashcroft and J. Wilkens, *Phys. Lett.* **14**, 285 (1965).
16. P. B. Allen and M. L. Cohen, *Phys. Rev. B* **1**, 1329 (1970).
17. For Na see A. D. B. Woods, B. N. Brockhouse, R. H. March, A. T. Stewart, and R. Bowers, *Phys. Rev.* **128**, 1112 (1962). For K see R. A. Cowley, A. D. B. Woods, and G. Dolling, *ibid.* **150**, 487 (1966).
18. E-Ni Foo and P. M. Platzman, *Phys. Rev. Lett.* **27**, 1568 (1971).
19. H. Scher and T. Holstein, *Phys. Rev.* **148**, 598 (1966).
20. P. Goy and G. Weisbuch, *Phys. Rev. Lett.* **25**, 225 (1970); S. J. Allen, to be published.
21. For Na, see M. E. Diedrich and J. Trivisonno, *J. Phys. Chem. Solids* **26**, 273 (1964).

ALKALI METAL PARAMETERS

METAL	a_0 (Å)	$N \times 10^{-22}$ (cm^{-3})	m_c/m_0	m^*/m_0 (SPECIFIC HEAT)	ω_p(CALC) eV	v_F (cm/s)	k_F cm^{-1}	r_s	g VALUE	E_F FREE ELECTRON	$\rho \times 10^{-6}\,\Omega\,\text{cm}$ (20°C)
Li	3.491 (4.2°K)	4.69		2.17	7.8	1.31×10^8	1.1×10^8	3.22	2.0023	4.72	8.55
Na	4.225 (4.2°K)	2.65	1.24	1.23	5.9	0.86×10^8	0.93×10^8	3.96	2.0015	3.12	4.3
K	5.225 (4.2°K)	1.40	1.21	1.23	4.25	0.71×10^8	0.746×10^8	4.87	1.9997	2.14	6.1
Rb	5.585 (4.2°K)	1.15	1.20	1.31	3.76			5.18	1.9984	1.82	11.6
Cs	6.045 (4.2°K)	0.905	1.44	1.58	3.46	$\approx .521 \times 10^8$	0.65×10^8	5.57	2.005	1.53	19.0

Author Index

Numbers in parentheses are reference numbers and indicate that an author's work is referred to, although his name is not cited in the text. Numbers in italics show the page on which the complete reference is listed.

A

Abrikosov, A. A., 274 (2), *294*
Adler, S. L., 42 (15), *48*, 77, 81, *81*
Aigrain, P., 120, *126*
Allen, P. B., 289 (16), *294*
Allis, W. P., 153 (1), 157 (1), *202*, 207 (10), *219*
Arrott, A., 160 (10), *202*
Ashcroft, N., 289, *294*
Azbel, M. Ya., 120 (3), *126*, 165 (17), 191 (36), *202*, 212 (12), *219*, 244, *271*

B

Baraff, G. A., 122 (17), *126*, 191 (34, 37), 197, 201, *202*, *203*, 252 (20, 21), 254 (21), 255 (21), *272*
Bardeen, J., 73 (10), *81*
Beaglehole, D., 67 (2), *81*
Benoit a la Guillanme, C., 99 (19), *118*
Bernstein, I., 172 (27), *202*
Bers, A., 153 (1), 157 (1), *202*
Blum, F. A., 84 (5, 9), 85 (9), *118*
Bohm, D., 12 (6), *19*, 49 (3), 54, *62*
Bowers, R., 120 (8), 121 (13), *126*, 127, 135, *150*, 163 (15), *202*, 290 (17), *294*
Brabiskin, A., 160 (9), *202*
Brinkman, W. F., 262 (24), *272*, 282 (6), 293 (6), *294*
Brockhouse, B. N., 290 (17), *294*
Brout, R., 49 (3), 54 (3), *62*, *63*
Brueckner, K. A., 54, *62*
Bruek, S. R. J., 99 (18), *118*
Buchsbaum, S. J., 120 (9), 121 (11), 125 (24), *126*, 142 (7), 143 (7), 146 (9), 147, *150*, 153 (1), 157 (1), 162 (14), 173 (28), 177, 191 (38), *202*, *203*

C

Carver, T. R., 204 (1), *219*
Case, K. M., 62 (12), *63*
Cohen, M. H., 49, *62*, 155 (2), 178 (2), *202*
Cohen, M. L., 289 (14, 16), *294*
Conte, S. D., 62, *63*
Cowley, R. A., 290 (17), *294*
Chambers, R. G., 255 (23), *272*
Cheng, Y. C., 122 (16), *126*, 247 (17), 252 (19), *272*
Clark, J. S., 122 (16), *126*, 252 (19), *272*

D

Damen, T. C., 84 (6), *117*
Davies, R. H., 84 (9), 85 (9), *118*
Debever, J. M., 99 (19), *118*
Debye, P. P., 11, *19*
DeGroot, S. R., 129 (3), *150*
Diedrich, M. E., 293 (21), *294*
Dimigen, H., 14 (9), *19*
Dolling, G., 290 (17), *294*
DuBois, D. F., 49 (3), 54 (3), *63*, 75 (12), 76 (12), *81*, 283 (7), 286 (7), *294*
Dunifer, G., 122, *126*, 136, *150*, 168, *202*, 205 (6), *219*, 264 (26), 268, 270, *272*
Dyson, F. J., 204, 209, *219*
Dzyaloshinski, I. E., 274 (2), *294*

E

Ehrenreich, H., 41 (14), 44 (14), *48*, 49, *62*, 79 (15), *81*
Einstein, A., 26, *48*
Engelsberg, S., 262 (24), *272*

F

Fal'kovskii, L. A., 5 (2), *19*
Fawcett, E., 120 (4), *126*

AUTHOR INDEX

Feher, G., 120 (5), *126*, 204, *219*
Ferrell, R. A., 59, *63*
Fischer, B., 67 (2), *81*
Fleury, P. A., 85 (12, 13), 114 (12, 13), 116 (13), *118*
Foo, E-Ni, 98 (18), *118*, 122 (15), *126*, 176 (19), 189 (32), 201 (19), *202*, 246 (15), *271*, 292 (18), *294*,
Foyt, A. G., 83 (2), *117*
Fredkin, D. R., 245 (14), 261 (14), 268 (14), 270 (14), *271*
Fried, B. D., 62, *63*, 106 (22), *118*
Friedel, J., 56, *63*
Fujita, E., 74 (11), *81*
Fukuda, N., 49 (3), 54 (3), *62*

G

Galt, J. K., 142 (7), 143 (7), *150*
Gantmakher, V. F., 196, 199, *203*
Gell-Mann, M., 54, *62*, 73 (10), *81*
Gerasimenko, V. I., 212 (12), *219*
Giriat, W., 99 (19), *118*
Gold, A. V., 221 (5), *271*
Goldstone, J., 49, *62*
Gorkov, L. P., 274 (2), *294*
Gorman, E., 160 (10), *202*
Gottfried, K., 49, *62*
Gould, R. W., 62 (13), *63*, 106 (22), *118*
Goy, P., 293 (20), *294*
Grimes, C. C., 121 (11), 122 (17), *126*, 128 (2), 146 (9), 147, *150*, 160, 162 (5), 163 (5), 167, *202*, 252 (20), 255 (20), *272*, 289, *294*
Griswold, T. W., 204, *219*
Groves, S. H., *118*

H

Harrison, M. J., 155 (2), 178 (2), *202*
Harrison, W. A., 155 (2), 178 (2), *202*
Hasegawa, A., 173 (28), *202*
Hasegawa, M., 75 (12), 76 (12), *81*
Hebel, L. C., 177, *202*
Hedin, L., 274 (4), 283 (4), 284 (4), 289, *294*
Heine, V., 289, *294*
Heitel, B., 160 (8), *202*
Heitler, W., 24 (1), 31 (1), *48*
Herring, C., 224 (9), *271*
Holstein, T., 292, *294*

Houck, J. R., 121 (13), *126*, 163 (15), *202*
Hubbard, J., 49 (3), 54 (3), *63*, 73 (9), *81*, 285, *294*
Hückel, E., 11, *19*

I

Isaacson, R. T., 5 (2), *19*

J

Jacobs, K. C., 255 (22), *272*
Jauch, J. M., 224 (8), *271*
Jones, H., 11 (5), *19*
Joyce, G., 60 (11), *63*

K

Kanazawa, H., 74 (11), *81*
Kaner, E. A., 119 (1), 120 (3), 121 (12), *125*, *126*, 130, *150*, 158 (1), 162 (13), 165 (17), 169 (22), 171 (23, 24), 191 (36), 196, 199, *202*, *203*
Keck, J. A., 287, *294*
Kelley, P. L., *118*
Kip, A. F., 120 (5), *126*, 167, 190 (33), *202*, 204, *219*, 289, *294*
Kittel, C., 204, *219*
Kivelson, M. G., 75 (12), 76 (12), *81*
Knapp, G. S., 263 (25), *272*
Koch, J. F., 190 (33), *202*
Kohn, W., 31 (8), *48*, 56, *63*
Konstantinov, O. V., 120, *126*
Kramers, H. C., 36, *48*
Kronig, R., 36, *48*
Kubo, R., 35, *48*
Kunz, C., 68, 69, 70, *81*

L

Lampe, M., 212 (9), *219*
Landau, L., 221, 232, *271*
Landau, L. D., 58 (8), *63*, *126*, 161 (11), *202*, 220, 232, *271*
Lang, W., 13 (7), 14, *19*
Langer, J., 56, *63*
Langmuir, I., 12 (6), *19*
Lax, B., 115 (26), *118*
Lax, M., 34 (10), *48*
Leder, L. B., 14 (9), *19*

Legendy, C., 120 (8), *126*, 127, 135, *150*
Leite, R. C. C., 84 (6), *117*
Lewis, R. B., 204 (1), *219*
Libchaber, A., 128 (2), *150*, 160, 162 (5), 163 (5), 199, *202*, *203*
Litshitz, I. M., 212 (12), *219*
Lundquist, S., 283, 289, *294*
Luttinger, J. M., 31 (8), *48*, 221, 232 (4), *271*

M

McWhorter, A. L., 28 (6), *48*, 84 (8), 88 (15), 104, 110, *117*, *118*
Malmberg, J. H., 62 (13), *63*
March, R. H., 290 (17), *294*
Marshall, N., 67 (2), *81*
Marton, L., 14 (9), *19*
May, W. G., 104, *118*
Mayer, H., 160 (8), *202*
Mendlowitz, H., 14 (9), *19*
Mermin, N. D., 122 (16), *126*, 247 (17), 252 (19), *272*
Migdal, A. B., 288, *294*
Misawa, S., 74 (11), *81*
Monod, P., 204 (2), *219*
Montgomery, D., 60 (11), *63*
Mooradian, A., 83, 84 (1, 2, 3, 5, 11), 85, 86, 88 (3, 15), 114 (23a), 117 (11), *117*, *118*
Mott, N. F., 11 (5), *19*

N

Nee, T. W., 192 (39), *203*
Ninham, B. W., 68, 75 (6), 76 (6), *81*
Nozières, P., 5 (3), *19*, 49 (3), 54 (3), *63*, 104 (21), *118*, 174 (29), *202*, 221, 222 (6), 229 (6), 232 (4), *271*, 274 (1), 275 (1), 276 (1), 278 (1), 280, *294*

O

O'Neil, T. M., 62 (13), *63*
Overhauser, A. W., 160, *202*, 206 (7), *219*

P

Patel, C. K. N., 84 (4, 10), 85 (12, 13), 96 (4, 16), 114 (4, 10, 12, 13, 24), 116 (4, 13), *117*, *118*

Pearson, G. A., 85 (13), 114 (13), 116 (13), *118*
Peercy, P. S., 184 (31), 196 (41), *202*, *203*
Perel, V. I., 120, *126*
Phelan, R. J., Jr., 99 (19), *118*
Philipp, H. R., 79, *81*
Pines, D., 2 (1), 5 (3), 12 (6), *19*, 26 (2), 48 (13), *48*, 49 (3), 54 (3), *62*, *63*, 104 (21), 107, *118*, 222 (6), 229 (6), *271*
Pippard, A. B., 120 (2), *125*, 158 (2)
Platzman, P. M., 27 (5), 28 (6), 29 (7), *48*, 84 (7), 97 (17), 109 (7), *117*, 121 (14), 122 (15, 17), 125 (24), *126*, 162 (14), 164 (16), 176 (19), 180 (16), 184 (31), 191 (38), 201 (19), *202*, *203*, 212 (9), *219*, 246 (15), 250 (18), 252 (20), 255 (20, 22), *271*, *272*, 282 (6), 292 (18), 293 (6), *294*
Powell, C. J., 68 (3, 6), 75 (6), 76 (6), *81*
Prange, R. E., 192 (39), *203*, 286 (11), 287 (11), 290, *294*

Q

Queisser, H. J., 67 (2), *81*
Quinn, J. J., 155 (2), 178 (2), *202*

R

Raether, H., 14 (8), *19*, 65 (1), 70 (1), *81*
Ramsey, N. F., 210, *219*
Rediker, R. H., 99 (19), *118*
Reuter, G. E., 133 (5), *150*
Reuter, G. E. H., 125 (23), *126*
Reuter, G. H., 191 (35), *202*
Rice, T. M., 245 (13), *271*, 274 (3, 5), 282 (6), 283 (5), 285, 286, 287 (3), 290, 293 (6), 294 (5), *294*
Rodriquez, S., 155 (2), 160 (7), 178 (2), *202*
Rohrlich, F., 224 (8), *271*
Rose, F., 120 (8), *126*, 127, 135, *150*
Rowland, T., 56, *63*
Rupp, L. W., Jr., 184 (31), 196 (41), *202*, *203*, 263 (25), *272*
Ruthemann, G., 13 (7), 14, *19*

S

Sachs, A., 286 (11), *294*
Sawada, K., 49 (3), 54 (3), *62*
Scher, H., 292, *294*
Schmidt, P. H., 184 (31), 196 (41), *202*, *203*, 263 (25), *272*

Schoenberg, D., 270 (28), *272*
Schrieffer, J. R., 107, *118*
Schultz, S., 122, *126*, 136, *150*, 168, 171 (25), 183, 187 (25), 201, *202*, 204 (2), 205 (6), *219*, 264 (26), 268, 270, *272*
Schumacher, R. T., 231 (10, 11), 271 (10), *271*, 287 (11a), *294*
Scott, J. F., 84 (6), *117*
Shak, J., 84 (6), *117*
Shaw, E. D., 84 (10), 114 (10), *117*
Siebenman, P. G., 160 (9), *202*
Silin, V. P., 122, *126*, 220, 246 (16), *272*
Silverstein, S., 285, *294*
Skobov, V. G., 119 (1), *125*, 130, *150*, 158 (1), 169 (22), 171 (23, 24), *202*
Skobov, V. I., 121 (12), *126*, 162 (13), *202*
Slichter, C. P., 231 (11), *271*
Slusher, R. E., 84 (4), 85 (13), 96 (4, 16), 99 (19), 114 (4, 12, 13, 24), 116 (4, 13), *117*, *118*
Smith, G. E., 177, *202*
Smith, N., 160 (9), *202*
Sondheimer, E. H., 125 (23), *126*, 133 (5), *150*, 191 (35), *202*
Spitzer, L., 161 (12), *202*
Stanford, J. L., 158 (4), *202*
Stern, E. A., 158 (3, 4), *202*
Stern, R. A., 67 (2), *81*
Stewart, A. T., 290 (17), *294*
Stiles, P. J., 270 (28), *272*
Stix, T., 121 (10), *126*, 172 (26), *202*
Stix, T. H., 58 (9), *63*
Stradling, R. A., 190 (33), *202*
Sueoka, O., 68, *81*
Swan, J. B., 68 (3), *81*
Swanson, N., 68, 75 (6), 76 (6), *81*

T

Teng, Y. Y., 67 (2), *81*
Tjoar, N., 97 (17), 98 (18), *118*
Tonks, L., 12 (6), *19*
Trivisonno, J., 293 (21), *294*
Tzoar, N., 27 (5), 29 (7), *48*

V

VanKampen, N. G., 62 (12), *63*
VanHove, L., 26, *48*
Varga, B. B., 86 (14), *118*
Vehse, W. L., 231 (10), 271 (10), *271*, 287 (11a), *294*
Vosko, S. H., 56, *63*
Vosko, S. J., 56, *63*

W

Walsh, W. M., 121 (14), 122 (15), *126*, 164 (16), 169 (21), 176 (19), 180 (16), 183, 184 (21), 186 (21), 201 (19), *202*, 246 (15), 264 (27), 269 (27), *271*, *272*
Walsh, W. M., Jr., 184 (31), 196 (41), *202*, *203*, 250 (18), 263 (25), *272*
Wannier, G. H., 34 (9), *48*
Watabe, M., 75 (12), 76 (12), *81*
Watanabe, H., 68, *81*
Weisbuch, G., 199, *203*, 293 (20), *294*
Werner, S. A., 160 (10), *202*
Wharton, C. B., 62 (13), *63*
Wigner, E., 73 (8), *81*
Wigner, E. P., 2, *19*, 212 (13), *219*, 223 (7), 233, *271*
Wilkens, J., 289, *294*
Williams, G. A., 5 (2), *19*, 144 (8), *150*
Wilson, A. R., 245 (14), 261 (14), 268 (14), 270 (14), *271*
Wiser, N., 42 (15), *48*, 77, *81*
Wolff, P. A., 84 (9), 85 (9, 13), 97 (17), 114 (13), 115 (25), 116 (13, 25), *118*, 122, *126*
Woods, A. D. B., 290 (17), *294*
Wright, G. B., 84 (3), 85, 86, 88, *117*, *118*

Y

Yafet, Y., 85 (12), 114, *118*, 206 (8), *219*

Z

Zubarev, D. N., 47 (16), *48*

Subject Index

A

A_0, 250, 285, 291–292
A_1, 251–252, 285, 291–292
A_2, 285, 291–292
Acoustic plasma waves, 84, 102–107, 109, 116
  degenerate plasmas, 104
  drafted plasmas, 106
  Landau damping of, 103, 105–106
  Maxwellian plasmas, 105
  two stream instability, 106–107
Alfven waves
  local dispersion relation, 142
  observation of, in Bi, 144–145
  phase velocity, 142
  vibrations on a stretched string, 143
Anomalous skin effect, 133, 198
Asymptotic current, 200
Azbel Kaner cyclotron resonance, 165–167, 245

B

B_0, 231, 270, 285, 291–292
B_1, 270, 285, 291–292
Band structure
  dielectric tensor, 41, 77–81
  light scattering, 111–117
Bernstein mode, 97–98, 173
  hybrid mode coupling, 97–98
Bloch equation, 211–212

C

Characteristic energy loss experiments, 13–14, 21, 23–26, 68–71, 100
Collision frequency, 18, 34
Compressibility of the electron gas, 222
Combination resonances, 261–262
Conductivity tensor, 33, 91, 123
  degenerate plasmas, arbitrary k and ω, 155
  longitudinal fields, 45
  long wavelength limit of, 176
  Maxwellian plasmas, 157
  transverse fields, 44–46
  uniform fields, 16, 132, 243
  zeros of, 247–249
Continuity equation
  current, 211
  interacting Fermions, 241
  spins, 211, 262–263
Correlation function
  density, 25–26, 107
  scattering experiments, 24–26
Coulomb interactions, 284–287
  quasi-particle effective mass, 285
Current sheets, 192–194
Cyclotron damping of helicons, 141
Cyclotron frequency, 15–18, 120
Cyclotron resonance, 139, 165–166
Cyclotron waves
  cutoff field, 182
  dispersion relation, 171, 173–174
    comparison to experiment, 180–190
    extraordinary wave, 177, 184–189
    intermediate wavelengths (numerical), 178–180
    long wavelength, 176
    ordinary wave, 177–183
    short wavelength, 177–178
  experimental results in
    Cu, 189
    K, 167, 182–187
    Na, 168, 186
  Maxwellian plasmas, 172–173
  standing wave modes, 180
  transmission experiments, 168
  zero current modes, 173–174
    electron motion, 175
    parallel propagation, 252–256

D

Debye length, 8–11
Density matrix, 232
  external fields, 234–236

interacting Fermions, 239–240
Wigner representation, 233
Dielectric function, 40–43, 95–97, 100–101
  band structure effects, 41, 77–81
  high frequency, 16–17, 57, 61
  inverse in a crystal, 42
  longitudinal, 46, 48, 54–55, 57, 60
  Maxwellian plasmas, 61
  singularities of, 198–199
  transverse, 46
Dielectric tensor, local limit, 132
Diffuse scattering, 197
Diffusion of electrons, 207–208
Doppler shifted cyclotron absorption
  arbitrary Fermi surface, 158
  spherical Fermi surface, 156, 198
Dysonian spin resonance, 206–213

E

Electromagnetic waves, 6
  magnetoplasmas, 94
Electron motion in a nonuniform field, 175, 200
Electron phonon interactions, 287–290
  Migdal's theorem, 288
  pseudopotential, 288
  quasi-particle mass, 291–292
Equations of motion in RPA, 50–51
Exchange corrections
  to RPA, 71–75
  spin–flip frequency, 240, 257
  susceptibility, 75

F

Fermi liquids
  entropy, 228
  Hartree–Fock, 73
  introductory remarks, 5–6
  specific heat, 228
  susceptibility, 229–231
Fermi momentum in Na and K, 270
Fermi–Thomas screening, 8, 11, 55–56
Fluctuation dissipation theorem, 47, 91
Four point function, 278–279
  integral equation, 280
  limiting forms, 280
  long range part, 281
Friedel oscillations, 56

G

g value
  metals, 206
  semiconductors, 113–114
Gantmacher–Kaner oscillations, 196, 199
Greens function, 275–276
  definition, 275
  free particle, 276
  spectral representation, 275

H

Hartree–Fock approximation, 54, 225, 277
Helicons, 17, 120–121
  cutoff field, 140–141
  cyclotron damping, 141, 158–160
  local dispersion relation, 131, 134–135
  magnetic Landau damping, 162–164
  nonlocal dispersion relation, 157
  nonparallel propagation, 161–164
  observation of, 127–128
  Overhauser ground state, 160–161
  phase velocity, 135
  polarization, 135
  vibrations on a stretched string, 143
Helicon phonon interactions
  dispersion relations in an anisotropic medium, 150
  dispersion relations in an isotropic medium, 147
  observation of, 147–149
Hybrid frequency, 96, 173
  coupling to Bernstein modes, 97–98

I

Induced charge, 38
Induced current, 43–47
Ion equation of motion, 146

J

Jellium model of a crystal, 104–105

L

Landau damping, 90, 96, 103, 105–106
Landau levels, 90, 115
Landau Raman scattering, 85, 115–117
Landau scattering function
  Coulomb interactions, 283–287
  definition of, 224

SUBJECT INDEX 303

four point function, relation to, 281
general form, 226
Hartree–Fock approximation, 224–225, 279
ladder diagrams, 289–290
Migdal theorem, 288–289
numerical estimates, 290–294
phonon effects, 287–290
pseudopotential, 288
spherical harmonic expansion, 227
sum rule, 281, 293–294
Landau–Silin transport equation
current transport, 240
isotropic exchange, 258–259
long wavelength solution, 246–247, 253, 261
parallel propagation, 254
spin transport, 236, 238, 257
Light scattering, 15, 27–31, 82–99, 107–117
acoustic plasmons, 102–104, 109
band structure effects, 111–117
effective mass approximation, 112–113
free electrons in a magnetic field, 115–116
Landau damping of plasmons, 89–90
Landau levels, 115–116
magnetic moment coupling, 113
magnetoplasmas, 91–99
multicomponent plasmas, 107
nonparabolic bands, 116
plasmon phonon coupling, 85–89
plasmons in semiconductors, 85–89
polarization of Landau level scattering, 115
spin density fluctuations, 113, 117
spin flip, 114
spin–orbit coupling, 111
stimulated Raman, 110
stimulated spin flip, 114

M

Magnetic Landau damping, 162–164
Magnetic moment, coupling to light, 113
Magnetoplasmas and light scattering, 90–99
Maxwell's equations, 123
Mean free path for plasmon emission, 66
Mirror field, 161

Multicomponent plasmas, 99–111
structure factor, 99–100

O

Onsager relations, 129
Overhauser ground state, 160–161
maximum velocity, 160
optical experiments, 160
position of helicon edge, 161

P

Paramagnetic spin waves, see Spin waves
Periodic boundary value problem, 193
Phase reversal, 134
Plasma dispersion function, 62
Plasma frequency, 11–15, 57
Plasma waves, 23, 85–88, 94, 102–107
band structure effects, 77–79
damping of, 75–76
dispersion of, 57, 65–68, 74
doped semiconductors, 14, 80–81
experimental results, 65, 71
Landau damping of, 58–59, 61–62, 70, 90
Maxwellian plasmas, 60–62
surface, 67
two-pair decay, 75–76
Plasmons, see Plasma waves
Poisson bracket, 233, 235
Polaritons, 95
Pseudopotential, 288

Q

Quasi-particles
effective mass
Azbel Kaner resonance, 245
electron electron effects, 285
electron phonon effects, 291–292
numerical values in Na and K, 269
translationally invariant system, 242
gyromagnetic ratio, 230
lifetime, 276, 280
renormalization constant, 280
scattering, 279–280
Quasi-static approximation, 94

R

Raman scattering, 23, 84, 86, 88, 97, 114
Random phase approximation (RPA), 49–54

exchange corrections to, 71–74
magnetoplasmas, 91
plasmon dispersion, 68
Relativistic energy momentum relation, 116
Relaxation time approximation, 152, 237–238, 243
Relaxation to local equilibrium, 243
Response functions, 33–47
   band structure effects, 41
   external field, 37–39
   induced charge, 38
   total field, 37–39, 44
   transverse fields, 43–44

S

Scattering experiments, see Light scattering
   electron beams, 13–14, 23–27
   general discussion, 20–23
   light scattering, 15, 27–31, 82–99, 107–117
   plasmon damping, 69
   plasmon dispersion, 68
Scattering
   from collective modes, 22–23, 27
   from noninteracting electron gas, 26–27
   from single particles, 22–23
   from stimulated Raman, 110–111
Screening
   degenerate plasmas, 11, 40, 55–56
   Maxwellian plasmas, 11, 40, 59–60
Self-energy, 276
Semimetals, 5, 143–144
Single particle scattering, 22–23, 27, 83–85, 89, 90, 111–117
Skin effect, anomalous, 133, 198
Specific heat of a Fermi liquid, 228
Specular reflection, 192, 195–196
Spin density fluctuations, 113, 117
Spin diffusion, 207, 211
Spin–flip scattering, 85, 114, 115
Spin lifetime, 205–206, 238
   exchange effects, 257
Spin–orbit coupling, 111
Spin resonance
   bounded medium, 212–219
   experiments in Na, 218–219
   non-interacting Dysonian, 206–213
Spin resonance frequency, 204–205
Spin transmission, 214, 216–217
Spin transport equation
   interacting Fermions, 257
   non-interacting Fermions, 236
   relaxation time approximation, 238
Spin waves, 6, 122, 256–270
   angular dependence of, 262–268
   determination of Fermi liquid parameters, 269–270
   experimental observation of, 266–268
   isotropic exchange, 258
   long wavelength limit, 260–263
Stimulated Raman scattering, 84, 110–111
Structure factor, 91–94, 96, 100
   definition of, 25–26
   dielectric function, 100
   light scattering, 47
   magnetoplasma, 91–94
   non-interacting, 27
   plasmon contribution, 57
   quasi-static approximation, 94
Susceptibility
   bounded medium, 213
   long wavelengths, 210
   magnitude in alkalis, 214
   tensor, 33, 123

T

Thomson scattering, 29, 115
Transmission experiments, 31–32, 136–137
   cyclotron resonance, 139
   helicons, 138
   spin resonance, 213–219, 265–266
Transmission spectrum
   metal, 190–191
   semiconductor, 190–191
Two band approximation, 113
Two-stream instability, 106–107

W

Weak coupling (conditions for)
   degenerate plasmas, 4
   Maxwellian plasmas, 3
Wigner solid, 2

/530.44P719W>C1/